Progress in Vaccinology

Progress in Vaccinology

Contraception Research for Today and the Nineties
Progress in Birth Control Vaccines

Progress in Vaccinology

Volume 1

G.P. Talwar, Editor

Contraception Research for Today and the Nineties

Progress in Birth Control Vaccines

With 99 Figures

Springer-Verlag
New York Berlin Heidelberg
London Paris Tokyo

Professor G.P. Talwar
National Institute of Immunology,
JNU Complex, Shahid Jeet Singh Marg,
New Delhi - 110 067, India

Library of Congress Cataloging-in-Publication Data
Contraception research for today and the nineties:
progress in birth control vaccines.
(Progress in vaccinology; v. 1)
Papers presented at the Symposium on
Contraception Research for Today and the Nineties,
Oct. 3-6, 1986 in New Delhi, India.
Includes index.
1. Contraception, Immunological – Congresses.
2. Pharmacology, Experimental – Congresses.
I. Talwar, G. P. II. Series. III. Symposium on
Contraception Research for Today and the Nineties
(1986 : New Delhi, India) [DNLM: 1. Contraception – methods – congresses. 2. Family Planning – methods – congresses. 3. Fertility – drug effects – congresses.
4. Vaccines – immunology – congresses. QV 177 P976 1986]
RG136.8.P76 1987 615'.766 87-12692

Softcover reprint of the hardcover 1st edition 1988

Typeset by Publishers Service, Bozeman, Montana.

9 8 7 6 5 4 3 2 1

ISBN-13: 978-1-4612-8331-7 e-ISBN-13: 978-1-4612-3746-4
DOI: 10.1007/978-1-4612-3746-4

Preface

The twentieth century will close with 5 billion people added to the current global population. Between 1980 and the year 2000, the total world population will increase from 4 billion to a little over 6 billion. There will be half as many more people on earth during these 20 years than the number accumulated since the origin of man to 1980. Overpopulation is particularly acute in economically developing countries, where contraception has become a social necessity.

Contraception Research for Today and the Nineties carries the proceedings of an international symposium convened in New Delhi in October, 1986, to review the status of current research in contraception. Major organizations supporting basic and applied research in contraception–The Population Council, World Health Organization (WHO), The Rockefeller Foundation, United States Agency for International Development (USAID), International Development Research Center of Canada (IDRC), National Institutes of Health (NIH), and the Indian Council of Medical Research (ICMR)–were represented by the heads of divisions who projected respective programs and strategies. Principal scientists responsible for many of the new leads participated.

The background and status of the most exciting current and emerging contraceptives are discussed in this volume. Included is the subdermal implant, Norplant®, which having satisfactorily passed the Phase IV clinical trials in several countries is on the verge of introduction in family-planning programs. The implant provides protection up to five years, unless removed by voluntary decision. The saga of 20 years of laboratory and clinical research, from the first mouse experiment to eventual effective device for human usage, is unfolded by the UFPA Population Award Laureate, S.J. Segal. The story of Norgestrel-IUD, whose efficacy is the highest recorded so far for a contraceptive, is told by the innovator of the device, T. Luukkainen and his colleagues. E. Coutinho writes on his novel idea which obviates the portal route of entry of the steroids, the vaginal pill. Lower concentrations are usable because of the direct local effects of the contraceptive steroids in the genital tract. The "morning-after" pill abrogating pregnancy is the new Roussel steroid which blocks the binding of progesterone to its receptor. E. Baulieu reviews its status and outlines further development.

The inherent link between brain and reproductive functions is well recognized. The contributions of the Hungarian School in delineating the neuroendocrine connections are presented by B. Flerkó. GnRH analogs as agonists and antagonists and the latest information on inhibin and other peptides of the reproductive tract are reviewed. An important recent advance has been the discovery and elucidation of structure of the peptide for inhibition of prolactin release. The cloning of the gene carrying information for this peptide and GnRH is described by J. Ramachandran, K. Nikolics, and P.H. Seeburg.

A large part of the symposium was devoted to immunological approaches to fertility control. The success of gestation demands nonrejection of the conceptus, which bears 50% genetic information of paternal origin. Intricate mechanisms involved in this process are being delineated. This has significant relevance to normal fertility and to problems of recurrent abortions. Five admirable chapters dealing with the immunobiology of pregnancy have been contributed by Y.W. Loke, P.M. Johnson, T.J. Gill, M.J.K. Harper and M.A. Jones, and J.A. Hill and D.J. Anderson. To this theme is added the chapter by G.A. Voisin, who draws attention to the facilitatory immune mechanisms in reproduction.

The feasibility of controlling fertility by antibodies and/or cell-mediated immune reactions has been amply demonstrated in experimental animals including subhuman primates. Cases of infertility due to immunological factors are recognized in clinical practice. Over the years enough data has accumulated to show that immunization against some key hormones of the reproductive tract leads to a reversible block of fertility. Toxicology studies have shown the lack of notable side effects. More than one birth control vaccine is under development. Four vaccines directed against the human chorionic gonadotropin (HCG) have entered the Phase I clinical trials. These are novel vaccines which seek to intercept one of the physiological processes in the body, in contrast to external pathogens to which vaccines have hitherto been directed. The symposium comprehensively reviewed progress in birth control vaccines. G.P. Talwar and O. Singh describe the vaccines against HCG and GnRH. R.B. Thau summarizes the experience of immunization with βoLH. N.R. Moudgal and colleagues review their studies in male bonnet monkeys on immunization with FSH. Relevant to these developments is the progress in cloning and expression of gonadotropin hormone genes reviewed in two papers by S.K. Jain, W.W. Chin, G.P. Talwar, and collaborators. Of concern are the possible immunopathological consequences of immunization, and N.R. Rose and colleagues describe results of their recent studies.

The next generation of birth control vaccines seeks to employ sperm antigens. The best characterized protein unique to sperm is $LDHC_4$, initially purified by E. Goldberg. He reviews current understanding on this enzyme. A number of other antigens of interest have been put in evidence by monoclonal and polyvalent antibodies; three chapters by S. Isojima and colleagues, T. Haneji and S.S. Koide, and R. Naz describe potential candidates. C.W. Bardin gives an overview of the processes regulating spermatogenesis. The importance of local immunity in the genital tract against sperm antigens is highlighted by N.J. Alexander. D.J. Anderson and J.A. Hill draw attention to the critical role of lymphokines, monokines, and cell-mediated immune reactions in fertility. J. Griffiths, N.A. Mitchison, and colleagues sketch the manner by which pertinent epitopes can be delineated using immunobiological and genetic engineering approaches. The volume closes with a philosophical note from E. Diczfalusy, a veteran who has influenced reproduction and fertility control research for many decades.

There will be a continuing need for safe, effective, and alternate methods of contraception. This volume describes new developments that will be open to general usage in the next two to five years, as well as those which are either in early clinical trials or in research domain. Prominent in the latter category are the birth control vaccines.

G.P. Talwar

Contents

Part III Neuroendocrine Regulation

Part IV Immunobiology of Pregnancy

Part V Birth Control Vaccines

Contributors

Nancy J. Alexander, Division of Reproductive Biology and Behavior, Oregon Regional Primate Research Center, Beaverton, Oregon, USA

Deborah J. Anderson, Fearing Laboratory, Department of Gynecology, Obstetrics and Reproductive Biology, Harvard Medical School, Boston, Massachusetts, USA

Nandita J. Arbatti, Institute for Research in Reproduction, Bombay, India

C. Wayne Bardin, The Population Council, New York, New York, USA

J. Barzelatto, Special Programme of Research, Development and Research Training in Human Reproduction, World Health Organization, Geneva, Switzerland

Etienne-Emile Baulieu, Université Paris-Sud, Lab-Hormones, Bicêtre, France

C. Lynne Burek, Department of Immunology and Infectious Diseases, The Johns Hopkins University School of Hygiene and Public Health, Baltimore, Maryland, USA

William W. Chin, Laboratory of Molecular Genetics, Department of Medicine, Brigham and Women's Hospital; Howard Hughes Medical Institute, Harvard Medical School, Boston, Massachusetts, USA

Elsimar Metzker Coutinho, Department of Maternal Health, Federal University of Bahia, Salvador, Bahia, Brazil

Egon Diczfalusy, Karolinska Institute, Reproductive Endocrinology Research Unit, Stockholm, Sweden

M.F. Fathalla, Special Programme of Research, Development and Research Training in Human Reproduction, World Health Organization, Geneva, Switzerland

Béla Flerkó, Department of Anatomy, University Medical School, Pécs, Hungary

Soheyla D. Gharib, Laboratory of Molecular Genetics, Department of Medicine, Brigham and Women's Hospital; Howard Hughes Medical Institute, Harvard Medical School, Boston, Massachusetts, USA

Thomas J. Gill III, Department of Pathology, School of Medicine, University of Pittsburgh, Pittsburgh, Pennsylvania, USA

Erwin Goldberg, Department of Biochemistry, Molecular Biology and Cell Biology, Northwestern University, Evanston, Illinois, USA

J. Griffiths, Imperial Cancer Research Fund, Tumour Immunology Unit, Department of Zoology, University College London, London, England

S.K. Gupta, Department of Immunopathology, Post-Graduate Institute of Medical Education and Research, Chandigarh, India

Tatsuji Haneji, Center for Biomedical Research, The Population Council, New York, New York, USA

Michael J.K. Harper, Department of Obstetrics and Gynecology, University of Texas Health and Science Center at San Antonio, San Antonio, Texas, USA

Maija Haukkamaa, Steroid Research Laboratory, Department of Medical Chemistry, University of Helsinki, Helsinki, Finland

Joseph A. Hill, Fearing Laboratory, Department of Gynecology, Obstetrics, and Reproductive Biology, Harvard Medical School, Boston, Massachusetts, USA

Shinzo Isojima, Department of Obstetrics and Gynecology, Hyogo Medical College, Nishinomiya, Japan

Swatantra K. Jain, National Institute of Immunology, New Delhi, India

Peter M. Johnson, Pregnancy Immunology Group, Department of Immunology, University of Liverpool, Liverpool, England

Marjorie A. Jones, Department of Chemistry, Illinois State University, Normal, Illinois, USA

Kinu Kameda, Department of Obstetrics and Gynecology, Hyogo Medical College, Nishinomiya, Japan

S.S. Koide, Center for Biomedical Research, The Population Council, New York, New York, USA

Koji Koyama, Department of Obstetrics and Gynecology, Hyogo Medical College, Nishinomiya, Japan

Pekka Lähteenmäki, Steroid Research Laboratory, Department of Medical Chemistry, University of Helsinki, Helsinki, Finland

Y.W. Loke, Division of Experimental and Cellular Pathology, Department of Pathology, University of Cambridge, Cambridge, England

Tapani Luukkainen, Steroid Research Laboratory, Department of Medical Chemistry, University of Helsinki, Helsinki, Finland

N.A. Mitchison, Imperial Cancer Research Fund, Tumour Immunology Unit, Department of Zoology, University College London, London, England

Sudhir B. Moodbidri, Institute for Research in Reproduction, Bombay, India

N.R. Moudgal, Centre for Advanced Research in Reproductive Biology, Department of Biochemistry, Indian Institute of Science, Bangalore, India

S. Mukerji, International Institute for Population Sciences, Bombay, India

G.S. Murthy, Centre for Advanced Research in Reproductive Biology, Department of Biochemistry, Indian Institute of Science, Bangalore, India

Rajesh K. Naz, Department of Obstetrics, Gynecology and Physiology, The George Washington University Medical Center, Washington, DC, USA

K. Nikolics, Department of Developmental Biology, Genentech, Inc., South San Francisco, California, USA

D.B.G. Oliveira, Imperial Cancer Research Fund, Tumour Immunology Unit, Department of Zoology, University College London, London, England

M.R.N. Prasad, Centre for Advanced Research in Reproductive Biology, Department of Biochemistry, Indian Institute of Science, Bangalore, India

J. Ramachandran, Department of Developmental Biology, Genentech, Inc., South San Francisco, California, USA

A.J. Rao, Centre for Advanced Research in Reproductive Biology, Department of Biochemistry, Indian Institute of Science, Bangalore, India

N. Ravindranath, Centre for Advanced Research in Reproductive Biology, Department of Biochemistry, Indian Institute of Science, Bangalore, India

Noel R. Rose, Department of Immunology and Infectious Diseases, The Johns Hopkins University School of Hygiene and Public Health, Baltimore, Maryland, USA

P. Salgame, Imperial Cancer Research Fund, Tumour Immunology Unit, Department of Zoology, University College London, London, England

Brij B. Saxena, Division of Reproductive Endocrinology, Department of Obstetrics and Gynecology, Cornell University Medical College, The New York Hospital, New York, New York, USA

P.H. Seeburg, Department of Developmental Biology, Genentech, Inc., South San Francisco, California, USA

Sheldon J. Segal, Population Division, The Rockefeller Foundation, New York, New York, USA

Shobha Sehgal, Department of Immunopathology, Post-Graduate Institute of Medical Education and Research, Chandigarh, India

Anil R. Sheth, Institute for Research in Reproduction, Bombay, India

Minoru Shigeta, Department of Obstetrics and Gynecology, Hyogo Medical College, Nishinomiya, Japan

Mukul Singh, Division of Reproductive Endocrinology, Department of Obstetrics and Gynecology, Cornell University Medical College, The New York Hospital, New York, New York, USA

Om Singh, National Institute of Immunology, New Delhi, India

Regine Sitruk-Ware, Department of Reproductive Endocrinology, Hôpital Necker, Paris, France

Julian P. Smith, Department of Immunology and Infectious Diseases, The Johns Hopkins University School of Hygiene and Public Health, Baltimore, Maryland, USA

G.P. Talwar, National Institute of Immunology, New Delhi, India

Rosemarie B. Thau, Center for Biomedical Research, The Population Council, New York, New York, USA

Juhani Toivonen, Steroid Research Laboratory, Department of Medical Chemistry, University of Helsinki, Helsinki, Finland

Shakti N. Upadhyay, National Institute of Immunology, New Delhi, India

Guy André Voisin, Centre d'Immuno-Pathologie et d'Immunologie Expérimentale, Hôpital Saint Antoine, Paris, France

Margaret E. Wierman, Laboratory of Molecular Genetics, Department of Medicine, Brigham Women's Hospital; Howard Hughes Medical Institute, Harvard Medical School, Boston, Massachusetts, USA

George Zeidenstein, The Population Council, New York, New York, USA

Part I
Population Control and Development

CHAPTER 1

Demographic Trends, Global and Indian Scene, and Implications

S. Mukerji

World population passed the 5 billion mark recently and India's population accounted for about 15% of that number. It is believed that world population will stabilize at about 10 billion around the year 2050, and at that time India is expected to have between 1.3 to 1.6 billion persons. Patterns of growth and ultimate size of the world population will depend heavily on what happens in Africa and Asia (Table 1.1). In 1985 when the world population was 4.8 billion, Asia and Africa accounted for 58.2% and 11.5% of the population, respectively. The medium variant of United Nations estimates and projections reported here suggests that by the year 2025, when the world population will be 8.2 billion, Asia and Africa will account, respectively, for 55.3% and 19% of the world population. Further, in mid-1985 Africa had 37% of its population below age 25 whereas Asia had 33% in the same age group. Level of urbanization was equally poor in the two areas, being 30% in Africa and 27% in Asia. The next 15 to 20 years will be crucial as the potential for higher fertility will be strong in this period.

In Asia, the two population giants, China and India, will continue to dominate the population scene. In both countries fairly strong population welfare and development programs have been in operation for the last several years. In China about 60% of eligible couples have adopted the one-child norm. The one-child policy is likely to continue for more years; the idea is to stabilize the population at about 1.2 billion. Only time will tell how far it will be successful. Some projections do indicate that India will have more people compared with China in early 21st century.

In India the official policy continues to be the adoption of the two-child family (Table 1.2). To achieve these targets vigorous efforts will be made to raise the mean age at marriage for women to 20 years, promote "two-child family limit," increase demand for contraception, improve program management and delivery of services, enhance child survival through increased maternal and child health immunization programs, oral rehydration, etc. The consensus among academicians and program officials is that the new strategy can be implemented in the present social and economic environment.

In spite of the fact that India has a fairly well-established population data collection system through censuses and Sample Registration Scheme (currently

Table 1.1. United Nations world population chart 1985: estimated vital statistics.

Country or area	Population midyear (thousands)			CBR	CDR	AGR	TFR	e_0^0	IMR	SR
	1985	2000	2025	1980–1985						
WORLD	4,836,645	6,121,813	8,205,765	27	11	1.7	3.5	60	78	101
Africa	554,928	871,817	1,616,515	46	17	2.9	6.3	49	112	99
Eastern Africa	166,404	272,244	536,859	49	18	3.1	6.8	47	120	98
Middle Africa	59,538	91,996	169,597	45	18	2.7	6.0	48	117	97
Northern Africa	122,960	175,563	260,767	39	12	2.6	5.6	57	100	102
Southern Africa	37,218	54,553	90,991	40	14	2.5	5.2	53	87	98
Western Africa	168,808	277,461	558,302	49	18	3.1	6.9	47	123	98
Americas	668,391	843,730	1,124,075	25	9	1.7	3.1	67	49	98
Latin America	404,806	546,395	778,662	32	8	2.3	4.1	64	62	100
Caribbean	31,959	40,985	57,837	27	8	1.5	3.3	64	65	100
Central America	105,099	149,059	222,589	35	8	2.7	4.8	65	57	101
Temperate South America	45,616	55,354	69,599	24	8	1.5	3.1	70	32	98
Tropical South America	222,133	300,987	428,636	32	8	2.4	4.1	63	69	100
North America	263,585	297,335	345,413	16	9	0.9	1.8	74	11	95
Asia	2,818,214	3,548,994	4,535,149	27	10	1.7	3.5	59	83	105
East Asia	1,249,755	1,475,034	1,721,382	19	7	1.2	2.3	68	36	105
South Asia	1,568,459	2,073,960	2,813,766	34	12	2.2	4.6	55	103	105
South Eastern Asia	399,626	519,500	688,477	32	11	2.1	4.1	57	73	99
Southern Asia	1,055,892	1,386,872	1,854,601	35	13	2.1	4.7	54	115	107
Western Asia	112,940	167,588	270,689	37	10	2.8	5.2	61	81	107
Europe	491,850	512,474	523,696	14	11	0.3	1.9	73	15	95
Oceania	24,644	30,062	37,822	21	8	1.5	2.6	68	31	101
USSR	278,618	314,736	368,234	19	9	1.0	2.4	71	25	90

CBR, crude birth rate; CDR, crude death rate; AGR, average growth rate (%); TFR, total fertility rate; e_0^0, life expectancy at birth for males and females; IMR, infant mortality rate; and SR, sex ratio as number of males per 100 females.

Table 1.2. The revised strategy for National Family Welfare Program: demographic goals.

Long-term goal: To reach zero-population growth by the year 2050, with an estimated population of about 1,300 million.
Medium-term goal: To reach net reproduction rate of unity by year 2000 with crude birth rate of 21, crude death rate of 9, and infant mortality rate below 60 per 1,000 live births.

covering nearly 6 million persons), we do not know the proportion of couples who have adopted the two-child family. Nor do we know the fertility and health conditions of the urban poor. There is a tendency to believe that high natural growth is mainly in rural areas and is more so in some of the northern states of India. Available data indicate a different picture. Between 1978 and 1985 it appears that crude birth rate (CBR) and age-specific fertility rates (ASFR) in

Table 1.3. Crude birth rate (CBR) in the urban areas of 22 states in 1978–1983 and change in age-specific fertility rate (ASFR) between 1978 and 1981 for 14 urban areas of major states.

	CBR		ASFR increased between 1978–1981
State	1978	1983	Age group
Andhra Pradesh	28.9	28.4	
Assam	23.5	24.5	15–19, 40–44, 45–49
Bihar	34.2 (1980)	32.1	
Gujarat	31.6	31.8	20–24, 25–29, 45–49
Haryana	29.8	29.7	15–19, 20–24, 25–29, 30–34, 35–39, 45–49
Himachal Pradesh	20.4	22.2	20–24, 30–34, 45–49
Jammu & Kashmir	21.4	25.0	15–19, 25–29, 40–44, 45–49
Karnataka	25.9	26.0	15–19, 25–29, 45–49
Kerala	24.8	24.8	20–24, 25–29, 45–49
Madhya Pradesh	30.3	32.0	20–24, 25–29, 30–34, 35–39, 40–44, 45–49
Maharashtra	23.9	26.9	15–19, 30–34, 40–44, 45–49
Manipur	23.0	23.9	
Meghalaya	15.3	17.2	
Nagaland	NA	18.6	
Orissa	30.5	30.5	20–24, 25–29, 30–34
Punjab	28.1	28.8	15–19, 20–24, 25–29, 40–44
Rajasthan	31.6	34.3	20–24, 25–29, 30–34, 35–39, 40–44, 45–49
Sikkim	26.6 (1981)	27.2	
Tamil Nadu	26.2	25.9	15–19
Tripura	21.2 (1980)	16.6	
Uttar Pradesh	33.0	32.8	25–29, 40–44, 45–49
West Bengal	19.6	21.7	
INDIA	27.8	28.0	

Source: Sample Registration Data.

both younger and older had gone up faster in urban areas compared with that of rural areas, and the so-called better states had a higher share in such increases (Table 1.3). This is likely to continue for a few more years. It is true that all indicators of fertility in urban areas continued to be lower than those in rural areas. But the possibility of higher natural growth in urban areas cannot be ruled out.

Both CBR and ASFR below age 30 can remain stable or even increase up to 1996 in spite of the increase in the couple protection rate. This will happen because we had one of the highest proportions of females in the 0 to 14 ages in 1971 compared with 1941, 1951, 1961, and 1981. For India the percentage of females in 0 to 14 ages in 1941, 1951, 1961, 1971, and 1981 were, respectively, 39, 39, 41, 42, and 40. In 1971 the lowest percentage (37.7%) was in Kerala and the highest (45.1%) was in Assam (Table 1.4). Owing to the high proportion of such females in 1961 and 1971, current fertility measures like CBR and ASFR in 1985 can be high, as those girls will be in the ages 25 to 39 and 15 to 29, respectively. Current measures of fertility may remain high until those cohorts go out of reproduction. The importance of promoting nonterminal methods also becomes apparent, as we do not expect a large number of sterilizations below age 25. By 1981 in almost all states the percentage of females below age 15 dropped below 40%. When these girls enter reproductive age, by the year 1996, current rates of fertility will show a declining trend. Eight out of 16 major states—namely, Assam, Gujarat, Haryana, Karnataka, Madhya Pradesh, Orissa, Rajasthan, and West Bengal—were well above the national average in this respect. Survivors of those girls were in the 15 to 29 years age group in 1985, where

Table 1.4. Percentage of females below age 15 among all females in the urban areas in three census years.

State	1961	1971	1981
Andhra Pradesh	39.9	40.4	38.0
Assam	46.5	45.1	NA
Bihar	42.2	42.8	41.8
Gujarat	41.1	40.5	36.6
Haryana	NA	43.1	38.1
Karnataka	42.3	41.4	38.2
Kerala	40.0	37.7	31.9
Madhya Pradesh	41.3	43.3	40.0
Maharashtra	40.2	40.0	37.4
Orissa	38.2	42.2	NA
Punjab	43.1	40.0	36.7
Rajasthan	42.3	43.0	41.2
Tamil Nadu	38.2	37.8	NA
Uttar Pradesh	42.7	43.3	41.5
West Bengal	40.6	39.6	33.9
INDIA	41.0	40.8	37.8

Source: Census of India and Mukerjee SB: *The Age Distribution of Indian Population. A Reconstruction for the States and Territories*, East West Centre, East West Population Institute, Honolulu, 1976.

Table 1.5. Literacy in women in the age groups 10–14 and 15–19 in 1981.

States	Percentage literate by age	
	10–14	15–19
Bihar	29.07	24.55
Gujarat	57.55	51.69
Haryana	43.16	38.56
Karnataka	48.13	46.09
Maharashtra	61.83	58.54
Punjab	64.75	59.61
Kerala	94.76	91.45
Madhya Pradesh	31.16	28.42
Rajasthan	21.37	20.80
Uttar Pradesh	29.82	27.87
INDIA	44.85	43.29

sterilization is not readily accepted. Thus, even if they adopt the small-family norm, the current rates of fertility like CBR and ASFR can go up. The spill-over effect is likely to continue up to 1996. Under this circumstance, a better indicator for fertility control will be the proportion of third or higher order births each year, which is falling in both rural and urban areas.

According to the seventh five-year plan document, the expected demographic indicators for 1990 are CBR 29.1, crude death rate 10.4, and infant mortality rate 87 per 1,000 live births. The available information as of 1984 (India's Population: Current Demographic Scenario, Ministry of Health & Family Welfare, 1985) shows the following:

1. Only three states—namely, Kerala, Tamil Nadu, and West Bengal—covering 18.7% of the population (according to 1981 census) had CBR either below or close to 29 per 1,000 population;
2. Nine out of 17 major states covering 36.9% of the population (according to 1981 census) had crude death rate below 11 per 1,000 population; and
3. Eight out of 17 major states covering 44.5% of the population (according to 1981 census) had infant mortality rate below 87 per 1,000 live births.

In the remaining six years, that is, up to 1990, we therefore expect that the death rate and infant mortality rate targets will be achieved, whereas the CBR target will continue to pose problems. There is very little reason to doubt that CBR will fall sharply after 1996.

One important reason for this optimism is that according to the 1981 census, about 44% of girls 10 to 14 years old and 15 to 19 years old were reported as "literates" (Table 1.5). Also, between 1951 and 1981 the rate of growth of "literates" among females was higher than those for males in all the important five-year age groups covering 5 to 29 years, for all states. From the table it can be seen that, except in the states Bihar, Madhya Pradesh, Rajasthan, and Uttar Pradesh, the prospect of literate couples in post-1981 marriages is good [for details see Mukerji (1)]. Those in the 10 to 14 age group in 1981 will enter reproductive age in 1986, and the 15 to 19 cohort will remain in highly fertile ages until 1991. Thus, in the post-1986 period we expect about 50% of the couples, both husband and wife, will be literate. This will have its effect as it is well known that literacy influences all demographic components favorably.

Beyond 1996 the interest in fertility and its control will decline. Mortality and distribution of population will occupy center stage, and finally it will be mortality that will hold the main interest. For the coming few years mortality–its relationship with social and economic variables–will attract national and international attention. Beyond the nineties the scenario of high infant and childhood mortality will be unimportant in most countries; then the interest will gradually shift toward the control of deaths due to genetic causes, environmental factors, social stress, and sheer old age.

We may soon find more persons in the over 60-year age group than in under 5-year age group. The emotional and material needs for such populations are different, with each country having its own social and cultural setup. What we have to offer to our senior citizens currently may be inadequate.

Some other areas that should receive attention are as follows:

1. The impact of nondemographic development programs like rural electrification, irrigation, employment schemes, etc. on marriages, fertility, mortality, and internal migration.
2. Impact of vocational and other training programs on income generation and the consequent changes in demographic indicators.

Reference

1. Mukerji S: Benefits we derived from present education. *Yojana* 1985;29(22):28–30.

CHAPTER 2

Population Control: Strategies and Experience

George Zeidenstein

In *The Discovery of India*, the best known book of independent India's first Prime Minister, Pandit Nehru expressed plainly a truth that was apparent to him during his nearly three-year-long internment in Ahmednagar Fort prison by the British authorities during World War II. Writing in 1944, he said:

> . . . [The] eastward sweep of technology, accompanied by education, sanitation, and better public health, is continuing and will cover many of the countries of Asia. Some of these countries, like India, far from needing a bigger population, would be better off with fewer people.

Then, in 1952, under the leadership of Prime Minister Nehru, India's government was the first in the world to adopt a national policy of limiting population growth by reducing birth rates. At that time, the population of India—in the post-Partition territory—was less than one half of what it is today.

Since India's historic 1952 decision, awareness has spread that rapid population growth impairs development performance in poorer countries. Everywhere in that so-called Third World there is striving to improve the well-being of the people, to achieve nations in which all are healthy, educated, and prosperous. But high rates of population growth, more than absolute numbers of people, are dragging down the chances for achievement. Now, among the less-developed countries, 70 governments support or endorse family-planning programs: 22 in Asia, 21 in Latin America, and 27 in Africa. More than 90% of all the people in those regions live under governments that support or endorse family planning or other programs to provide contraceptive services and information. Resources devoted to contraceptive service programs have risen, too. The global figure now approximates $2 billion annually of which about $500 million is contributed by donors.

With this accumulated experience and with much contemporary work in progress, what can be said about population-control strategies and experience? Human reproduction and its modern manifestation in rapid population growth is inevitably a subject beset by controversy. It touches on intimate family behavior at one extreme and on broad issues of national development and international relations at the other. Both are areas where moral and ideological differences cannot be avoided. In each case, however, there is a core of scientific knowledge representing the reality around which moral debate can center. Reproductive biology is such a core–a steadily growing understanding of the natural processes of reproduction and of the means by which fertility can be modified according to individual volition. Empirically established effects of rapid population growth on social, economic, and political systems constitute the other core of basic knowledge.

What are these effects on human society? National economies in developing countries, already struggling to find productive employment for their present work force, face decades more of unprecedented growth in numbers of new labor-force entrants. Take as a simple illustration the crucial young adult age group of 20 to 40 years, the group that plays such a large role in determining economic growth and political stability. Over the period 1960–1980, in the developing countries, this age group increased by 355 million; from 1980–2000 it will grow by 635 million. In India, the corresponding 1960–1980 increase was 64 million and that for 1980–2000 is likely to be 131 million. The developing countries' increase alone will be greater than the present combined work forces of Europe, the United States, and the Soviet Union. The greater part of this expansion will take place in urban areas, and the giant cities of the future (in India, Bombay and Calcutta are already among them and Delhi will be added by the year 2000) will face ever greater difficulties of sustenance and governance.

In human experience, growth of this magnitude is inconsistent with substantial progress in reducing the numbers or even the proportions of people in extreme poverty. Physical and intellectual impairment of children born into poverty will continue, reinforcing economic differentials within societies. Other less drastic but still important human costs are likely to include degradation of public amenities and physical environment and steadily increasing government regulation of social life.

At best, the costs of rapid population growth are opportunities foregone. Material advance is slower than it could be; improvements in the quality of life fall short of what would be possible without the pressure of relentlessly expanding numbers.

Although most population growth effects are felt at the national level or below, there are also international consequences. Narrowing the gap between rich and poor nations is more difficult. Progress toward an integrated world economy is slowed or even reversed.

There are those who dispute the nature or seriousness of these consequences–who see a world of smoothly clearing markets and sovereign consumers only waiting to be released from government interference. To them, population growth happens because it is what people want. But that is not the real world, or not much of it. In the real world, people can rarely achieve what they want simply through individual action. When they try, nevertheless, to do so, the aggregate outcome may be bad for all. Population growth is a prime illustration of this simple truth.

Where, then, do we stand as a world community in dealing with this problem? In 1974, when government representatives met at the World Population Conference in Bucharest, it was not yet clear that the world had passed through its peak period of population growth. Evidence of fertility decline was scant; the cause for alarm about global demographic prospects was very clear. In retrospect, the polarization at Bucharest between those seeing the solution to the population problem in economic development and those favoring more narrowly focused family-planning policies can be seen as a side issue. Increasingly, governments have moved on both fronts.

By 1984, when governments met again at the International Conference on Population in Mexico City, and still today, the signs of significant progress in reducing fertility are widespread. Two decades ago women experiencing then-prevailing average world levels of fertility had five children during their reproductive span; today that average is 3.5. In the developing countries the fertility decline has been from six children to about four, a drop of one-third. Since declining fertility has been partially offset by admirable progress in reducing mortality, reduction in overall population growth rates has lagged behind, but the downward trend in population growth, too, is now unmistakable.

The chief mechanism of fertility decline has been a dramatic rise in contraceptive practice. During the past two decades, in the developing countries, the proportion of married women of reproductive age estimated to be using contraception has increased fourfold. Nearly 40% now do so. Credit for this achievement lies both with progress in social and economic development—leading couples to lower their family size goals—and with vigorous efforts by governments and nongovernmental organizations to distribute modern means of contraception and to promote their use.

Altogether, there is reason for some satisfaction on the part of the world community. The first United Nations projection of world population growth over the rest of this century was made in 1963. It forecast a substantial fall in fertility over those four decades. If that projection is taken as a target trajectory, half of it is now in the past and we are on course. The most recent UN projections (1983) have found little reason to change the population growth path predicted 20 years earlier or its implied pace of fertility decline. For developing countries, the most recent projections foresee fertility in the year 2000 averaging 3.0 children per woman. By 2025 this number is projected to have dropped further to 2.4 children, approaching replacement-level fertility.

Despite accomplishments to date, however, there is no cause for complacency. The population growth of the last several decades and the future growth anticipated in current UN projections—a world total of 6 billion in the year 2000, 8 billion in 2025, perhaps 11 billion in 2100—pose enormous problems of economic and social adjustment. These problems will be grave even if population growth can be held to the projected levels.

A disproportionate part of the explanation for the overall progress thus far in fertility reduction in the developing countries is the dramatic demographic change that has taken place in China, essentially in the last 15 years. Many regions of the world have yet to experience any significant fertility decline. This is true of virtually all of sub-Saharan Africa; much of West and South Asia; and several countries in Latin America. The same would have been said about several large Indian states until just a few years ago. The combined population size of countries in which fertility decline has not really begun is about one billion – about one third of the total Third World population.

Is this present pattern of increasingly uneven incidence of fertility decline simply a lag phenomenon that will remedy itself as more and more governments come to view population issues seriously and take appropriate action? I do not believe so. Some of these high-fertility countries have a long-standing government concern about rapid population growth and commitment to providing family-planning services, yet with little effect. In others, including many of the poorest countries, continued rapid population growth has yet to be recognized as an impediment to development. In the long run, as demographic pressures continue to build, effective government responses *are* likely to be generated. But much may have been lost in the delay, including the gradualism and voluntarism of earlier and more orderly action.

What, then, can be done to initiate fertility decline in the regions yet to experience it or where the declines to date have been inadequate? In attempting to answer this question I have sifted through the experience of programs in many countries by looking to the cases, historical and contemporary, where demographic transition has taken place or is now proceeding. The lessons that I draw from this experience are not perhaps the obvious ones, but taken together I think they point to several needed strategies for population policy and programs in the world of today and tomorrow.

The demographic history of the developed countries suggests that socioeconomic development brings with it economic pressures and institutional arrangements that undermine rationales for high fertility and that steadily reduce desired family size. Close examination of the conditions prevailing at the onset of fertility decline, however, fails to show distinct threshold levels of development variables. The level of average income, rate of child survival, extent of schooling, and degree of economic and social equality and inequality by class and gender are all important for fertility behavior, but their importance is not uniform across societies. Exceptions to general knowledge on the determinants of fertility are too easily found – instances where fertility fell despite "unfavorable" socioeconomic conditions or where it failed to fall when the socioeconomic conditions indicated that it should have.

The problem is not that fertility behavior is mysterious or inaccessible to rational analysis. It is that its determinants need to be understood in specific settings: social, economic, institutional, and cultural. The course of fertility over time needs to be related to changes in those settings.

Nevertheless, there are transferable insights that can be learned from the course of developing countries that have successfully lowered fertility levels. A number of such insights are available from the experience of contemporary societies. Certain conditions are more conducive than others to fertility decline. Recent analyses done in India and elsewhere suggest that conditions conducive to fertility decline include a small number of development indicators: high female literacy, low infant mortality, and high contraceptive use. High female literacy and low infant and child mortality are outcomes of public policies supporting social development; and high contraceptive use is fostered by public policies supporting effective family planning programs.

Education is one of the most important elements of social change, and female education is one of the most robust predictors of change in reproductive behavior. Better educated couples do have smaller families. This happens because improvements in female education enhances their status within and outside their families and increases their exposure to the information and ideas disseminated through printed material. In addition, staying longer at school increases their age at marriage. Thus, emphasis on education of females at least up through primary levels is an important aspect of population-related policies in a number of societies – in addition, of course, to its inherent value for society.

Decreasing infant and child mortality can create demand for family planning services even before desired family size has changed. For example, surveys conducted in the 1970s have shown that couples in India want three to four children; they mean three to four living children rather than three to four live births. Under high mortality conditions, say, in the most populous state of Uttar Pradesh with an infant mortality rate of close to 170, five to six births may be required to produce three to four living children. Even before lowering desired family size, a decrease in infant and child mortality would eliminate the need for five to six live births. A time lag between the change in mortality and fertility behavior is inevitable. Nevertheless, there is no example in contemporary societies where fertility decline has preceded declines in infant and child mortality. Thus, among high mortality societies, an emphasis on infant and child survival should be an important element of population-related policies, as well as a goal to be pursued for its own sake.

About *effective family planning programs* three important elements need to be considered. These are program decentralization, user perspective, and means of fertility regulation.

Program Decentralization

In a situation often characterized by deficient demand for fertility control, centralized service delivery is usually unable to effect change; it is like pushing on a string. One remedy for this problem is to bring program decision-making into direct contact with those whose needs are ultimately to be served or whose preferences in matters of reproduction it is hoped to influence.

This is not a policy direction with a particular ideological bent. It is a lesson that can be drawn from the historical experience of much of Western Europe and of Japan, and from the recent record of China and Indonesia. In these radically different settings, selective devolution of authority and accountability to the community level is perhaps the only common factor.

We live in a pluralistic world and one in which government intrusiveness in people's lives can be as great an impediment to progress as the absence of collective organization. I am not advocating any standardized program recipe in the population case, let alone one that calls for closer government supervision. I am suggesting, however, the need in each setting to explore how best to take account of population policy objectives in the setup of community organization and government administration. Options for decentralization are too often overlooked.

Related to the often excessive centralization of program effort, posing a serious impediment to successful policy implementation, it is too often the case that contraceptive service programs seem to have been designed with more attention to the convenience of the providers than to that of the users of the service. This problem leads me to my second suggestion about delivering contraceptive services.

User Perspective

Designing and operating contraceptive service programs is no easy matter—even after decentralization is accomplished. In large countries, huge numbers of workers are to be recruited, trained, deployed, and supervised; super-shiploads of supplies are to be received or produced, processed, stored, transported, and accounted for. It is easy to understand why the programs often seem to have solution of management concerns uppermost among their objectives. But understandable or not, contraceptive service programs that fail to pay adequate attention to the perspectives of their users, that is, their customers, too often fail to achieve what is most wanted: high rates of contraceptive acceptance followed by high rates of contraceptive prevalence. This failure is not only demographically costly; it causes immediate economic waste as well.

The thinking about contraceptive service programs in governments and in professional population organizations is generally dominated by the perspectives and attitudes of managers, scientists, and population professionals. But, as we all know, achievement of sustained use of contraception—the objective of both the user and the program manager—depends on the perspectives and responses of individual users.

It is likely to be challenging to implement even a strongly held conviction about the need for greater attention to user perspectives. But in my opinion, efforts toward that end are vital. The five elements or principles to be worked on are these:

1. availability, choice, and self-selection;
2. responsiveness;
3. opportunities for learning;
4. integration in the life-experiences of users; and
5. women as decision-makers at all levels.

I will explain briefly what I mean about each of these.

Availability, Choice, and Self-Selection. The widest possible range of contraceptives needs to be made equally available and accessible, without excessive cost to program or user, and keeping in mind that costs as perceived by clients are likely to include a wide range of economic, social, cultural, and geographic factors such as travel-time, inconvenience, physical discomfort, indignity, and indiscretion. Among other reasons, such availability and choice would make switching from one contraceptive to a different one easier.

Responsiveness. In the last analysis, program clients are the judges. If the service provided fails to meet their perceived needs and wants, fails to address their perceived problems, they are not likely to continue using it or to motivate nonusers to begin. Increased program responsiveness will translate into greater acceptance and sustained contraceptive use. To achieve it, there must be effective mechanisms for feedback and for program modification.

Opportunities for Learning. Contraceptive users communicate among themselves. Program elements need to be provided to raise the level and quality of relevant information within the community. Opportunities for sustained contraceptive use will be improved as people know more about their bodies and the available contraceptives. There are obvious connections between learning opportunities, self-selection, and responsiveness. The network of communicators is enlarged and strengthened if it includes more than the points of views of demographically defined target groups. Feedback to programmers from such a learning group of clients is likely to be more useful than it would otherwise be.

Integration Into the Life Experiences of Users. Women are the vast majority of the individual users of contraceptives. Only if the actual life experiences of the affected women in the particular program are understood and taken into account is it likely that the program can adequately serve their perceived wants and needs and address their perceived problems. Contraceptive service programs can be integrated into broader aspects of women's lives. For example, the Working Women's Forum in Madras helps 13,000 poor, urban women to secure credit for their small businesses; it also promotes family planning among them and secures other desired services for them as well.

Women as Decision-Makers. Given the existing state of contraceptive development, women are likely to remain for some time the principal users of contraceptive service programs. Therefore, many more women in decision-making roles at every level are necessary before programs can attain optimal user perspective. A program cannot possibly comprehend women's views about availability, choice, self-selection, program responsiveness, learning opportunities, or integration into women's life experiences without women in vital decision-making roles, at every level. It is challenging to devise ways and means to select, recruit, train, support, and supervise women in these jobs because it is difficult to find enough women with required backgrounds. But if decision-makers continue to wait until enough fully qualified women turn up, not enough progress is likely to be made. Much more substantial affirmative action is needed.

Means of Fertility Regulation

Much of the fertility decline that has occurred in Europe, Japan, and the United States took place before the advent of modern contraception. Before the industrial revolution in the West and in Japan, fertility was adjusted to collective interest mainly through the regulation of marriage. A substantial minority of the population was, in effect, excluded from having children. Infanticide, too, was a widespread practice. With the greater geographic mobility and expanding labor demands of industrial society, marriage controls tended to break down, to be replaced by traditional methods of contraception within marriage. These methods at the individual level were often unreliable, but in aggregate were quite effective. Fertility in both the United States and France, for example, had fallen to an average of 2.3 children per woman by around 1930–decades before the appearance of the pill or the intrauterine device.

Today, even though traditional contraceptive methods continue to play a significant role in securing Europe's current low levels of fertility, modern contraception predominates on a world-wide basis. And it is almost a truism to note that limitations of each of the available modern contraceptives require that research and development on additional methods be strengthened and accelerated. There are several promising new methods near the final stages of development; other leads that require more research–such as the contraceptive vaccine–are extraordinarily exciting. All of these are discussed in this volume. The Population Council and its distinguished colleagues in the International Committee for Contraception Research, including Dr. G. Pran Talwar, are devoted to carrying this work forward. We face funding constraints that are increasingly worrisome, because governments and private foundations allocate too little for this vital work. Nevertheless, we will continue with our greatest energy to do all that can be done with available funds and to persuade donors to allocate larger amounts.

CHAPTER 3

WHO Special Programme in Human Reproduction and Contraceptive Needs of Developing Countries

M.F. Fathalla and J. Barzelatto

The Special Programme of Research, Development, and Research Training in Human Reproduction is the main instrument of the World Health Organization for promoting the international research and development effort required by developing countries for the successful implementation of their health strategies in the area of fertility regulation.

The Special Programme, established in 1972, addresses three main areas of research needs in developing countries: (a) new and improved methods of fertility regulation, (b) improved performance of currently available methods, and (c) promotion of national self-reliance to undertake research in family planning.

Development of New and Improved Methods of Fertility Regulation

There is an unmistakable need for new methods of fertility regulation:

- methods that may be accepted by those couples who do not find present methods acceptable or who cannot continue their use, taking into consideration the diversity and complexity of the sociocultural background in developing countries;
- methods that are relatively more safe, taking into consideration the health profile of developing country populations and the ability of health services to help those who may have problems;
- methods that can have a more extended availability, taking into consideration the status of the health service infrastructure in developing countries.

It has been the experience in several family-planning programs that whenever a new method is introduced, a new layer of acceptors is added.

Strategy

The WHO Special Programme recognizes the various potential approaches for fertility regulation. Three main criteria guide the Programme in selecting the approaches to support (a) the need and potential demand for the new method in developing countries, (b) feasibility, including scientific assessment, as well as cost and time estimates, and (c) the extent of involvement of other research and development agencies and industry, and the extent to which an input from the Programme can push the development process forward.

At present the following task forces of the WHO Special Programme are actively involved in the development of new and improved methods of fertility regulation:

Task Force on Long-Acting Systemic Agents of Fertility Regulation
Task Force on Postovulatory Methods of Fertility Regulation
Task Force on Vaccines for Fertility Regulation
Task Force on Methods for the Regulation of Male Fertility
Task Force on Methods for the National Regulation of Fertility
Task Force on Plants for Fertility Regulation

New Methods Under Development

A number of methods of fertility regulation in the Programme are now at an advanced stage of development:

Once-a-Month Injectables. Following extensive pharmacokinetic studies and a Phase III clinical trial, two types of new monthly injectable contraceptives containing both estrogen and progestogen (HRP.102 and Cycloprovera) are being evaluated. Both provide significantly improved bleeding patterns compared with those induced by progestogen-only formulations. Large-scale clinical testing is expected to be completed early in 1987. The Programme is the only agency that is engaged in the development of once-a-month injectable formulations. This method was developed with the collaboration of scientists from four developed and seven developing countries.

Vaginal Rings. A vaginal delivery system that provides constant blood levels of levonorgestrel (LNG) by releasing 20 μg/24 h of the drug during a period of at least 3 months has been developed. Its main advantages are easy self-administration and self-removal, and fertility regulation with a minimal steroid load, the lowest of all existing steroidal contraceptives. There is also no need to remove the ring at any time during its active life. Prototype equipment has been designed and constructed for large-scale manufacturing of the devices, and a Product Licence Application was submitted in June 1985. Phase III trials are run with the collaboration of scientists from eight developed and 11 developing countries.

Prostaglandins. In international comparative trials in 25 centers, involving more than 25,000 volunteers, the Programme has established which prostaglandin derivatives, dose regimens, formulations, and routes of administration are the most suitable in different clinical situations. The Programme has also generated a large body of clinical pharmacological data for the registration of several products and has negotiated agreements with three manufacturers for the supply of their products for distribution to the public sector at concessionary prices.

Antiprogestins. In collaboration with a pharmaceutical company, the Programme has developed a sequential method for the medical termination of early pregnancy, consisting of the administration of a powerful antiprogestin, followed by a prostaglandin analog in such small quantities that the occurrence of gastrointestinal side effects is minimized but a high success rate is still obtained.

Other new methods in the pipeline include, among others:

Synthesized New Long-Acting Progestins. These provide contraceptive efficacy for 3 to 4 months in doses seven to ten times less than that employed in the currently available injectable preparations. Work has been done in collaboration with the US National Institutes of Health (NIH).

Biodegradable Implant (CAPRONOR). This implant, developed in collaboration with NIH, is a single device that releases LNG for about 12 to 18 months and is retrievable during its active life, after which it biodegrades in a few years.

Progesterone-Releasing Vaginal Ring. This device releases the natural hormone progesterone at a rate of about 5 to 10 mg/24 h. The ring is intended for postpartum use and is being developed jointly with the International Committee for Contraceptive Research (ICCR) of the Population Council.

Antifertility Vaccine. The Programme has always recognized that a safe, effective, and reversible birth control vaccine would be a greatly needed addition to the currently available fertility-regulating methods and an attractive proposition to family-planning programs, a view that is shared by the Indian government, which puts birth control vaccines as a top priority area for national research.

From the early studies of the task force, four promising leads emerged: β-subunit of human chorionic gonadotropin (βHCG), nonhormonal placental protein PP5, lactic dehydrogenase isoenzyme LDH-C4, and the zona pellucida. In view of financial constraints, the task force concentrated on the most promising one, the βHCG. The Programme has now developed an almost totally synthetic birth control vaccine consisting of a synthetic peptide corresponding to the C-terminal 109-145 amino acid region of the β-subunit of HCG conjugated to diphtheria toxoid as the carrier and administered with muramyl dipeptide adjuvant in a squalene-arlacel vehicle. After a full decade of in-depth study on its efficacy and safety in monkeys, this new vaccine has been brought now to the stage of clinical testing, after permission from both the Australian Drug Regulatory Authority and the Food and Drug Administration of the United States.

The Programme has supported the early animal experimentation with vaccine development in the All India Institute of Medical Sciences. It is most gratifying to see the progress that Indian scientists have made in this field. Dr. G. Talwar is also, since 1985, a member of the Programme's Task Force on Vaccines for Fertility Regulation.

Improving the Performance of Currently Available Methods for Fertility Regulation

Strategy

As the Programme works to promote the development of new and improved methods of fertility regulation, it recognizes that the development process is a very long one, that the need for fertility regulation in developing countries is of great urgency, and that time delays carry heavy penalties for health and development in terms of unchecked population growth. Thus best use must be made of what is currently available.

The strategy of the Programme for improving the performance of current methods is based on recognizing three main targets for research:

1. The method—by generating more information, through epidemiological research, on safety and efficacy in the setup of developing countries;
2. The user—by generating more information, through social research, on the determinants of fertility regulation;
3. The health delivery service—by generating more information through health systems research to improve the availability, accessibility, and efficiency of services.

The following two task forces of the WHO Special Programme are actively involved: Task Force on Safety and Efficacy of Fertility Regulating Methods and Task Force on Behavioural and Social Determinants of Fertility Regulation.

Health systems research is presently done in collaboration with WHO regional offices.

Safety and Efficacy of Methods of Fertility Regulation

Rationale

The great majority of published safety studies have been conducted in developed countries, and corresponding information from developing countries is scanty. Since people from developing countries may differ markedly, with respect to genetic, environmental, nutritional, and other risk factors for disease, and since risk-benefit ratios are also very different, it is of questionable validity to extrapolate conclusions about contraceptive safety from developed to developing countries. Moreover, developing countries may dramatically differ from each other with respect to risk factors for disease. Hence, there are compelling arguments to obtain comparative data for every important method from a variety of developing and developed countries.

As regards efficacy, several competitive products are being offered with different claims and prices to developing country governments. International comparative trials organized by the Programme help to provide answers to such questions.

WHO is in a unique position to fulfill this task of generating and providing information on the safety and efficacy of methods of fertility regulation in view of its mandate in the health field, its impartiality, its capability to undertake global and inter-regional studies, plus its traditional strengths in the area of epidemiological studies.

The Programme's Work

The Programme's work in this area has been extensive and intensive. Reference can be made to the Programme's annual reports and in various publications.

Examples of relevant projects include assessing the effectiveness and side effects of different formulations of currently available oral contraceptives (OCs); assessing the effects of various OCs on lactation; assessing interactions between OC use and diseases common in developing countries; assessing the risk of cardiovascular hazards; assessing the relationship between long-term use of OCs and the development of neoplasia; comparative studies of various intrauterine devices (IUDs); insertion of IUDs after pregnancy termination; assessing the risk of ectopic pregnancy with IUD use; assessing the risk of pelvic inflammatory disease (PID) with IUD use; assessment of the cervical mucus (ovulation) method of natural family planning in a large-scale five-country study; study of the relationship of depot medroxyprogesterone acetate (DMPA) use and the development of neoplasia; the relationship between injectable contraceptives (DMPA and norethisterone enanthate [NET-EN]) and the risk of PID; studies on the return of fertility after long-term use of injectable contraceptives; and on the long-term development of children exposed to DMPA in utero or to DMPA via breastmilk.

Indian Projects

The following are some examples of significant studies carried out by Indian scientists in collaboration with the Programme in this area:

Comparative study of the biochemical effects of combined estrogen/progestogen OCs containing less than 50 μg estrogen
Vitamin supplements to women using OCs
Interactions between OCs and malaria
Comparative study of a number of IUDs inserted immediately following therapeutic termination of pregnancy
Case-control study on the relative risk of ectopic pregnancy and pelvic inflammatory disease associated with IUD use
Phase III comparative evaluation of the contraceptive effectiveness of NET-EN given every two or three months
Return of fertility following discontinuation of injectable contraceptives
Multicentered clinical evaluation of the use-effectiveness of the ovulation method
Objective assessment of menstrual blood loss following tubal occulation
Psychosomatic sequelae of female sterilization
Comparative evaluation of surgical sterilization procedures

Comparison of flexible versus rigid cannulas in the early termination of pregnancy
Studies on sterilization and abortion

Psychosocial Research

The Programme, in its efforts to improve the effective utilization of presently available methods of fertility regulation, promotes research aimed at a better understanding of the user's perspective. User preferences for methods have been investigated in several countries.

In 1985 the Programme redirected its social science activities to address a major concern for many national programs, the poor continuation rates of modern methods and the incomplete understanding of the role of contraception in human reproduction behavior. In this context, the role of communities in facilitating or impeding access to methods, the changing status of women, and the costs and benefits of contraception emerged as main research priorities.

Indian Projects

The following are examples of social science research carried out by Indian scientists in collaboration with the Programme:

Psychosocial factors affecting continuation/discontinuation of IUDs and pills
Acceptability of male fertility regulating methods in India (field study)
Patterns and perceptions of menstrual bleeding
Acceptability of methods in a free-choice situation.

Health Systems Research

Health systems research aims at improving the effective utilization of presently available methods of fertility regulation.

The use of nonphysician personnel in the delivery of family-planning services is one example of relevant research with policy and service implications. Large-scale collaborative studies have been conducted on family formation patterns and health. Large-scale introductory field trials of injectable contraceptives involve more than 25,000 subjects.

Indian Projects

The following are examples of health systems research carried out by Indian scientists in collaboration with the Programme.

Participation of rural practitioners in the delivery of services
Introduction of NET-EN in the national family planning program
Use of services for the termination of pregnancies
Role of health delivery services on acceptance of family planning

The Promotion of National Self-Reliance for Research in Family Planning

Strategy

A major objective of the Programme is to strengthen the capabilities in developing countries to undertake research relevant to their needs. While it is generally accepted that there is a major shortage in manpower and facilities required for research in human reproduction, the need is by far the greatest in developing countries. The strategy of the Programme has been to strengthen human resources, to strengthen material resources, and, to promote participation of developing countries in the global research effort. The responsible body in the Programme for this area is the Committee on Resources for Research.

Strengthening of Human Resources

During the period 1972–1985, the Programme awarded research training grants to 842 scientists from 59 countries and visiting scientist grants to 152 senior scientists from 36 countries. More than 900 of these grants were awarded to scientists from developing countries (102 to Indian scientists) (see Table 3.1). In addition, "re-entry grants" have been awarded to 79 scientists from 31 countries. The total value of these grants is approximately $13 million.

In addition, the Programme held 115 training courses in collaboration with Ministries of Health in 30 countries. The Programme has also organized 34 seminars, 45 workshops, and 31 symposia, altogether 110 major scientific events in 38 countries. The proceedings of these events are frequently published. Particularly important symposia are also published in the language of the host country.

Strengthening of Material Resources

The grand total of equipment and supplies ordered through WHO on behalf of developing country centers funded by the Programme since its inception is

Table 3.1. Research training and visiting scientists grants (1972–1985).

WHO Region	Countries	Research trainees	Visiting scientists
African	10	106	11
Latin American	14	136	25
Eastern Mediterranean	6	80	10
European	16	61	33
Southeast Asian	6	186	39
Western Pacific	9	273	34
Total	61	842	152

slightly above $25 million. Furthermore, since 1976 the Programme distributed reagents to the centers and institutions strengthened for a total cost of $3 million. Hence, since its inception, the Programme has spent $28 million of its total income of $185 milion (i.e., 15%) to provide equipment, reagents, and consumable supplies to institutions predominantly in developing countries. Small supplies have been received by 28 Indian institutions.

As part of its institutional support, a program has been established for the provision of matched reagents, together with an associated external quality assessment scheme. Reagents are currently distributed to some 80 laboratories in 40 countries, and a further 90 centers participate in the external quality assessment scheme.

National reagent programs are also now being established in a number of countries, including Cuba, Mexico, India, the People's Republic of China, and the USSR. The materials produced by these programs will be standardized against WHO calibration materials.

The Programme is also in the process of establishing a bank of reagents for immunoassay hormones.

Participation in the Global Research Effort

Task Force Steering Committees and Scientists

From its inception, the Programme undertook efforts to organize multicentered research teams for collaborative research and development. This concept

Table 3.2. Task-force scientists 1986.

	No. of scientists		No. of their countries		
	Total number	From developing countries	Total countries	Developing countries	Women scientists
Long-acting systemic agents	91	47	31	17	28
Vaccines	28	8	12	6	2
Postovulatory methods	42	24	18	9	10
Plants	60	47	16	11	9
Methods for the regulation of male fertility	103	45	31	16	11
Methods for the natural regulation of fertility	11	4	7	4	4
Safety and efficacy of fertility-regulating methods	102	60	32	21	36
Behavioral and social determinants of fertility regulation	60	58	27	23	22

Table 3.3. Designated WHO collaborating centers for research in human reproduction.

WHO Region	City (year of designation)
African	Ibadan (1972)
	Lusaka (1975)
	Nairobi (1984)
American	Cali (1980)
	Havana (1976)
	Mexico City (1972)
	Salvador, Brazil (1972)
	Santiago, Chile (1973)
Eastern Mediterranean	Alexandria (1974)
European	Leningrad (1978)
	London (1974)
	Ljubljana (1972)
	Moscow (1973)
	Stockholm (1970)
	Szeged (1972)
Southeast Asian	Bangkok, Chulalongkorn (1977)
	Bangkok, Siriraj (1972)
	Bombay (1972)
	Chandigarn (1972)
	Karachi (1977)
	New Delhi (1973)
Western Pacific	Manila (1972)
	Seoula (1973)
	Shanghai (1983)
	Singapore (1972)

was developed into a unique multinational and multidisciplinary collaborating mechanism, the task forces and their steering committee.

Developing-country scientists play a major role in task force activities (Table 3.2). In 1985, 57% of steering committee members came from developing countries. The Programme is privileged to have five Indian scientists presently as members of Steering Committees for Research and Development. Indian scientists have carried out 113 task force projects in the past 14 years.

Collaborating Centers for Research (CCR)

The Programme has developed and strengthened a global network of centers to conduct coordinated clinical trials using common protocols and carefully standardized diagnostic techniques and laboratory methods. The number of CCRs is presently 25, 19 (75%) of which are in developing countries (Table 3.3). Three of the centers are in India: The Institute of Research in Reproduction in Bombay, the Department of Obstetrics and Gynaecology of the Postgraduate Institute of Medical Education and Research in Chandigarh, and the All India Institute of Medical Sciences in New Delhi.

Apart from the CCRs, more than 120 nondesignated centers have participated in the Programme's multicentered clinical trials. The Programme has so far conducted more than 180 multicentered clinical trials involving approximately 185,000 subjects in 150 centers in 53 countries.

Publications

The work of the Programme in promoting the participation of developing-country scientists in the global research effort can be seen in the number of papers authored by scientists from developing countries. There are more than 6,000 publications describing the results of various studies supported by the Programme. Developing-country scientists contributed about one third of the papers in the period 1972–1979 and one-half of the papers in the years 1984–1985.

Collaboration with ICMR

The collaboration of the Programme with the Indian Council of Medical Research has been outstanding. Collaborative projects are carried out; combined task force meetings are held; scientific meetings are jointly organized; consultations, formal and informal, are always ongoing.

The Future

> The United Nations estimates that the world population will continue to grow for another 110 years . . . it is projected to stabilize at 10.5 billion by the end of the next century. NINETY-FIVE PERCENT of this future growth of the global population will occur in the developing countries of the world . . . The continued momentum of population growth in developing countries will lead to a doubling by the year 2025 in the number of women in the reproductive ages, currently estimated at 873 million.
>
> –R.M. Salas
> Mexico City, 1984

Contraceptive needs of developing countries will continue to increase dramatically. The challenge is great to governments, international organizations, the scientific community, and to all concerned about the future and welfare of mankind. The WHO Special Programme of Research, Development and Research Training in Human Reproduction will continue to collaborate with all concerned, to serve the health needs of populations in developing countries in the area of fertility regulation.

Part II
Recently Developed Contraceptive Methods at Phase II/III/IV

CHAPTER 4

Steroid Implants for Long-Term Contraception

Sheldon J. Segal

Overview

Subdermal implants of Silastic® containing levonorgestrel (LNG) can release the synthetic steroid at a relatively steady rate for five years or longer. The objective of our program over the past 20 years has been to utilize this principle to develop a new form of long acting contraception. In the initial phases of the work, the release rate from Silastic of various synthetic progestins was studied, and the resultant blood levels determined. While some progestins pass through Silastic too rapidly to be practical, approximately 90% of an initial load of 35 mg of LNG within a Silastic capsule remains after one year of subdermal placement in women. About 50% remains after five years. Blood levels above 0.2 ng/ml LNG are maintained. The original version of the contraceptive implant, named Norplant®, utilizes six capsules, each 3 cm in length; daily release of LNG is approximately 40 μg. A second version, made of homogenized Silastic-steroid solid rods, achieves for three years similar performance characteristics with two implants of 4 cm each. With both versions, contraceptive effectiveness exceeds that of other reversible contraceptives and approaches the high level of pregnancy protection expected with surgical sterilization. A pregnancy rate of 0.3 is recorded for women using Norplant for one year. This rate is maintained for years two and three. Contraceptive activity of Norplant is provided through ovulation inhibition in about 50% of cycles and reduced sperm penetration through the progestin-thickened cervical mucus. Bleeding disturbances are the main reason for discontinuation of use. Approximately 10% of users terminate use in year one. The aggregate discontinuation rate for bleeding after three years is 16%. Following Norplant use, women experience a rapid restoration of fertility. At the end of one year, 80% of women who discontinue in order to have a planned pregnancy are successful. Subdermal contraceptive implants provide a new, effective and safe form of long-acting, reversible contraception.

Introduction

Hormonal contraception began in the 1950s as an ovulation-inhibiting procedure based on the oral administration of an estrogen-progestin combination used three weeks out of every month (1). Throughout the 1960s and early 1970s a series of different progestins appeared, usually in combination with the same original estrogen, ethinyl estradiol or its 3-methyl ether. Through the years, the dose of each constituent was reduced drastically so that it became possible eventually for women to achieve effective contraception using combination pills with only 5% of the quantity of steroid contained in the original oral contraceptives (2). One modification in the dosage schedule, the so-called sequential pills, involved the administration of the two hormones not in combination, but in sequence, as the name implies (3). In the mid-1960s Martinez-Manautou and his colleagues in Mexico demonstrated that the estrogenic component could be dispensed with entirely and contraception can be achieved with the daily, oral administration of a progestin only (4). This led to the so-called low-dose "minipill" which, while feasible, tended to have a higher rate of contraceptive failure than combination pills. The principle of progestin-only contraception opened the possibility of a number of new delivery systems based on the slow and continuous release of progestins from the synthetic, rubber-like polymer, Silastic (5).

Evolution of Subdermal Implants

The development of a long-acting form of contraception that takes advantage of the tissue compatability of Silastic began in our labortories some 20 years ago with the demonstration that many steroid substances can be released from Silastic at a slow and steady rate (6). Lipophilic compounds, like steroids, are soluble in, and pass through, a solid matrix of polydimethylsiloxane. Thus, a supply of crystalline steroid within a Silastic capsule will pass through the polymer wall at a controlled rate. The steroid can also be released from a solid rod composed of a mixture of liquid polymer and crystalline steroid, which is homogenized and then prompted to solidify by adding an appropriate catalyst. Silastic was also deemed suitable as a basis for a contraceptive modality because of its record of medical safety. Silastic tubing has been used for subdermal hydrocephalus drains, and a medical-grade preparation of the polymer is used in various reconstructive or cosmetic surgical procedures (7).

The laboratory work began in 1965. By 1968 the first clinical trials had begun (8,9,10), and three years later, a series of studies were undertaken to choose the best progestin candidate for use in a long-acting, low-dose, continuous-administration contraceptive. Levonorgestrel was selected as the compound of choice for the first subdermal contraceptive implant system because of its high level of effectiveness in preventing pregnancy (11). Also, it is released at the slow rate of about 10% of the original load per annum (7). This means that it would take about five years for 50% of the original dose of levonorgestrel to be released, which portended a very long duration of contraceptive efficacy.

The subdermal contraceptive system Norplant has been developed as a hollow capsule 30 mm in length and 2.4 mm in diameter filled with about 36 to 40 mg

Table 4.1. Plasma levonorgestrel (LNG) levels with Norplant

Month of treatment	n	LNG, ng/ml	SD	p
1-12	27	0.35	0.13	
13-24	23	0.31	0.11	ns
25-36	30	0.34	0.11	ns
37-48	8	0.32	0.08	ns
49-60	42	0.29	0.11	.05

of crystalline levonorgestrel. The capsules are placed under the skin of the upper arm with a 10-gauge trocar, using a local anesthetic. The steroid is released at a slow, controlled rate of about 6 μg per capsule per day. A set of six implants, therefore, results in a release rate of about 36 to 40 μg/day, which is sufficient to achieve an antifertility effect (12). A second system, Norplant-2, has also been developed, consisting of a solid rod of homogenized 50% Silastic-50% steroid mixture covered with ultrathin Silastic tubing to hold the entire rod in place as steroid dissolution proceeds (13).

Efficacy and Safety

The effectiveness of Norplant has been evaluated in long-term studies that have progressed for well over five years. Table 4.1 shows that plasma levonorgestrel remains at a relatively constant level of 0.3 ng/ml over a five-year period, declining just slightly in the fifth year. Plasma levonorgestrel levels high enough to protect against pregnancy are maintained two to three years beyond the recommended five-year period of effectiveness (14). This may be potentially important should the user delay returning for a replacement set of implants.

In one five-year study, only 11 pregnancies occurred in 2380 woman-years of use, which translates into a pregnancy rate well under 1 pregnancy per 100 woman-years of use (Table 4.2). Life-table analysis of another five-year study

Table 4.2. Five-year study (cumulative) of Norplant efficacy.

Years of use	Pregnancies	Woman-years	Rate
1	3	894	0.3
2	4	1459	0.3
3	6	1878	0.3
4	10	2187	0.5
5	11	2380	0.5

Table 4.3. Norplant net cumulative continuation rates per 100 acceptors N=992.

	Month				
Rate or number	12	24	36	48	60
Pregnancy	0.3	0.5	0.7	0.7	0.7
Menstrual problems	8.6	12.9	14.4	15.2	16.6
Medical	4.7	8.2	10.2	12.0	12.8
Continuation	79.4	58.3	49.0	43.2	39.3
No. at risk in month	787	331	174	92	51
Woman-months of use	10,601	16,334	18,785	20,147	20,823

involving nearly 1000 women indicates that the level of protection against pregnancy year by year approaches the effectiveness of surgical contraception (Table 4.3). In this clinical trial, the 50% discontinuation-of-use rate was reached at about three years.

The main reason for discontinuation of Norplant is menstrual problems, most commonly frequent irregular bleeding (Table 4.4). About 5% of women discontinue the method for this reason during the first year of use, an additional 1.5% do so during the second year, and an additional 1% or less during the third year. Amenorrhea is rarely a cause for discontinuation (15).

Extensive safety studies carried out with Norplant indicate no unfavorable changes in liver function, lipid metabolism, or blood pressure; no increase in the risk of ectopic pregnancy; and no increase in total menstrual blood loss (16). In fact, Norplant tends to increase hemoglobin levels in women who had low levels before using the implants.

Since the implants can be removed at any time, this contraceptive method is voluntarily reversible, and there has been sufficient experience to ensure that fertility recurs promptly after discontinuation. According to one study of women who discontinued the use of either Norplant or an intrauterine device because

Table 4.4. Rate of discontinuation of Norplant because of menstrual disturbances.

	Termination rate/ 100 acceptors		
Menstrual disturbance	Year 1	Year 2	Year 3
Frequent, irregular bleeding	4.9	6.6	7.1
Heavy menstrual flow	0.5	1.4	1.9
Prolonged bleeding	2.8	4.1	4.8
Dysmenorrhea	0.0	0.0	0.0
Amenorrhea	0.7	0.9	0.9
Spotting	0.5	0.8	0.8
Other bleeding	0.0	0.2	0.5
Total	9.5	13.9	16.0

Table 4.5. Restoration of fertility after removal of Norplant implants.

Month	% of women who became pregnant after discontinuing contraception	
	Norplant system	TCu 200
1	21.6	11.3
3	43.1	33.0
6	67.3	47.8
12	76.6	73.9
No. of women	53	29

they wished to become pregnant, approximately 75% of women in both groups became pregnant within one year (Table 4.5).

Much of the data discussed here comes from studies of the 6-capsule Norplant system. However, clinical trials of the Norplant-2 system show essentially the same level of performance over a three-year period of observation.

Norplant has been approved for general use in Finland, Sweden, Thailand, Indonesia, Ecuador, Colombia, and the Dominican Republic. Experience with the method has been obtained in 31 countries including China, Brazil, Egypt, Nigeria, Kenya, and the United States. Results from field studies in several countries confirm the very high level of effectiveness that had been observed in the initial clinical trials (Table 4.6). The continuation rates are even higher than those reported in the clinical trials, ranging around 90% after the first year of use. This generally favorable experience has continued in years two, three, and even four and five in some of these field trials.

Conclusion

The high efficacy of Norplant makes this contraceptive method an important reversible alternative to sterilization. It affords long-acting protection, which means that one visit to the clinic can replace virtually thousands of days of pill

Table 4.6. First-year results of field studies of effectiveness and continuation of Norplant implants.

Country	Pregnancy rates	Continuation	n
Colombia	0.0	93.0	300
Ecuador	0.0	87.4	283
Egypt	0.8	89.6	250
Indonesia	0.5	95.3	813
Thailand	0.3	88.4	704

taking. Norplant can be used for birth spacing because the implants can be removed at any time and fertility restored without impairment. It is also a suitable option for immediate postabortal use and when intrauterine devices or estrogens are contraindicated. This route of administration should thus prove to be an important adjunct to existing modalities for effective and reversible contraception.

References

1. Pincus G, Rock J, Garcia CR, Rice-Wray E, Paniagua M, Rodriguez I, Pedras R: Fertility control with oral medication: *Amer J Obstet Gynecol* 1958;75:1333–1346.
2. Lachnit-Fixson U: Development and clinical evaluation of a new triphasic oral contraceptive. In Proceedings of 5th international meeting on fertility control, Genoa, Italy, March 1–3, 1984. ed. Sciarra JJ, Pescetto G, Martini L, and De Cecco, L.;353–362
3. Goldzieher JW, Livingstone NB, Martinez-Manautou J, Moses LE, Rice-Wray E: Use of sequential estrogen and progestin to inhibit fertility. *West J Surg Obstet Gynec* 1963;71:188.
4. Rudel HW, Martinez-Manautou J, Maqueo-Topeto M: The role of progestogens in the hormonal control of fertility. *Fertility and Sterility* 1965;16:158–169.
5. Dziuk PJ, Cook B: Passage of steroids through silicone rubber. *Endocrinology* 1966;78:208–211.
6. Segal SJ, Croxatto HB: Single administration of hormones for long-term control of reproductive function. Presentation at the XXIII Meeting of American Fertility Society, April 14–16, 1967, Washington D.C.
7. Segal SJ: Contraceptive subdermal implants in Mishell DR Jr (ed): Advances in Fertility Research. New York, Raven Press, 1982;1:117–127
8. Croxatto H, Diaz S, Vera R, Etchart M, Atria P: Fertility control in women with a progestin released in microquantities from subcutaneous capsules. *Am J Obstet Gynecol* 1969;105:1135–1139.
9. Coutinho EM, Mattos CER, Sant'Anna ARS, Adeodato Filho J, Silva MC, Tatum HJ: Long-term contraception by subcutaneous silastic capsules containing megestrol acetate. *Contraception* 1970;2:313–317.
10. Tejuja S: Use of subcutaneous silastic capsules for long-term steroid contraception. *Am J Obstet Gynecol* 1970;107:54–958.
11. Coutinho EM, Segal SJ: Contraception with long acting subdermal implants: I. An effective and acceptable modality in international clinical trials: ICCR. *Contraception* 1978;18:315–333.
12. Weiner E, Johansson EDB: Plasma level of d-Norgestrel, estradiol and progesterone during treatment with silastic implants containing d-Norgestrel. *Contraception* 1976;14:81–92.
13. Robertson DN: Implantable levonorgestrel rod systems: In vivo release rates and clinical effects in Zatuchni GI, Goldsmith A, Shelton JD, Sciarra JJ (eds): Long-Acting Contraceptive Delivery Systems. Philadelphia, Harper and Row, 1983, pp 133–144
14. Croxatto HB, Diaz S, Miranda P, Elamsson K, Johansson EDB: Plasma levels of levonorgestrel in women during long-term use of Norplant. *Contraception* 1980;22: 583–596.

15. Diaz S, Pavaz M, Miranda P, Herreros C, Johansson EDB, Croxatto HB: Bleeding patterns, outcome of pregnancies, and levonorgestrel plasma levels associated with method failure in Norplant implant users. *Contraception* 1986;33:347–356.
16. Croxatto HB, Diaz S, Robertson DN, Pavez M: Clinical chemistry in women treated with levonorgestrel implants (Norplant) or a TCu 200 IUD. *Contraception* 1983;27: 281–288.

CHAPTER 5

The Levonorgestrel-Releasing IUD

Tapani Luukkainen, Maija Haukkamaa,
Pekka Lähteenmäki, and Juhani Toivonen

The progestin-only contraceptive method, the minipill, has few general side effects, but there are problems with efficacy and bleeding control. These problems have affected the acceptance of the method, and the need for daily motivation leads to high discontinuation.

The progestin-only method can be improved by using parenteral administration and sustained steroid release with semipermeable membranes. In this way the release of progestin for many years from a single administration can be obtained. The constant release eliminates unnecessarily high or low blood concentrations of a contraceptive steroid. Sustained release of levonorgestrel from subdermal implants is used in Norplant® (giving reliable contraception for five years) and in the levonorgestrel-releasing intrauterine device (LNG-IUD). Experience with the LNG-IUD is summarized here.

Structure

For intrauterine release, levonorgestrel (LNG) is mixed with a silicone rubber, Silastic® (382 Medical Grade Elastomer, Dow Corning Corporation, Medical Products Division, Midland, Michigan, USA). This mixture, containing 50% steroid by weight, is used to make sleeves around the vertical stem of a carrier IUD, a plain Nova-T (Fig. 5.1). The sleeve is covered with Silastic® tubing. LNG-releasing devices were manufactured by Leiras Pharmaceuticals, Turku, Finland. Various models have been tested. Models releasing 50 to 75 μg/d suppressed ovulation. This high range of release could be maintained only for less than a year. Models releasing 10 to 12.5 μg/d did not show improved bleeding control and were considered to be less effective in long-term use. Models releasing 20 and 30 μg/d were tested for five years (9). Because no significant differences between these release rates were observed in clinical performance or in bleeding control, a 20 μg/d-releasing device was selected for further evaluation. These devices release about 60% of the original steroid load during the first five years of use (9). The in vitro release after 7 years of use has been 10 μg/d (1). The recommended active life of the device is five years, and the devices should

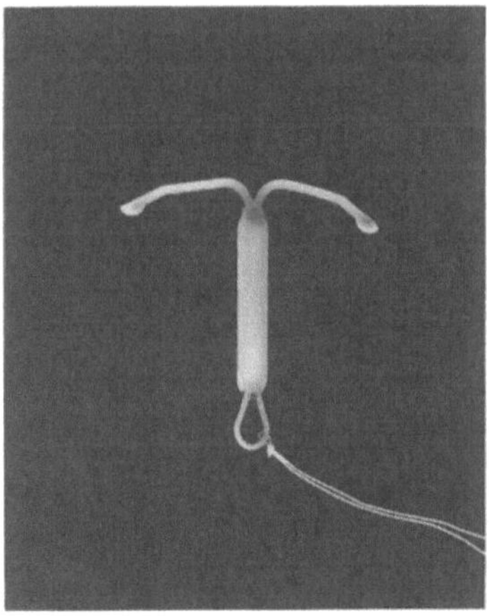

Fig. 5.1. The levonorgestrel-releasing IUD.

be removed during the sixth year of use; if desired, a new device can be inserted at the removal.

Plasma Concentrations of Levonorgestrel

The interindividual variations in the plasma concentrations of LNG are clear (Fig. 5.2). The individual plasma concentrations are fairly stable. The mean concentrations of LNG during seven years of use are shown in Figure 5.3. There is a significant difference between plasma concentrations associated with devices releasing 20 μg or 30 μg/d. The mean plasma concentrations decreased slightly during the first three years and were fairly stable from the fourth to the seventh year of use.

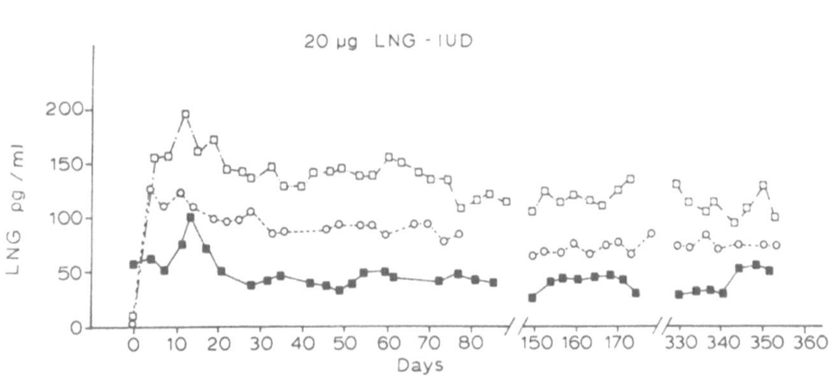

Fig. 5.2. Plasma levonorgestrel (LNG) levels in three women during the first year of use of a 20 μg/d-releasing LNG-IUD.

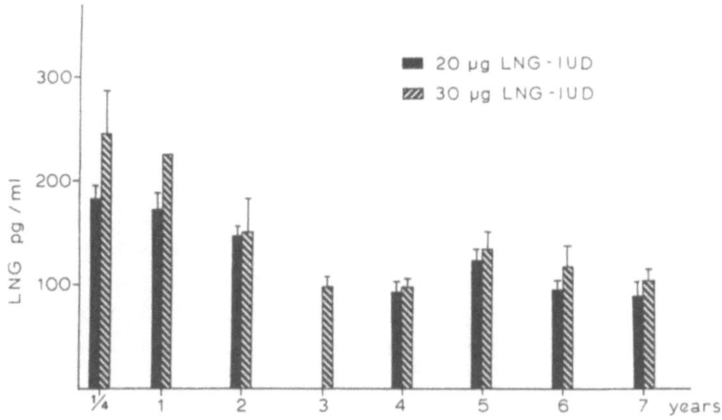

Fig. 5.3. Mean (± SEM) plasma levonorgestrel (LNG) concentrations over seven years of use of 20 and 30 μg/d-releasing LNG-IUDs.

Systemic Effects of LNG-IUDs

Randomized comparative trials between LNG-IUDs and copper-releasing IUDs have given information on the metabolic effects of LNG-IUDs in large numbers of long-term users of the device. These studies have shown that the weight gain of LNG-IUD and copper-IUD users is the same (2,14). In the same study, a slight decrease in both systolic and diastolic blood pressure was found after one year of use in users of the LNG-IUD and of the copper-releasing IUD. An analysis of body weight and blood pressure in more than a thousand users of an LNG-IUD did not reveal any significant differences in the mean change or in the pattern of changes between women using an LNG-IUD or a copper-releasing Nova-T (2). The effect of a LNG-IUD on lipoproteins was evaluated after two years (14) and five years of use (1), but no difference was found between users of an LNG-IUD or of a Nova-T. The serum liver enzyme levels were also the same in women using an LNG-IUD or a Nova-T over five years (1). Women using an LNG-IUD significantly more often reported an increase in headache, nervousness, and acne than women using a copper IUD, when asked at the one-year control visit (14). Low circulating LNG levels could offer an explanation for the lack of observable alterations in biochemical studies.

Tissue Concentrations of Levonorgestrel

The concentration of LNG in the endometrium of women using the LNG-IUD is very high, in the order of 470 to 1,500 ng per gram of tissue wet weight (12). This high local steroid concentration explains the strong progestin action seen in microscopic examination of the endometrium (13). When compared with myometrial or Fallopian tube tissue a great concentration gradient was seen, as only 1.8 to 2.4 ng of LNG per gram of tissue wet weight were determined in these tissues. Fat tissue contained similar amounts of LNG as myometrium and Fallopian tube. Fat tissue concentrations of LNG were correlated with plasma levels (12). The study showed that fat tissue is able to concentrate the nonpolar progestin LNG during sustained release as well as during oral administration. The high local concentration of LNG in the endometrium leads to strong suppression and atrophy, which are reflected in scanty menstrual bleedings or even amenorrhea in some subjects. This effect is completely reversible. It reflects the local effect of LNG in the endometrium without suppression of ovarian function. During oral administration of LNG, 250 μg/d, the endometrial steroid concentration does not differ from that in the adjacent tissues, and effects on the endometrium are correspondingly much weaker (12).

Endometrial Morphology

Intrauterine administration of LNG with the LNG-IUD leads to strong suppression of the endometrium by the progestin. The endometrial glands become atrophic and the stroma decidualized (13,17). The mucosa is thin, the stroma is swollen, and the endometrial glands are sparse and atrophic, with an inactive epithelium. These histologic changes are uniform as soon as after 1 month of use of the LNG-IUD, and the picture is not affected by the phase of the menstrual cycle. In addition to the changes mentioned, stromal leukocytic infiltrates, sometimes containing plasma cells, and usually accompanied by small or large foci of necrosis and scarring, are prominent in samples taken after long-term use of the LNG-IUD. All these histologic changes are reversible even after seven years of use of the LNG-IUD, both in amenorrheic women as well as in those who have been menstruating. Removal of the LNG-IUD results in normal menstruation within 30 days. Endometrial biopsies taken three months after removal of the LNG-IUD show a normal proliferative or secretory histological pattern (17). Return of fertility is not impaired and achievement of pregnancy after removal of the IUD is not affected (11).

Excretion of Levonorgestrel in Breast Milk

To study if infants are exposed to LNG ingested in breast milk, LNG-IUDs were inserted in 70 breast-feeding amenorrheic women six weeks after delivery. Milk and plasma samples were collected for 12 weeks, and the concentrations of LNG were determined by radioimmunoassay. Small but measurable amounts of LNG were detected in breast milk (4). The plasma to milk ratio of LNG was 100:25. The results showed that even with a low maternal steroid dose of 10 to 30 μg/d of LNG, the infants are exposed to the progestin. However, this small dose was only 0.1% of the maternal daily dose and obviously does not have any harmful effects on the infants (5). The weight gain of infants, their blood chemistry, and the time of breast feeding were not significantly affected when the children of mothers with a copper-releasing IUD were compared with those of mothers with LNG-IUDs (5). The estrogen levels were low in infants at six weeks after delivery and thereafter, leading to low progestin receptor concentrations. Thus the negligible amounts of progestins ingested in breast milk have no detectable effects on child health. Accordingly, the LNG-IUD can be used during the breast-feeding period but better performance is achieved with copper-releasing IUDs as regards the bleeding patterns during the first three months.

Ovarian Function

The daily release of 20 μg of LNG from the IUD does not result in serum levels high enough to inhibit ovulation completely. Indeed, using serum progesterone elevations above 5 ng/mL during the second half of the menstrual cycle as an indicator of ovulation, about 50% of women were regarded as being ovulatory during the first year of LNG-IUD use (15). However, interindividual variation between women is seen in plasma progesterone concentrations. Some women have a high progesterone concentration, indicating normal luteal function, whereas low plasma progesterone concentrations in some women indicate inadequate luteal function. We have measured plasma progesterone concentrations in 86 women after one or two years of LNG-IUD use. When values were arranged according to the stage of the menstrual cycle, we found on only three occasions a progesterone rise before day 14, indicating that in those cases, spotting during the luteal phase was regarded as menstruation. In 25% of the luteal phase samples no progesterone rise was seen, indicating anovulation. In samples obtained from amenorrheic women, in about 50% no progesterone rise was seen (Fig. 5.4).

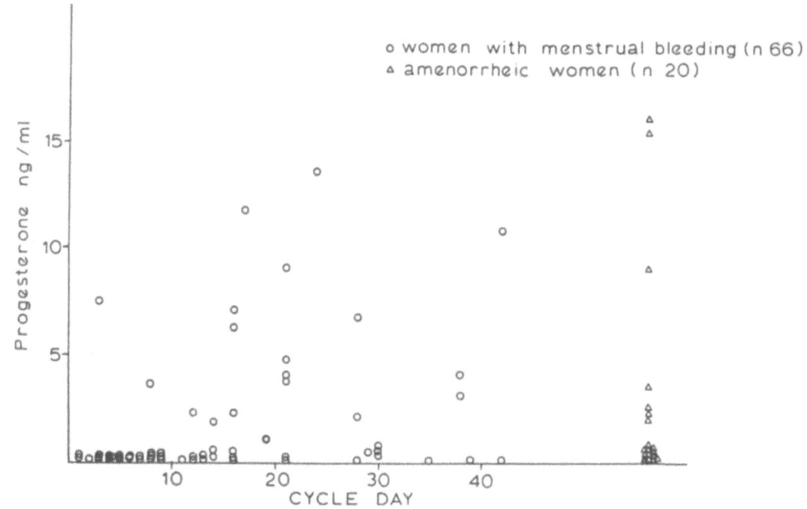

Fig. 5.4. Individual plasma progesterone concentrations in menstruating and amenorrheic women using an LNG-IUD with a daily release of 20 μg.

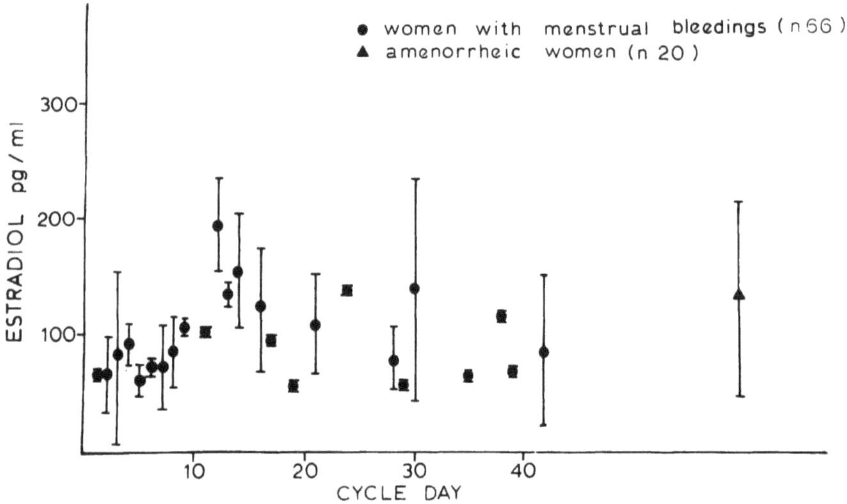

Fig. 5.5. Mean (± SD) plasma estradiol concentrations during the menstrual cycle and in amenorrhea in women using a LNG-IUD.

The avoidance of estrogen deficiency is an important prerequisite for a long-acting contraceptive method in order to prevent osteoporosis. Figure 5.5 shows mean plasma estradiol concentrations in 86 women after one or two years of LNG-IUD use. The mean level for amenorrheic women is depicted on the right of the figure. Other mean concentrations are arranged according to the stage of the menstrual cycle. It can be seen that the values for menstruating women fall into the range typical of normal menstrual cycles. Importantly, no individual menopausal estradiol values were seen. Estradiol values in amenorrheic women fall within the same range. Table 5.1 gives mean estradiol concentrations as a function of time of LNG-IUD use. There are no changes in the mean concentrations between one and seven years of use. The higher mean concentration at three months of use is due to very high concentrations of up to 900 pg/mL in five women. These values are due to follicular cyst formation in these women.

Table 5.1. Mean plasma estradiol (E_2) concentrations (pg/mL) in women using an LNG-IUD (20 μg/d).

	3 Months	1 Year	2 Years
E_2	168.4	104.6	128.4
N	33	53	42
	5 Years	6 Years	7 Years
E_2	109.6	105.6	111.6
N	32	27	34

Table 5.2. Mean plasma estradiol (E_2) and levonorgestrel (LNG) concentrations in menstruating and amenorrheic women using an LNG-IUD.

	Menstruating	Amenorrheic
E_2	103.9	132.7
N	66	20
LNG	175.2	179.9
N	62	20

Table 5.2 shows that there are no differences in plasma estradiol concentrations between menstruating and amenorrheic LNG-IUD users. It can be concluded that the LNG-IUD may cause disturbances in luteal function, but, excluding the possibility of follicular cyst formation at the beginning of the treatment, folliculogenesis, and hence estradiol production, continues normally. The use of an LNG-IUD should therefore not predispose the subject to osteoporosis.

Bleeding

As mentioned above, the use of an LNG-IUD results in a high LNG concentration in the endometrium, which is morphologically reflected by suppression and atrophy of this tissue. The average pattern of bleeding is characterized by an increasing number of days of spotting during the initial three to four months of LNG-IUD use, after which a plateau of scanty bleeding, both in number of days and amount of blood loss, is reached. Figure 5.6 shows this effect compared with use of a copper-releasing IUD, Nova-T, of the same size and shape.

The reduction in the amount and duration of menstrual bleeding leads to a high incidence of oligomenorrhea or amenorrhea. In our study with a five-year follow-up of LNG-IUD and Nova-T users, removals of the LNG-IUD because of oligomenorrhea took place during the initial 24 months, especially at the begin-

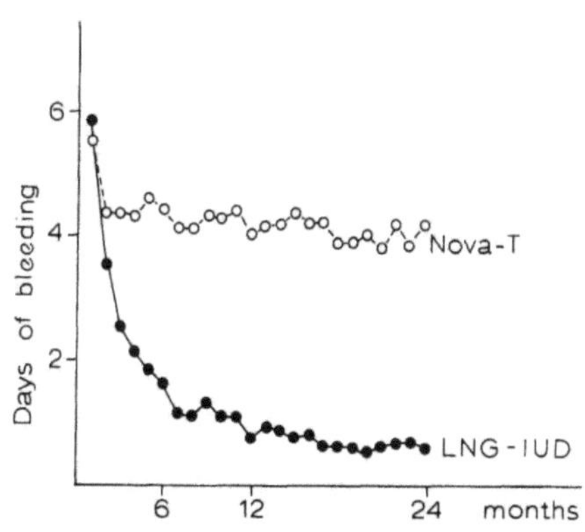

Fig. 5.6. Days of bleeding during the two initial years of Nova-T and LNG-IUD use.

ning of the second year (9). We could clearly see that the high removal rate for this reason was due to inadequate counseling of the users and doctors, because at that time there was not enough known about the nature of oligomenorrhea during LNG-IUD use and/or the effectiveness of the device. When we were able to provide accurate information, there were very few removals because of oligomenorrhea. Amenorrhea is not caused by low serum estradiol, but by local LNG dominance, making the endometrium insensitive to estradiol.

Scanty bleedings are a health benefit for LNG-IUD users. There is no body iron depletion, as reflected by slightly increased serum ferritin concentrations after one year of use (6). Another health benefit is the relief of menstrual pain, which takes place in 40% of LNG-IUD users, according to their subjective evaluation. This may be connected with atrophic endometrium, resulting in a low concentration of endometrial prostaglandins.

Percentual changes in menstrual characteristics and the blood hemoglobin concentrations of Nova-T and LNG-IUD users at one year of use are summarized in Figure 5.7. Excluding hemoglobin determination, the information was obtained by interview of users. Hemoglobin levels have been found to rise slightly but significantly (2,19).

Effectiveness and Continuation

The LNG-IUD is a highly effective device in preventing unplanned pregnancies. In a multicenter clinical trial, two women out of 755 acceptors became pregnant during the first year, giving a net rate of 0.3 per 100 (19). In another study the cumulative net rate at five years was 0.4 per 100 (9). In this study (9) the only unplanned pregnancy occurred during the first year of use of the LNG-IUD. The Nova-T was used as a comparative device and the cumulative net rate of unplanned pregnancies with this device was 5.5 at five years.

The continuation rate appeared to be very similar to that seen with the copper-releasing IUD throughout a five-year follow-up period (9). In this study oligomenorrhea created some confusion, and removals because of this reason decreased the continuation rate, as has been discussed earlier. Another reason for termination, and where a difference between the LNG-IUD and the copper-releasing IUD was observed, was removals for hormonal reasons. However, no discontinuation because of these reasons took place after the third year of use. It is not possible to evaluate if these side effects become less common or better tolerated during use or if those suffering them discontinue at an earlier stage of treatment.

Preliminary results on acceptance of LNG-IUD use show that after three months' use, 75% of women are satisfied and 5.8% are not (16). The respective figures for Nova-T users are 56.7% and 10%. Although these figures cover only the three initial months, they are interesting because bleeding problems with LNG-IUD use are most common during that period. Some data show that more than 80% were very confident in the device and no one was unsure.

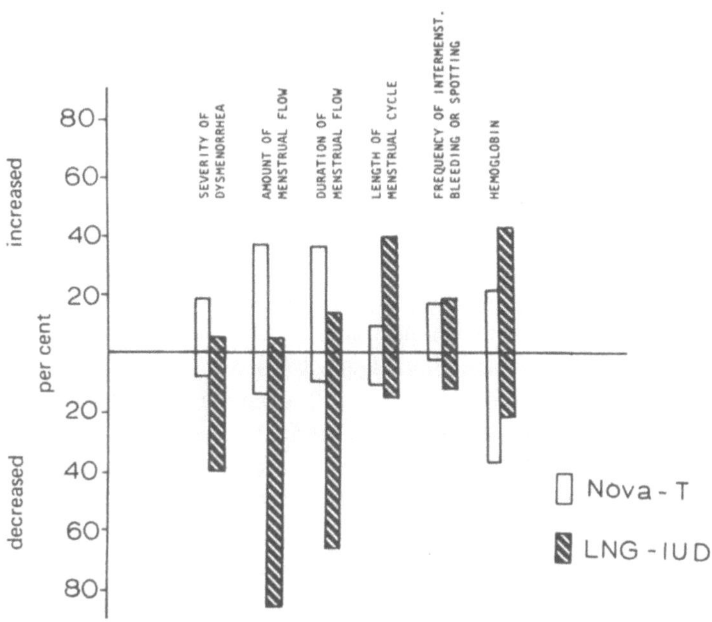

Fig. 5.7. Percentage changes in menstrual parameters and blood hemoglobin levels in Nova-T and LNG-IUD users during the first year of use.

Selection of Patients

The LNG-IUD is representative of a method of long-term contraception for parous women. Continuation is best in women aged more than 25 years, and particularly in those aged more than 30. It is best suited to women with the planned number of children who need long-term, reliable, and reversible contraception. The spotting after insertion, lasting two or three months, necessitates good motivation for continuation. If headache (migraine), acne, or hirsutism are already problems for the patient, the LNG-IUD cannot be recommended because the condition may be worsened. The indications for LNG-IUD use are listed in Table 5.3. Special indications for LNG-IUD use are related to two major characteristics of the device: reduction of menstrual blood loss and high effectiveness. In addition to contraception, those women with heavy bleedings could avoid operative treatment of hypermenorrhea.

Counseling of Patients and Health Personnel

Before insertion of an LNG-IUD the patient should be well informed of the general characteristics of the IUD and possible side effects. Special attention should be given to the possible occurrence of amenorrhea. The patient has to

Table 5.3. Indications for use of an LNG-IUD.

- Age over 30 and at least over 25, parous women
- Age over 25, cannot take pills, nonparous
- Problems with other contraceptive methods
- Heavy menstrual bleedings and/or dysmenorrhea
- Heavy bleeding during use of a copper-releasing IUD
- Surgical sterilization not yet acceptable
- Previous pregnancy during use of a copper-releasing IUD

know that menstrual bleeding will be scanty and that oligomenorrhea may occur, which does not mean that ovarian function is affected. Lack of menstrual bleeding is not a sign of pregnancy. Blood hemoglobin is increased, and the scanty bleedings mean better health of the patients. The initial irregular spotting lasting for three months must be explained beforehand to the patients. At the same time it should be made clear that bleeding will become scanty after this initial spotting phase and that heavy bleedings never occur during use of an LNG-IUD. In addition, menstrual pain usually disappears.

After insertion, during the first few months there may be episodes of headache, swelling, breast tenderness, and skin problems. These hormonal side effects usually disappear in the long run. Other side effects besides initial spotting at the beginning are rare. The IUD may be expelled, as with other IUDs. If normal menstrual bleeding returns, it most probably indicates expulsion of the IUD, and the woman should contact the clinic.

The patients should also be told that normal menstruation and fertility returns after removal of the LNG-IUD. After five years of use the contraceptive efficacy of the LNG-IUD cannot be guaranteed and the IUD should be replaced by another one. In general, the patients should attend the clinic three months after insertion and then annually.

Conclusions

The effectiveness of the LNG-IUD is exceptionally high. A Pearl Index of 0.1 has been reached in all three major studies (2,9,19). There have been three extrauterine pregnancies in all users of an LNG-IUD releasing 20 μg/d, giving a rate of 0.4 ectopic pregnancies per 1,000 women-years. The rate in users of a copper-releasing IUD is 1.4 ectopic pregnancies per 1,000 women-years in Scandinavia (18). High rates of ectopic pregnancy have been found with low activity progestin IUDs such as an LNG-IUD releasing only 2 μg/d or Progestasert® releasing 65 μg of progesterone daily, which has been estimated to be equivalent to 2 μg of LNG (3,8). The LNG-IUD seems to be a most effective reversible contraceptive method. The pregnancy rate is not related to parity or age. The position of the device in the uterine cavity seems not to be critical for contraceptive effectiveness, because a device also releasing 20 μg of LNG daily is as effective if situated in the cervical canal (7,10). This effectiveness, regardless of the position of the device in the uterine cavity, may explain the uniform low pregnancy rate in every clinic.

The LNG-IUD can be used by women who cannot use oral contraceptives because of ethinyl estradiol-related metabolic or subjective side effects. The strong suppression of uterine bleeding makes the use of the LNG-IUD beneficial in women who have hypermenorrhea or who have bleeding problems with a conventional IUD. The reduction of menstrual blood loss is a health benefit, particularly in countries with a high incidence of anemia and dietary iron depletion.

The effect on menstrual bleeding during the use of an LNG-IUD resembles that with oral contraceptives, which also give protection against gonococcal or chlamydial acute salpingitis (20). However, clinical trials have failed to demonstrate better protection against pelvic inflammatory disease with an LNG-IUD than with modern copper-releasing IUDs (2,19). It could be that a tailless LNG-IUD would perform like oral contraceptives in this respect, but the use of an IUD without a tail may lead to more problems than it solves.

The safety of long-term use has been evaluated by careful microscopic studies of the endometrium and examination of cervico-vaginal smears. These studies have not revealed any alarming findings. The lack of alterations in lipoproteins, in serum chemistry, in electrolytes, or in blood pressure gives a reassurance of the safety of this method. During long-term use, ovarian estrogen production is not suppressed even if women have oligomenorrhea. Hence, osteoporosis is not a cause of concern during use. The oligomenorrhea is of local origin and is a result of intrauterine release of progestin, which makes the endometrium insensitive to estrogen. This action is also responsible for the strong reduction of menstrual blood loss. The local action is reversible, and after removal of the LNG-IUD normal menstruation and fertility quickly returns.

References

1. Allonen H: Personal report, 1985.
2. Allonen H, Haukkamaa M, Luukkainen T, et al: Effective contraception with levonorgestrel-releasing intrauterine device. 12-month European multicenter study. To be published 1987.
3. Diczfalusy E: New developments in oral, injectable and implantable contraceptives, vaginal rings and intrauterine devices. A review. *Contraception* 1986; 33:7–22.
4. Heikkilä M, Haukkamaa M, Luukkainen T: Levonorgestrel in milk and plasma of breast-feeding women with a levonorgestrel-releasing IUD. *Contraception* 1982;25: 41–49.
5. Heikkilä M, Luukkainen T: Duration of breast-feeding and development of children after insertion of a levonorgestrel-releasing intrauterine contraceptive device. *Contraception* 1982;25:279–292.
6. Heikkilä M, Nylander P, Luukkainen T: Body iron stores and patterns of bleeding after insertion of a levonorgestrel- or copper-releasing intrauterine contraceptive device. *Contraception* 1982;26:465–474.

7. Kurunmäki H, Toivonen J, Lähteenmäki PLA, et al: Intracervical release of levonorgestrel for contraception. *Contraception* 1981;23:473–485.
8. Larsen S, Hanssen MK, Jacobsen JC, et al: Comparison between two IUDs: Progestasert and Cu T 200. *Contracept Deliv Syst* 1981;2:281–286.
9. Luukkainen T, Allonen H, Haukkamaa M, et al: Five years experience with levonorgestrel releasing IUDs. *Contraception* 1986;33:139–148.
10. Lähteenmäki P, Luukkainen T, Ratsula K, et al: Levonorgestrel-releasing intracervical device in Zatuchni GI, Goldsmith A, Sciarra JJ (eds): Intrauterine Contraception. Philadelphia, Harper & Row, 1985, pp 238–242.
11. Nilsson CG: Fertility after discontinuation of levonorgestrel-releasing intrauterine devices. *Contraception* 1982;25:273–278.
12. Nilsson CG, Haukkamaa M, Vierola H, et al: Tissue concentrations of levonorgestrel in women using a levonorgestrel-releasing IUD. *Clin Endocrinol* 1982;17:529–536.
13. Nilsson CG, Luukkainen T, Arko H: Endometrial morphology of women using a D-norgestrel-releasing intrauterine device. *Fertil Steril* 1978;29:397–401.
14. Nilsson CG, Luukkainen T, Diaz J, et al: Clinical performance of a new levonorgestrel-releasing intrauterine device. A randomized comparison with a Nova T copper device. *Contraception* 1982;25:345–356.
15. Nilsson CG, Lähteenmäki P, Luukkainen T: Patterns of ovulation and bleeding with a low levonorgestrel-releasing intrauterine device. *Contraception* 1980;21:155–164.
16. Shain R, Ratsula K, Luukkainen T, et al: to be published, 1987.
17. Silverberg SG, Haukkamaa M, Arko H, et al: Endometrial morphology during long-term use of levonorgestrel-releasing IUD. *Int J Gynecol Pathol* 1986;5:235–241.
18. Sivin I: IUD-associated ectopic pregnancies, 1974 to 1984, in Zatuchni GI, Goldsmith J, Sciarra J (eds): Intrauterine Contraception. Philadelphia, Harper & Row, 1985, pp 340–353.
19. Sivin I, Alvarez F, Diaz J, et al: Intrauterine contraception with copper and with levonorgestrel: A randomized study of the T Cu 380 Ag and levonorgestrel 20 mcg/day devices. *Contraception* 1984;30:443–456.
20. Wølner-Hanssen P, Svensson P, Mårdh P-A, et al: Laparoscopic findings and contraceptive use in women with signs and symptoms suggestive of acute salpingitis. *Obstet Gynecol* 1985;66:233–238.

CHAPTER 6

Antiprogestin Ru 486: A Contragestive Agent

Etienne-Emile Baulieu

Overview

The steroidal derivative RU 486 [17β-hydroxy-11β-(4-dimethylaminophenyl 1)-17α-(prop-1-ynyl)-estra-4,9-dien-3-one] is the first potent antiprogestin to be useful medically. Acting reversibly at the molecular level of receptor binding, RU 486 interrupts progesterone action, while allowing endocrine functions to return quickly to normal afterwards. However, target-cell dynamics that depend on a continuity of progesterone action is irreversibly disrupted. In cycled women, RU 486 acts during the luteal phase in the endometrium, provoking bleeding and, via a decrease in luteinizing hormone (LH), resulting in a secondary luteolysis. In pregnant women it affects mainly the decidualized mucosa, increases myometrial contractility and maturation of the cervix, leading to termination of pregnancy. Luteolysis, brought on by secondary alteration or detachment of the trophoblast, is secondary to the decrease in human chorionic gonadotropin (HCG). Clinical studies thus far indicate that RU 486 can be a very efficient and safe contragestive agent, especially for the medical termination of early pregnancy and as a postcoital menses inducer or menstrual regulator. The failures observed when RU 486 is given alone may be overcome when efficient form(s) of RU 486 are administered and/or when uterotonics are used in combination with the RU 486. A small amount of prostaglandin, which by itself is inefficient, when given at the end of RU 486 treatment yields highly satisfactory results in women with up to 8 weeks of ammenorhea. Treatment with the antiprogestin RU 486 does not expose women to a hormonal drug continuously, and no significant systemic side effects have been observed, including those that seemed possible because of the compound's antiglucocorticosteroid activity. Based upon physiological and molecular understanding of RU 486 actions, this antiprogestin may be a second generation agent to achieve safe and effective control of human fertility.

Introduction

Progesterone (*progestare*) is named after its supporting effect in pregnancy and is believed to be essential in mammals (14). In the 1950s Gregory Pincus championed hormonal *contraception*, which was essentially based on the inhibition of ovulation (38). Here I shall describe RU 486, a compound designed to achieve "contragestion" in women through its antiprogestin activity (3,24,42). The term *contragestion* (for contra-gestation) is proposed to cover all modalities of fertility control interfering in the establishment or continuation of pregnancy. Similarly, Segal and Tietze (44) and Segal and Atkinson (45) have suggested a "contragestational pill" or "contragestational agents" as possible means of fertility control in women.

In 1970 Carl Djerassi defined "as an important example of future contraceptive methodology in the female, a 'once-a-month' pill with luteolytic or abortifacient properties, or both" (18). He further noted:

> Such an agent has at least four advantages over agents now being used. 1) Administration of one pill a month is clearly more convenient than daily administration of pills. This is true, both for major fertility control programs in developing countries and for highly motivated individuals in advanced countries. 2) Periodic short-term administration of a drug may be expected to give rise to fewer long-term side effects, primarily because the agent is intended to act more specifically on a well-defined biological process. 3). . . it does not matter whether the woman is pregnant or not. 4) Ideally, the agent might be active any time during the first 8 weeks after fertilization, so that it could also act as an 'abortifacient.'

As Djerassi said, such a compound "may well turn out to be a steroid." However, while he praised the properties of prostaglandins (5,53), the concept of antiprogestin was not mentioned. Indeed, at that time, the key biological component had not yet been revealed. It is only in 1970 that work at INSERM U 33 gave the actual definitive description of the intracellular progesterone receptor in the uterus (32). It mediates progesterone action and is the molecular target for antiprogestin.

It took years before a progesterone antagonist was developed. Not all the available scientific resources were used, as the trend in most drug companies was not in favor of contraceptive research at that time. In the seventies the biopharmaceutical fashion did not lie with steroids, even at Roussel Uclaf (Paris, France) where some of the best steroid chemists in the world worked. Edouard Sakiz, who was then its President, and myself, acting as a consultant, insisted on keeping that potential alive and on using it. There were other difficulties also; some individuals did not believe in the commercial future of fertility control, and some were negatively influenced by their sociological class and/or religious background.

Molecular and Physiological Background

Progesterone is produced and active only in temporally distinct periods, and its receptor is only found in specific organs engaged in the reproductive processes. Indeed, of all steroid receptors, the progesterone receptor is certainly the one that has the most limited distribution in the body. In addition, there are variations of receptor's concentration in potential target cells, making them sensitive to the hormone only at appropriate times (31,32) (Fig. 6.1).

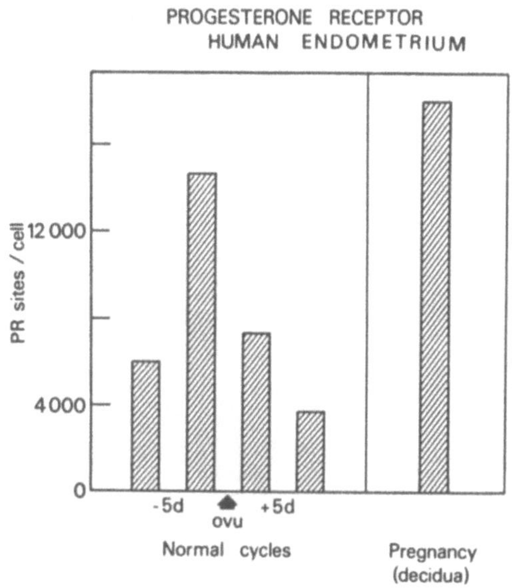

Fig. 6.1. Progesterone receptor in endometrium (decidua): the main physiological-molecular target of progesterone, progestins, and antiprogestin derivatives. During the menstrual cycle, there is an increase of progesterone receptor in endometrium before ovulation and a decrease during the luteal phase. In the decidua (obtained on the occasion of voluntary early pregnancy interruptions), there is more receptor (almost totally occupied by progesterone). Bars represent the sum of "cytosol plus nuclear" fractions of the receptor (31) in endometrium during the 5th-9th days, 10th-14^{n}, 15^{n}-19^{n}, and 20th-28th days of the cycle, and decidua of 7th-8th weeks of amenorrhea.

Precluding progesterone action at the receptor level seemed to us more promising than decreasing progesterone biosynthesis or progesterone transport. It is known that progesterone, besides being produced by the corpus luteum and the placenta cells and then implicated in pregnancy, is also an intermediary step in adrenal steroidogenesis. Thus an inhibition of progesterone synthesis (36) may lack the necessary physiological specificity. On the other hand, very convincing experiments have demonstrated that antiprogesterone antibody can interfere with the development of gestation (16,54). However, permanent immunization of women against progesterone is obviously not a sound alternative.

The progesterone receptor, naturally designed for the flat and rather "simple" progesterone structure, was found to loosely bind a number of steroids of different structure (32,47). In most cases, these compounds did not have any assayable progesterone-like or antiprogesterone activities. In the mid-1970s we found, again in our laboratory, (a) that a steroid hormone antagonist, an antiestrogen in this instance, can indeed be fully antagonist, without any agonist properties: this was the triphenylethylene tamoxifen in the chick oviduct system (49) (Fig. 6.2), and (b) that 4-hydroxytamoxifen, a derivative with high affinity for the estrogen receptor, even more than the natural agonist estradiol, could be a perfect antagonist (7). (These observations were made against the background of the

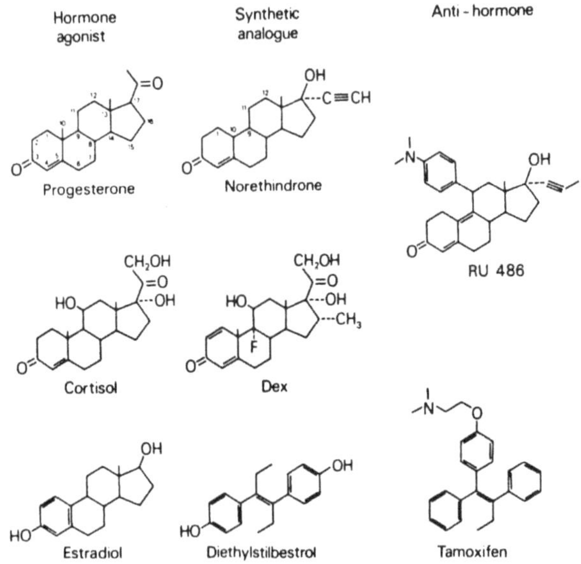

Fig. 6.2. Progestins, glucocorticosteroids, and RU 486. Estrogens and antiestrogen tamoxifen. Norethindrone, a 19-norsteroid, is a synthetic progestin. Note the main structural characteristics of RU 486: the presence of 11β-extra cycle and the third cycle of tamoxifen, which are situated at a position corresponding to the β-side of the steroid overall plane, just off the C11 carbon. "RU 486" is a short for RU 38486, the number used internally in Roussel Uclaf. Diethylstilbestrol is a synthetic agonist of estradiol.

theory prevailing at that time suggesting that an antagonist would have low affinity, with a rapid dissociation rate from the receptor.) Although the two phenyl rings of the antiestrogens are analogs to the A and D cycles of steroids (27), the third lies in a region corresponding just near and outside carbons C10, 11, and 12 of steroids (Fig. 6.2). It was then logical to try similar substitution on the steroid skeleton itself. In fact, the Roussel chemists (51) found a new method to produce 11β-substituted steroids, and several compounds now have been obtained. The RU 486 is a norethindrone derivative with an 11β-substitution. It is striking that norethindrone is a synthetic progestagen used originally in oral contraception (38) and that RU 486, the first developed antiprogestin, can be considered as one of its parent compounds.

In the 1970s work with steroid derivatives in general and in the field of reproductive medicine in particular was not very popular among drug companies for a series of reasons, including the cost of toxicity studies necessary for contraceptive development and use. At Roussel, the search for antiprogesterone was actually included in a wider program of antagonist synthesis, including research for antiglucocorticosteroids (glucocorticosteroids have a 11β-hydroxyl group), and we knew about the spill-over binding of progestagens and glucocorticosteroids by the glucocorticosteroid and the progesterone receptors (1).

Table 6.1. RU 486: binding to receptors.

	Primate	Rodent	Avian
Receptors			
Progestin	+++[a]	++++	0
Glucocorticosteroid	++++	++++	++++
Androgen		+	
Estrogen	0	0	0
Mineralocorticosteroid		0	
Plasma Proteins			
Specific (CBG[b], SBP[b])	0	0	0
α_1-mucoid	++ (human)	(0)	(0)

[a] +++ is ~ affinity of the corresponding natural hormone.
[b] CBG, corticosteroid binding globulin; SBP, sex steroid binding plasma protein.

Molecular Mechanism

We do not know very much except for hormone binding (Table 6.1), about the mechanism of action of antagonists. In the absence of knowledge of the steroid binding-site structure and functioning, the global hypothesis is that an antagonist should exclude the agonistic progesterone from binding and favor a receptor conformation unable to trigger hormonal action. This implies that the receptor does not have the appropriate conformation for interaction with specific regulatory segments of DNA and/or other chromatin components, as indicated in A of Figure 6.3. In spite of some work with progesterone (39) and glucocorticosteroid (8,22,33) systems, this concept has not been demonstrated. Recent work at INSERM, by Françoise Cadepond and André and Ghislaine Groyer, suggests another possibility that is represented in B of Figure 6.3. The "8S form" of the receptor, a native hetero-oligomer with no biological activity, is stabilized by RU 486. The DNA binding site of the receptor interacts with a heat-shock protein (hsp 90) (11,28), and when the receptor binds an antagonist, hsp 90 is not released and the receptor cannot act (4). Work in progress will provide further data to assess this very recent hypothesis.

Physiopathological and Pharmacological Effects of RU 486

In a large number of cellular and animal mammalian systems, the antiprogesterone activity of RU 486 has been demonstrated. This includes effects on endometrium, myometrium, implantation, pregnancy development, specific protein synthesis, cell growth, and behavior (3). However I will only summarize the effects of RU 486 in pregnant and nonpregnant women.

In the case of pregnancy, the *decidua*, which contains high progesterone receptor levels (31) (Fig. 6.1), is a primary target of RU 486. Its alteration is rapid,

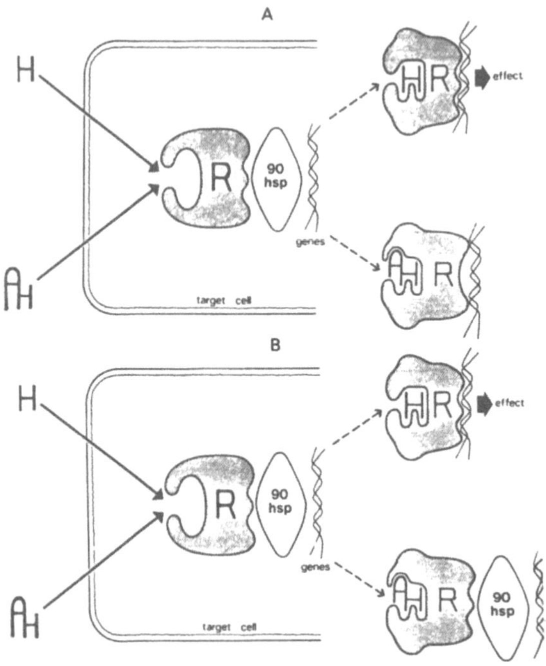

Fig. 6.3. Steroid hormones and antihormones at the receptor level. Steroidal hormone (H) and antihormone (AH) penetrate into the target cells and reach their corresponding receptors originally present in the nucleus (20). **A**: Hormone (agonist) provokes the "transformation-activation" of the receptor under the intracellular conditions; this includes separation from the heat-shock protein hsp 90, and therefore the DNA-binding site of the receptor can interact with regulatory segments of the gene and trigger action. However, AH stabilizes a conformation that does not interact "well" with the gene and does not provoke a response. Therefore there is a difference in DNA binding of hormone- and antihormone-receptor complexes. **B**: On this panel is presented a different possibility recently proposed at INSERM U 33 (see text): the binding of antihormone to the receptor stabilizes the receptor-hsp 90 complexes, which are unable to bind to DNA and to trigger the response. In any case, competition between hormone and antihormone for the hormone binding site of the receptor is the basic physical mechanism for explaining reversible antihormonal activity of antihormone.

with bleeding (vascular endothelium is damaged) and increased prostaglandin production (25,30). Detailed studies suggest that when RU 486 is given, HCG does not increase as it should at this period of pregnancy (25). Thereafter HCG decreases rapidly when the embryo is detached (Fig. 6.4). It has been demonstrated in pregnant rats that RU 486 provokes separation of the embryo from the uterus earlier than that of castration. This is a predictable result since progesterone, secreted by the corpus luteum, does not decrease as rapidly after ovariectomy as it does when acted upon by RU 486. The decrease of HCG causes

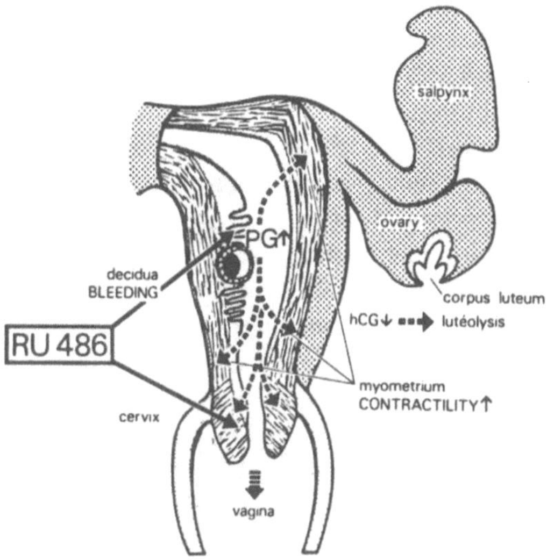

Fig. 6.4. A representation of the effects of RU 486 in early pregnancy. The drawing indicates direct action on the endometrium and cervix, the secondary increase of PGs, which stimulate myometrium and contribute to cervix maturation. Luteolysis is secondary to the decrease of HCG provoked by detachment/alteration of the trophoblast.

secondary and irreversible luteolytic effect. Antiprogesterone activity increases contractility of the myometrium, as does the increase in prostaglandins (37). Evacuation of the uterus is also favored by the effect on the cervix, which is softened and dilated. The mechanism may be direct by antiprogesterone action on the cervix and indirect by prostaglandin activity.

During the cycle, when endometrium is under progesterone influence, RU 486 provokes bleeding. Independently, the compound decreases the LH output during the luteal phase, even though there is tendency toward frequent secretory episodes. This effect was not predicted for an antiprogestin, but early results (24) have now been confirmed by several groups (21,43,53,55). Concomitantly, a decrease in follicle-stimulating hormone (FSH) is also recorded. Indeed, FSH and LH decrease when RU 486 is given to estrogen-treated postmenopausal women (21). We do not know the mechanism of this central nervous system effect of RU 486. LH decrease is dose-dependent (43) and likely to be involved in indirect luteolysis, which occurs more frequently when RU 486 is administered late in the cycle, as though, during the first postovulatory days, sensitivity of the corpus luteum was decreased either by receptor occupancy after the large mid-cycle peak or by down-regulation.

When RU 486 was administered before the expected date of menses but after fertilization, data indicate a decrease in LH under RU 486, which may precede the decrease in newly secreted HCG. A direct effect of RU 486 on corpus luteum steroidogenesis and/or life span has not been demonstrated, even though it is not totally excluded. Unpublished experiments with animal and human luteal cells may indicate some antagonistic effects of RU 486 in vitro, especially in cases of large concentrations, but it is improbable that a direct luteolytic effect is operational in vivo in most instances.

The paradoxical effect of RU 486 on LH secretion and the observed changes in receptor concentration and DNA-polymerase activity in human endometrium (21) have evoked the possibility of *progesterone-like* action of RU 486, especially when administered after or with estrogen and in absence of progesterone. Although there is no known clinical consequences of such observations, they should be borne in mind. The membrane progesterone receptor of *Xenopus laevis* oocytes (2) mediates definitively an agonist activity of RU 486 (but this is a totally different biological system), and the intracellular progesterone receptors of chick (40) and marsupial (46) do not bind the compound.

In human beings, and indeed in all tested animals, RU 486 is *neither an antiestrogen nor an estrogen*. There has been argument that antiprogesterone may favor endogenous estrogenic activity by removing the opposing hormone effect. Indeed, when RU 486 interrupts pregnancy or cycle, it is clear that the consequence is to abolish simultaneously the secretion of progesterone *and* of placental and/or ovarian estrogen. RU 486 has some affinity for the androgen receptor and some antiandrogenic activity (29,37), probably causal for gynecomastia in long-term treated patients with Cushing's disease (35). However, at doses and for period of time used in fertility control, no effect is expected. Also, some discrete but complex effects on water and mineral metabolism have been described, with apparently secondary increase in circulating aldosterone in dog studies (48). The RU 486 does not have an affinity for the rat mineralocorticosteroid receptor, and no related effects have been observed clinically when it was administered for fertility control; nevertheless, further studies should be undertaken. In fact, no side effects have been observed that would have indicated mandatory research to evaluate their risk, and the relatively long-term (weeks) exposure of cancer patients and patients with Cushing's disease (35,41) to RU 486 has been very reassuring.

The *antiglucocorticosteroid* activity of RU 486 deserves some discussion. As expected from binding to receptor and innumerable cell-free, cellular and animal experiments, RU 486 is also an antiglucocorticosteroid, including in humans (6,19). It has been clearly observed that there are effects, both "central" on the control system of adrenal steroids, at the hypothalamus and pituitary levels, and at the "periphery," where cortisol exerts its actions. Centrally, antiglucocorticosteroid activity at the receptor level leads to an increase in the corticotropin-releasing factor (CRF)-ACTH system (associated pro-opiomelanocortin peptides, and also vasopressin augment). This effect is not only dose-dependent, but also time-dependent, parallelly to the endogenous diurnal rhythm. Thus, administration in the evening and early night provokes an increase in cortisol 8 to 12 hours later, amplifying the morning peak. On the contrary, administration in the morning does not influence cortisol in the following afternoon (19). It is therefore evident that when administering RU 486 for fertility control, since

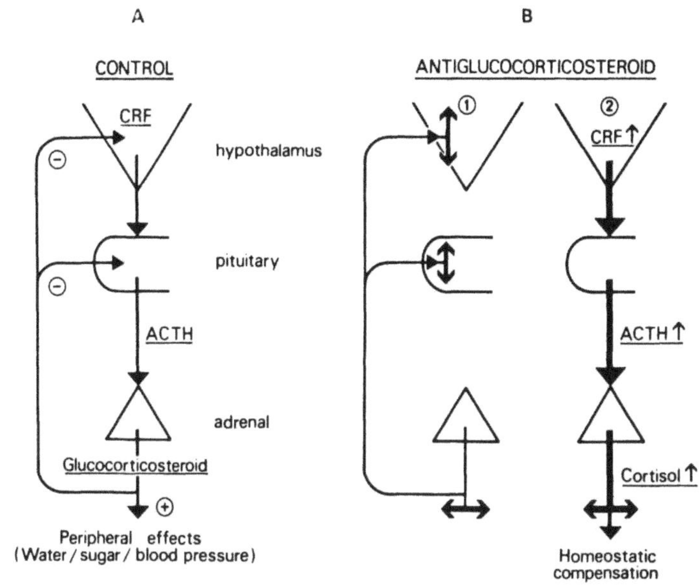

Fig. 6.5. The hypothalamus-pituitary-adrenal system: negative feed-back control by glucocorticosteroid and effect of antiglucocorticosteroid. Normally, as indicated as **A**, cortisol negatively feed-back control corticotropin-releasing factor (CRF) and ACTH productions, while it exerts its periphereal effects on various metabolic parameters. **B**: In case of antiglucocorticosteroid acting at the receptor level, the feed-back mechanism is decreased, and CRF and ACTH are increased and lead to an augmentation of cortisol secretion. This increased cortisol overcomes the antiglucocorticosteroid activity at the periphereal level and reestablishes the glucocorticosteroid function, thanks to this homeostatic mechanism.

there is no demonstrated differences of activity in the sex steroids related to diurnal rhythm, it is advantageous to administer it in the late evening. The subsequent increase in endogenous glucocorticosteroid will eventually compensate, by natural homeostatic mechanism, the peripheral antiglucocorticosteroid activity of the compound, and then preclude signs of hypoadrenalism (Fig. 6.5). On the contrary, if RU 486 is administered early in the morning, there is, at least in principle, more risk that its antiglucocorticosteroid activity is not compensated by the endogenous corticosteroid. In any case, it remains of great interest to study further the quantitative and kinetic parameters of the hypothalamo-pituitary-adrenal system, particularly in situations of stress. It is also of interest to note that the increase of cortisol may serve as a marker for proper administration of RU 486.

The *metabolism* of RU 486 has been studied in human beings. The main points are that after *oral* intake, the bioavailability is relatively low, thus responsible for large intake of the product contrasting with high affinity to receptor. Several metabolites have been described (17); none of them demonstrate any unexpected activity, some are inactive, and some have weak RU 486-like activity.

The half-life of the compound is about 20 hours in nonpregnant or pregnant women. Since the endocrine parameters of the sex and adrenal hormonal systems return to normal, it is clear that, as with all steroids, *reversibility* is part of the mode of action of RU 486. The reproductive events have their own temporal characteristics, and a reversible effect can influence definitively the functioning and survival of hormone-dependent cells (as those of the endometrium or decidua and secondarily the corpus luteum) and thus provide an ad hoc effect in interrupting cycle or pregnancy. By the same token, the body is only exposed *temporarily* to the foreign drug.

Further studies will indicate whether other forms of RU 486 administration–i.e., through the vagina mucosa, the skin, or by injection–give better quantitative or kinetic characteristics to its activity.

Clinical Studies

Present Status

Data available on more than 1,500 patients have led to the proposed use of orally administered antiprogestin RU 486 for the following:

Induction of Labor After Intrauterine Fetal Death. RU 486 helps delivery of dead fetuses (9). Preliminary results indicate that RU 486 may be useful in the delivery of abnormal fetuses, or in cases of pathological situations (toxemia), even if high doses or associated therapy are necessary owing to lateness in pregnancies. Comparisons with already available techniques should continue.

Cervix Maturation. RU 486 can be used to prepare endo-uterine interventions, including mechanical interruption of pregnancy, due to its opening effects on the cervix. It may also facilitate second trimester abortion, which is currently performed with prostaglandins (49).

After the Expected Day of the Menses. Systematic trials for early interruption of pregnancy have been performed in 15 countries, within programs conducted by Roussel Uclaf, World Health Organization, Population Council, and several individuals. In one recent trial (13), 100 women received RU 486 alone *within 10 days after the expected time of menses* and had successful pregnancy interruption in 85% cases. The used doses were 400 mg (in four days), 600 mg (in four days), or 800 mg (in two days). Since then, a single 600-mg dose (~10 mg/kg) of RU 486 administered at 8 PM has given comparable results in 95 women with *up to 41 days of amenorrhea* (~93% successes: French trial conducted by André Ulmann and Catherine Doubois) (Table 6.2).

Administration of these relatively high doses [in contrast with the first trials, which were arbitrarily and prudently limited for fear of signs of cortisol insufficiency (24)] is well tolerated, and the results confirm Csapo's concept (15,18) that brief interruption of progesterone action during pregnancy leads to irreversible damage. However, RU 486 alone displays only ~70% success during the

Table 6.2. RU 486 alone, 600 mg once.[a]

Number		Failure		
Up to 41 days of amenorrhea	Success	Total	Incomplete expulsion	Ongoing pregnancy
95	88 (92.6%)	7 (7.4%)	2 (2.1%)	5[b] (5.3%)

[a] French multicenter trial, September 1986.
[b] One ectopic pregnancy.

seventh and eighth weeks of amenorrhea and even less later (admittedly with doses that could be insufficient). If this global evaluation is based on more than 1,000 cases, we are still short of detailed comparisons with reference to varied doses and regimens of administration, age of pregnancy, and hormonal levels. It is however now clear that antiprogestin RU 486 alone, even at relatively large doses, may not be efficient. When we analyze the results according to the duration of amenorrhea (Table 6.3), we note that 1) with increasing length of pregnancy, the rate of successes decreases and 2) among the failures, the number of ongoing pregnancies increases less than that of incomplete expulsions. This suggests that ongoing pregnancies may be a result of quantitatively insufficient exposure to RU 486 to counteract progesterone action at the decidualized endometrium level, and naturally progesterone concentration increases with time. Incomplete evacuation frequently may be due to a limiting "post-RU 486" mechanism, particularly an insufficient increase of prostaglandin synthesis, and since the conceptus becomes more difficult to eliminate with progressing pregnancy, pharmacological supplementation is indicated to stimulate enough myometrium contraction. Indeed, other unknown factors also may be implicated in resistance to antiprogesterone effect. It has been shown that increased by RU 486 alone, myometrium contractions are further vigorously provoked by the addition of a small amount of exogenous prostaglandin analog, an effect improving the rate of success in pregnancy interruption (50). This has been demonstrated by the intramuscular injection of 16-phenoxy-tetranor-PGE_2-methyl-sulfonamide (Sulprostone) at the end of four days administration of RU 486, and very recently with pessaries of 16,16-dimethyl-trans-Δ_2-PGE_1. Further trials, using orally or vaginally administered prostaglandin, are to be conducted (52). It is then advisable, at this time point, to preferentially give a *single* dose of RU 486 in the *evening*, on the basis of 600 mg experience (a larger or smaller dose may be envisaged according to future results at different stages of pregnancy). Then, *prostaglandin* should be administered *afterwards*, perhaps (and ideally for practical reasons) on the next morning, or, more probably, a little later, for instance just *when bleeding* has started (on the second and third day). While improving the rate of success, the post-RU 486 administration of a small dose of prostaglandin clearly avoids the digestive, painful, and vascular effects of

Table 6.3. RU 486 alone: successes and failures in September 1986.[a]

		Failures		
Weeks of amenorrhea	Successes (%)	Total %	Incomplete expulsion (%)	Ongoing pregnancy (%)
7-8[b]	70	30	50	50
6[b]	88	12	30	70
5	93	7	20	80

[a] A simplified account of the data obtained in more than 1,500 patients.
[b] Preliminary trials (~150 cases) in patients with six to eight weeks of amenorrhea indicate ~95% successes with the successive RU 486 plus prostaglandin administration (see text and references 10 and 50).

larger amounts of prostaglandin. A concern with antiprogestin-interrupted pregnancy is the possibility of excessive and/or prolonged bleeding, which may hamper its use in developing countries. It is of great importance to perform large trials with RU 486 and complementary prostaglandin.

Before the Expected Day of the Menses. RU 486 can be used as a menstrual regulator, in case of unexpected intercourse. This may also be called late postcoital contraception and is of particular interest in teenagers, or women more than 35 years old. Both reasoning and available data suggest that the drug should be taken at the end of the current cycle and may be administered either according to the once-600-mg regimen, or for several days (e.g., four) at moderate doses (e.g., 100–200 mg/d). The main reason for the several day administration is that one has to be certain that the antiprogestin is given after ovulation has taken place. Globally, whenever there is no pregnancy, and just cyclic corpus luteum, or, on the contrary, fertilization, menstrual bleeding will occur in more than 98% of total patients. An HCG test applied one week later will be useful, particularly if there is no bleeding. In case of anovulatory cycle, no bleeding is expected, and the pregnancy test will be negative.

Presumably because of the long half-life of RU 486 and involvement of progesterone in follicle development, retardation of the next cycle ovulation is occasionally observed after administration at the end of the cycle. Patients should be informed of possible menstrual irregularity. This spill-over effect of the product in the next cycle is the basic difficulty for using the product as a once-a-month menstrual regulator, since cycle disturbances lead to irregular ovulation.

In summary, RU 486 has clear indications in fertility control. This does not mean that for the four indications described above, the best formulation has been found, nor that all complications have been discovered. The results, however, point to its potentiality of becoming a widely used second-generation drug for fertility control.

Other Applications

From the previous discussion, it is clear that improvements have to be looked for, but this should not preclude the present use of RU 486.

The vaginal and percutaneous routes of administration should be tested. From monkey experiments, it has been shown that injected RU 486 may be more effective than oral administration (23). Although the development of a derivative with no antiglucocorticosteroid activity may be of great interest, a shorter-acting derivative that could repeatedly be used once a month would be even more suitable. We have already discussed the related use of prostaglandin, and there is a number of possibilities in terms of molecules, regimens, and routes of administration. Also, it is not ruled out that even in the use of RU 486 as a menses regulator or menses-inducer, the systematic addition of prostaglandin may improve results. Another possibility, in these last indications, would be the adjunctive use of antigonadotropin-releasing hormone (to help get rid of the corpus luteum function). For these indications, more studies on embryo-toxicity will be useful, although there is currently no detected danger.

In the reproductive field, we may mention other possibilities.* The use RU 486 in ectopic pregnancies has not given very encouraging results so far.

Antiprogestins may also be useful to help normal delivery, but, again, specific toxicity studies have to be continued, including the watching of newborns for a long period of time.

After birth, the onset of lactation may be delayed by progesterone, and RU 486 could be tested for triggering it in the postpartum period.

We also know, from primate experiments (26,34), that given during the first part of the cycle, RU 486 *retards ovulation*. This was also observed in humans (25) and the long-acting injected product leads to long-term anovulation (26,34). In humans the subsequent hypoestrogenism of such chronic anovulation may not be suitable.

The *central* effect of the compound on regulatory peptides of both the sex and the adrenal functions (6,19,23,24,55) may be the background for trials exploring RU 486 effect on premenstrual syndrome, on peri- and postmenopausal hot flushes, and on the corresponding behavioral troubles.

Finally, RU 486 may be also used as adjuvant therapy in progesterone-sensitive *tumors*, and indeed recent trials have been started in breast cancers (41).

Why "Contragestion"? (Fig. 6.6)

Fertility control in general and interruption of pregnancy in particular are and will remain subjects of debate. Problems are naturally not only biological, but highly cultural and social, not to mention metaphysical. In any case, practice has

*We cite, only for the record, the use of RU 486 to regularize reproductive cycles in farm animals.

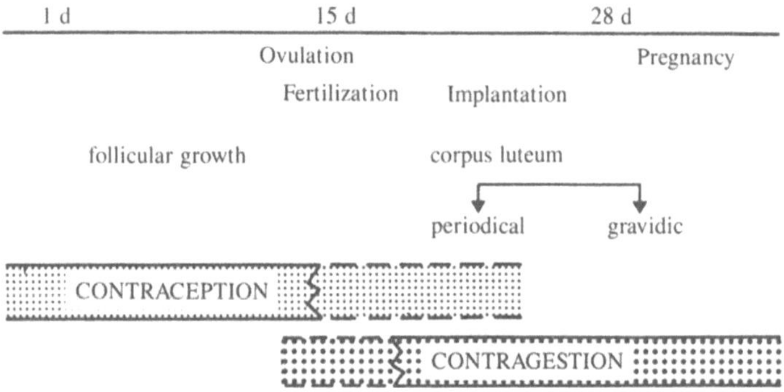

Fig. 6.6. On the basis of the classical sequence on events for ovulation, fertilization, implantation, menses, and beginning of pregnancy, we have indicated what are and what become contraception and contragestion. Contraception, originally, has been used to designate the lack of fertilization, essentially by suppression of ovulation. In fact, contraception is also used in "postcoital contraception" and for qualifying IUD functioning. Contragestion, in the broad sense of the term, may include what is known as contraception. However, initially, and because it has been originally used to designate early interruption of pregnancy, it still keeps the connotation of "postovulatory" methods: the use of antiprogestin, and also that of IUD and anti-HCG vaccine, essentially relates to contragestion.

coined a wrong meaning to the word "contraception" (contra-conception), since conception cannot be equated to fertilization as commonly believed. Conception is a continuing process and is, in fact, difficult to describe temporally. Certainly life does not start with fertilization, and human life is not defined by the events occurring at the time of fertilization. In addition, the word "abortion" has been used in a very traumatic context, not only psychologically (and this is understandable), but also physically, because of the implicated mechanical maneuvers. Indeed, IUDs today, and possibly antipregnancy vaccines tomorrow, are also, in large part, postovulatory methods to control reproduction. To use terminology not charged with emotional connotation and, at the same time, not hide the biological reality, we have proposed contragestion, with the hope to change the abortion debate, thanks to medical progress. Clearly, antiprogestins are not the only contragestive agents. However, their biological characteristics bring about two series of consequences. First, they give responsibility to the women and to the couples, since the decision of using an antiprogestin has to be made each time. It is also clear that for acceptability, it has to be compared with vacuum aspiration, which has approximately the same rate of success in early pregnancies: in developed countries doctors may still prefer aspiration, but, as Marc Bygdeman says, women may not agree and may prefer a medical method. In the developing world, a nonsurgical, noninstrumental method would be more convenient, but it too should be managed under strict medical supervision, at least for some time. Also, contragestion does not currently fit in any legal context (12), whereas contraception has been legally extended to include postcoital contraception and the use of IUD. In many medical and legal instances, pregnancy is defined as starting at implantation time, that is, one week after fertilization (10). Indeed, this also has become somewhat disputable because of the practice of in vitro fertilization (IVF). With quasi-unanimous agreement, "preembryos" may be kept up to 15 days after fertilization, and then be occasionally discharged. Therefore, any limitation in time applying to a postovulatory method may be argued. Very early use of RU 486 around the date of expected menses may be envisaged in the context of well-known irregularity of normal cycles, and then considered in the same spirit as is contraception, justifying the concept of menstrual regulation. Nature spontaneously gets rid of a majority of fertilized ova. Human beings should be given, beyond the biological considerations, the opportunity to make a safe and medically efficient decision. Education, responsibility, freedom, and medical care are key words of the modern times. Contragestion is a humanistic attitude.

References

1. Baulieu EE, Raynaud JP: A "progesterone graph" method for measuring binding systems. *Eur J Biochem* 1970;13:293–304.
2. Baulieu EE, Schorderet-Slatkine S: Steroid and peptide control mechanisms in membrane of Xenopus laevis oocytes resuming meiotic division, in Porter R, Whelan J (eds): *Molecular Biology of Egg Maturation – Ciba Foundation Symposium 98*. Bath, Pitman Press, 1983, pp 137–158.
3. Baulieu EE, Segal SJ (eds): *The Antiprogestin Steroid RU 486 and Human Fertility Control*. New York, Plenum Press, 1985.
4. Baulieu EE: Steroid hormone antagonist at the receptor level: a mechanism implicating hsp 90: *UCLA Symposia on Molecular and Cellular Biology*, 1987, in press.
5. Bergstrom S, Diczfalusy E, Borell U, et al: Prostaglandins in fertility control. *Science* 1972;175:1280–1287.
6. Bertagna X, Bertagna C, Luton JP, et al: The new steroid analog RU 486 inhibits glucocorticoid action in man. *J Clin Endocrinol Metab* 1984;59:25–28.
7. Binart N, Catelli MG, Geynet C, et al: Monohydroxytamoxifen: an antioestrogen with high affinity for the chick oviduct oestrogen receptor. *Biochem Biophys Res Commun* 1979;91:812–818.
8. Bourgeois S, Pfahl M, Baulieu EE: DNA binding properties of glucocorticosteroid receptors bound to the steroid antagonist RU 486. *EMBO J* 1984;3:751–755.
9. Cabrol D, Bouvier d'Yvoire M, Mermet E, et al: Induction of labour with mifepristone after intrauterine fetal death. *Lancet* 1985;8462:1019.
10. Cameron IT, Michie AF, Baird DT: Therapeutic abortion in early pregnancy with antiprogestogen RU 486 alone or in combination with prostaglandin analogue (Gemeprost). 1987; to be published.
11. Catelli MG, Binart N, Jung-Testas I, et al: The common 90kD protein component of non-transformed "8S" steroid receptors is a heat-shock protein. *EMBO J* 1985;4: 3131–3135.
12. Cook RJ: Legal abortion: limits and contribution to human life, in Porter R, O'Connor M (eds): *Abortion: Medical Progress and Social Implications. Ciba Foundation Symposium 115*. London, Pitman Press, 1985, pp 211–227.

13. Couzinet B, Le Strat N, Ulmann A, et al: Termination of early pregnancy by the progesterone antagonist RU 486. *N Engl J Med* 1986;315:1565–1570.
14. Csapo AI: Antiprogesterone in fertility control, in Zatuchni GI, Sciarra JJ, Speidel JJ (eds): *Pregnancy Termination: Procedures, Safety, and New Developments*. Hagerstown, Harper & Row, 1979, pp 16–34.
15. Csapo AI, Pulkkinen MO, Wiest WG: Effects of luteectomy and progesterone replacement therapy in early pregnant patients. *Am J Obstet Gynecol* 1973;115:759– 765.
16. Csapo A, Dray F, Erdos T: The biological effects of injected antibodies to estradiol-17β and to progesterone in pregnant rats. *Endocrinology* 1975;97:603–614.
17. Deraedt R, Bonnat C, Busigny M, et al: Pharmacokinetics of RU 486, in Baulieu EE, Segal SJ (eds): *The Antiprogestin Steroid RU 486 and Human Fertility Control*. New York, Plenum Press, 1985, pp 103–122.
18. Djerassi C: Birth control after 1984. *Science* 1970;169:941–951.
19. Gaillard RC, Riondel A, Muller MF, et al: RU 486: a steroid with antiglucocorticosteroid activity that only disinhibits the human pituitary-adrenal system at a specific time of day. *Proc Natl Acad Sci USA* 1984;81:3879–3882.
20. Gasc JM, Ennis BW, Baulieu EE, et al: Récepteur de la progestérone dans l'oviducte de poulet: double révélation par immunohistochimie avec des anticorps antirécépteur et par autoradiographie à l'aide d'un progestagène tritié. *CR Acad Sci Paris* 1983;297:477–482.
21. Gravanis A, Schaison G, George M, et al: Endometrial and pituitary responses to the steroidal antiprogestin RU 486 in postmenopausal women. *J Clin Endocrinol Metab* 1985;60:156–163.
22. Groyer A, Le Bouc Y, Joab I, et al: Chick oviduct glucocorticosteroid receptor. Specific binding of RU 486 and immunological studies with antibodies to chick oviduct progesterone receptor. *Eur J Biochem* 1985;149:445–451.
23. Healy DL, Hodgen GD: Non-human primate studies with RU 486, in Baulieu EE, Segal SJ (eds): *The Antiprogestin Steroid RU 486 and Human Fertility Controls*. New York, Plenum Press, 1985, pp 127–140.
24. Herrmann W, Wyss R, Riondel A, et al: Effet d'un stéroide anti-progestérone chez la femme: interruption du cycle menstruel et de la grossesse au début. *CR Acad Sci Paris* 1982;294:933–938.
25. Herrmann WL, Schindler AM, Wyss R, et al: Effects of the antiprogestérone RU 486 in early pregnancy and during the menstrual cycle, in Baulieu EE, Segal SJ (eds): *The Antiprogestin Steroid RU 486 and Human Fertility Control*. New York, Plenum Press, 1985, pp 179–198.
26. Hodgen G: 1985, personal communication.
27. Hospital M, Busetta M, Bucourt R, et al: X-ray crystallography of estrogens and their binding to receptor sites. *Mol Pharmacol* 1972;8:438–445.
28. Joab I, Radanyi C, Renoir JM, et al: Immunological evidence for a common non hormone-binding component in "non-transformed" chick oviduct receptors of four steroid hormones. *Nature* 1984;308:850–853.
29. Jung I, Baulieu EE: Inhibition of glucocorticosteroid action in cultured L-929 mouse fibroblasts by RU 486, a non anti-glucocorticosteroid of high affinity for the glucocorticosteroid receptor. *Exp Cell Res* 1983;147:177–182.
30. Kelly RW, Healy DL, Cameron MJ, et al: The stimulation of prostaglandin production by two antiprogesterone steroids in human endometrial cells. *J Clin Endocrinol Metab* 1986;62:1116–1123.

31. Levy C, Robel P, Gautray JP, et al: Estradiol and progesterone receptors in human endometrium: normal and abnormal menstrual cycles and early pregnancy. *Am J Obstet Gynecol* 1980;136:646–651.
32. Milgrom E, Atger M, Baulieu EE: Progesterone in uterus and plasma. IV. Progesterone receptor(s) in guinea pig uterus cytosol. *Steroids* 1970;16:741–754.
33. Moguilewsky M, Philibert D: RU 38486: potent antiglucocorticoid activity correlated with strong binding to the cytosolic glucocorticoid receptor followed by an impaired activation. *J Ster Biochem* 1984;20:271–276.
34. Nadler RD, Roth-Meyer C, Baulieu EE: Behavioral and endocrine consequences of long-term antiprogesterone (RU 486) administration to cynomolgus monkeys: preliminary results, in Baulieu EE, Segal SJ (eds): *The Antiprogestin Steroid RU 486 and Human Fertility Control*. New York, Plenum Press, 1985, pp 169–177.
35. Nieman LK, Healy DL, Spitz IM, et al: Use of a single dose of the antiprogesterone steroid RU 486 for induction of menstruation in normal women, in Baulieu EE, Segal SJ (eds): *The Antiprogestin Steroid RU 486 and Human Fertility Control*. New York, Plenum Press, 1985, pp 279–284.
36. Odlind V, Birgerson L: 1986, this volume.
37. Philibert D, Moguilewsky M, Mary I, et al: Pharmacological profile of RU 486 in animals, in Baulieu EE, Segal SJ (eds): *The Antiprogestin Steroid RU 486 and Human Fertility Control*. New York, Plenum Press, 1985, pp 49–68.
38. Pincus G: *The Control of Fertility*. New York, Academic Press, 1965.
39. Rauch M, Loosfelt H, Philibert D, et al: Mechanism of action of an antiprogesterone, RU 486 in the rabbit endometrium. Effect of RU 486 on the progesterone receptor and on the expression of the uteroglobin gene. *Eur J Biochem* 1985;148:213–218.
40. Renoir JM: 1986, unpublished.
41. Romieu G, Maudelonde T, Ulmann A, et al: The antiprogestin RU 486 in advanced breast cancer: preliminary clinical trial. *Bull Cancer* 1987, in press.
42. Sakiz E, Euvrard C, Baulieu EE: The antiprogesterone activity of RU 486, a contragestive agent in the human, in LaBrie F, Proulx L (eds): Endocrinology. Amsterdam, Excerpta Medica, 1984, pp 239–242.
43. Schaison G, George M, Lestrat N, et al: RU 486 in women with normal or anovulatory cycles, in Baulieu EE, Segal SJ (eds): *The Antiprogestin Steroid RU 486 and Human Fertility Control*. New York, Plenum Press, 1985, pp 271–278.
44. Segal SJ, Teitze C: Contraceptive technology: current and prospective methods. *Rep Popul Fam Plann* 1971;1:1–24.
45. Segal SJ, Atkinson LE: Systemic contragestational agent, in Osofsky HJ, Osofsky JD (eds): *The Abortion Experience*. Hagerstown, Harper & Row, 1973, pp 400–414.
46. Short R: 1986, personal communication.
47. Smith HE, Smith RG, Toft DO, et al: Binding of steroids to progesterone receptor proteins in chick oviduct and human uterus. *J Biol Chem* 1974;249:5924–5932.
48. Spitz IM, Wade CE, Krieger DT, et al: Effect of RU 486 on the pituitary-adrenal axis in the dog, in Baulieu EE, Segal SJ (eds): *The Antiprogestin Steroid RU 486 and Human Fertility Control*. New York, Plenum Press, 1985, pp 315–330.
49. Sutherland RL, Mester J, Baulieu EE: Tamoxifen is a potent "pure" anti-oestrogen in chick oviduct. *Nature* 1977;267:434–435.
50. Swahn ML, Cekan S, Wang G, et al: Pharmacokinetic and clinical studies of RU 486 for fertility regulation, in Baulieu EE, Segal SJ (eds): *The Antiprogestin Steroid RU 486 and Human Fertility Control*. New York, Plenum Press, 1985, pp 249–258.

51. Teutsch G: Analogues of RU 486 for the mapping of the progestin receptor: synthetic and structural aspects, in Baulieu EE, Segal SJ (eds): *The Antiprogestin Steroid RU 486 and Human Fertility Control*. New York, Plenum Press, 1985, pp 27–47.
52. Van Look P: 1986, personal communication.
53. Wilks JW: Pregnancy interception with a combination of prostaglandins: studies in monkeys. *Science* 1983;221:1407–1409.
54. Wright LJ, Feinstein A, Heap RB, et al: Progesterone monoclonal antibody blocks pregnancy in mice. *Nature* 1982;295:415–417.
55. Yen SCC: 1986, personal communication.

CHAPTER 7

Vaginal Pills: Systemic Contraception Induced by Vaginal Administration of Contraceptive Pills

Elsimar Metzker Coutinho

The most widely accepted contraceptive is undoubtedly "the pill." Highly effective because it inhibits ovulation, the pill remains the first choice for most women. In its low estrogen formulation, which was introduced a decade ago, it is well tolerated and causes few undesirable side effects. In fact, epidemiological evidence indicates that pill use may reduce the incidence of endometrial and ovarian cancer, an added bonus for those who rely on its contraceptive protection for long periods of time (21). The pill is easily available and may be self-administered. It is therefore very close to being an ideal contraceptive, leaving other methods of contraception as alternatives to be considered only when a formal contraindication obviates the use of oral contraceptives. Attempts to improve the qualities of the pill and reduce its few limitations have been and should be further encouraged, in order to make its use even more widely accepted.

High effectiveness is probably the best quality of the pill. Nevertheless, it lasts only a short time and for this reason the pill has to be administered daily. To prolong effectiveness of the pill is one major aim of new methods. On the other hand, while the pill is well tolerated and almost devoid of risks for young women, it may cause serious trouble for those over 35 years of age. Both poor tolerance and cardiovascular side effects appear to be associated with the impact of the pill on the liver, and for this reason, attempts to administer the same steroids of the pill by other routes became a major endeavor of new methods (18).

Administration of contraceptive steroids by the vaginal route was first attempted by Mishell Jr., Lumkin, and Stone in the early seventies (13) using a vaginal ring made of silastic rubber impregnated with medroxyprogesterone. These authors have shown that the progestogen released from the ring was readily absorbed by the vaginal mucosa, reaching the bloodstream and the pituitary gland at levels sufficiently high to inhibit ovulation. Medroxyprogesterone rings developed by Mishell et al were intended as a product, but development was discontinued probably because the drug was not approved as a contraceptive in the United States. The method was held in abeyance for a few years but studies were resumed by Mishell himself and others, using several other progestrogens or combinations of progestogens with estrogens (14,19,22).

Development of vaginal rings has taken two different approaches. In one approach the ring releases enough hormone to prevent ovulation. These rings are intended to be used intermittently but may be used continuously for up to six months. In the most widely tested regimen, the ring is left in the vagina for three weeks and then removed for one week to bring on withdrawal bleeding. A second approach is a ring that releases low levels of progestogen only. Prevention of pregnancy is achieved through local effects on cervical mucus and the endometrium, not by ovulation inhibition. The ring is worn continuously for three months and does not need to be removed monthly.

A large multicenter trial suggests that combined vaginal rings are as effective in preventing pregnancy as low-dose oral contraceptives. More than 1,100 women using the rings in two sizes were compared with 533 women using the low-dose combined oral contraceptive, Nordette®, consisting of 0.15 mg of levonorgestrel and 0.3 mg of ethinyl estradiol. The life-table pregnancy rates at one year were similar for all three groups: 1.8% of women using the 50-mm ring or oral contraceptives and 0.7% of women using the 58-mm ring (19).

In this trial, similar percentages of women stopped using the two methods because of side effects; at one year about 23% of women using rings and 19% of women using oral contraceptives. Side effects differed. Vaginal problems, primarily discharge, caused 5% to 7% of ring users to discontinue the method but none of the oral contraceptive users. Medical examination showed that vaginal problems, particularly vaginitis, leukorrhea, and other discharge, were significantly more common among women using rings than among oral contraceptive users. Other side effects, such as headache, nausea, and dizziness, were less common among ring users.

Despite the estrogen, breakthrough bleeding remains a problem. About 7% to 8% of women discontinued the rings because of menstrual problems compared with 4% of oral contraceptive users. After the first month of use, women using the rings experienced an average of about five days of bleeding and spotting every month. Roughly 10% of the women had four or more days of breakthrough bleeding or spotting during the three weeks in which the ring was in place.

Other side effects of the estradiol-levonorgestrel ring appear to be minor. Blood pressure, hemoglobin, and hematocrit, blood clotting factors, liver function, and carbohydrate metabolism are not significantly affected. Plasma levels of high-density lipoprotein cholesterol are significantly lower. This may be considered an adverse effect but the clinical significance of the change is not certain.

Despite all these attractive features, vaginal rings still carry few but significant drawbacks. The ring may cause erosion and irritation of the vaginal wall. It may retain menstrual residues absorbed during breakthrough bleeding and occasionally develop unpleasant odor. It may interfere with coitus. In addition, some women (and men) object to the continuous presence of a foreign body in the vagina for aesthetic, hygienic, or even cultural reasons. To overcome these objections to the use of the vaginal ring, but with a view to the preservation of the advantages of the vaginal route of administration, we proposed daily administration of a vaginal pill containing a progestogen or a progestogen-estrogen combination, as originally prescribed for oral contraceptive use, as an alternative to the ring.

The pill is inserted manually in the vagina daily for 21 days. After an interval of eight days allowed for withdrawal bleeding, insertion is resumed. If the patient does not wish to bleed, continuous administration is recommended. Several steroids and steroid combinations have been tested as vaginal pills.

dl-Norgestrel and Ethinyl Estradiol

A pill containing 0.5 mg of *dl*-norgestrel and 0.05 mg of ethinyl estradiol, a widely used oral contraceptive, was the first combination tested as a vaginal pill. The first study with this combination, carried out in a small number of women, revealed that vaginal administration inhibits ovulation in all subjects (7). Both progesterone and estradiol levels remained low throughout the period of administration, as shown in Figure 7.1.

Comparative studies between oral and vaginal administration of this combined pill containing 50 μg *dl*-norgestrel and 35 μg ethinyl estradiol revealed that plasma levels of *l*-norgestrel rise at a slower pace and to a lower peak value after vaginal administration than after ingestion. After four hours, the two curves are parallel; 24 hours after vaginal insertion of the pill, the mean *l*-norgestrel plasma concentration is between 0.3 and 0.4 ng/mL, which is less than one half the level observed at 24 hours after oral administration. At any day in a 21-day treatment cycle, the plasma level after vaginal administration is lower than after oral administration, the difference being more pronounced during the first two weeks. Despite the relatively low plasma levels of *l*-norgestrel, patients using the vaginal pill did not ovulate, as indicated by the low levels of both progesterone and estradiol throughout the treatment cycle (Fig. 2) (2).

In a larger study, 124 women were enrolled. Of these women, 117 were between 15 and 35 years of age. Sixty-four of the women had never used oral contraceptives. The remaining had used oral contraceptives but had discontinued their use because of side effects such as nausea or weight gain (10). The women

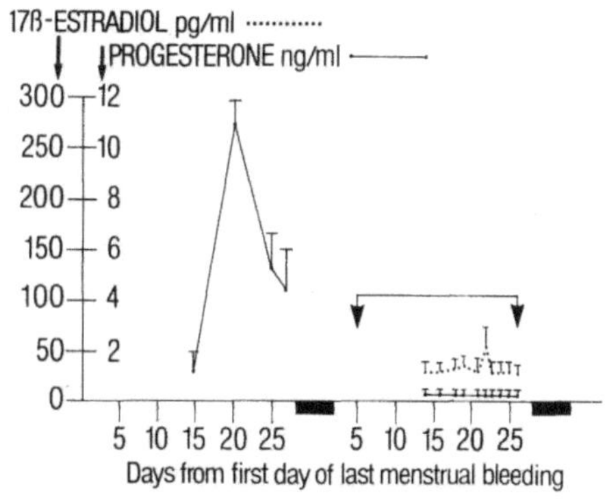

Fig. 7.1. Points represented are plasma levels plus SEM, n=12. Solid bar indicates days of menstrual bleeding. ↓——↓ indicates period of daily vaginal use of tablets (days 5–26).

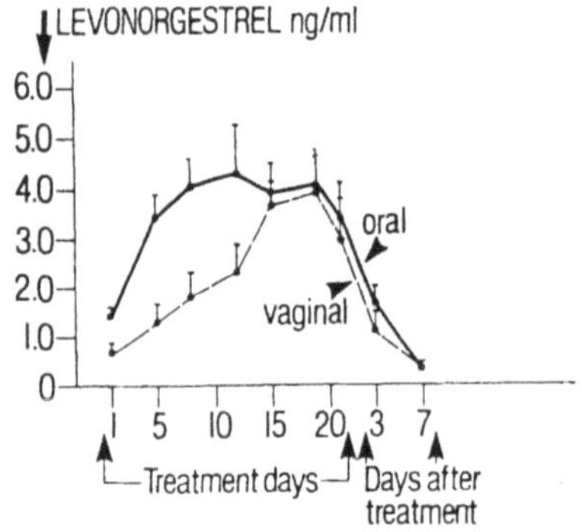

Fig. 7.2. Plasma levels of levonorgestrel in women treated with a daily pill, containing 500 μg of *dl*-norgestrel and 50 μg of ethinyl estradiol either orally or vaginally.[2]

enrolled in this study used the vaginal pill for a minimum of six and a maximum of 20 months. A total of 1,438 woman-months were recorded. No pregnancies occurred. Bleeding usually developed three to five days following insertion of the last pill and lasted in over 90% of cycles three to five days. In only 2% of the cycles did bleeding last longer than five days. The interval between withdrawal bleeding events was 26 to 30 days in 86% of cycles (Fig. 7.3). Figure 7.3 shows the bleeding pattern of a patient who used the vaginal pill for 20 months. Figure 7.4 shows the bleeding intervals and duration of withdrawal bleeding in all subjects. Amenorrhea, defined as an interval longer than 45 days, was reported twice by one subject and once by two subjects. Breakthrough bleeding occurred in six subjects and spotting in only four.

Weight increase greater than 2 kg occurred in 27 patients (22%), whereas weight loss occurred in 11 (9%). In 86 subjects (69%) there was no weight change. Blood pressure remained unchanged in 118 patients throughout the treatment period. In five subjects a moderate ($<$20 mm Hg) increase in systolic pressure was recorded. Headache was reported by 12 subjects and nausea by eight. Acne developed in two subjects. Nervousness was reported by eight subjects; mastalgia by eight subjects. Varicose veins developed or became worse during treatment in three women. Dizziness was reported by six women. Vaginal discharge requiring treatment developed in 26 subjects. In nine of them, the pathogenic agent was identified as *Trichomonas vaginalis*. Other agents identified in vaginal discharge occurring or flaring up during treatment were *Candida albicans*, 5; *Neisseria gonorrhoeae*, 2; *Escherichia coli*, 4; *Staphylococcus epidermidis*, 2; *Streptococcus β-hemolyticus*, 1; *Proteus mirabilis*, 1; and *Proteus vulgaris*, 2. Other pelvic and vaginal conditions detected during treatment are listed in Table 7.1.

Forty-four women discontinued the vaginal pill before completing one year of use, whereas the remaining 80 women (64%) opted for continuation. Only seven women discontinued for medical reasons. Two discontinued because of mastalgia; two others because of vaginal irritation. Vaginal discharge, unreponsive to treatment and requiring discontinuation, developed in three women. Six women decided to conceive. The remaining 31 women, although satisfied with the performance of the vaginal pill, discontinued because they felt daily vaginal

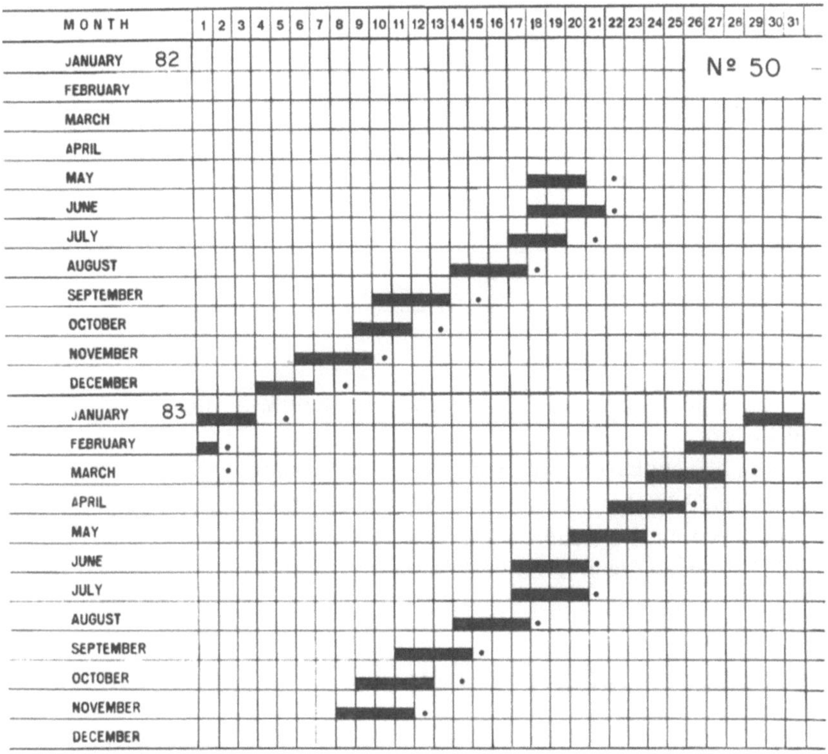

Fig. 7.3. The bleeding pattern of a patient who used the vaginal pill for 20 months.

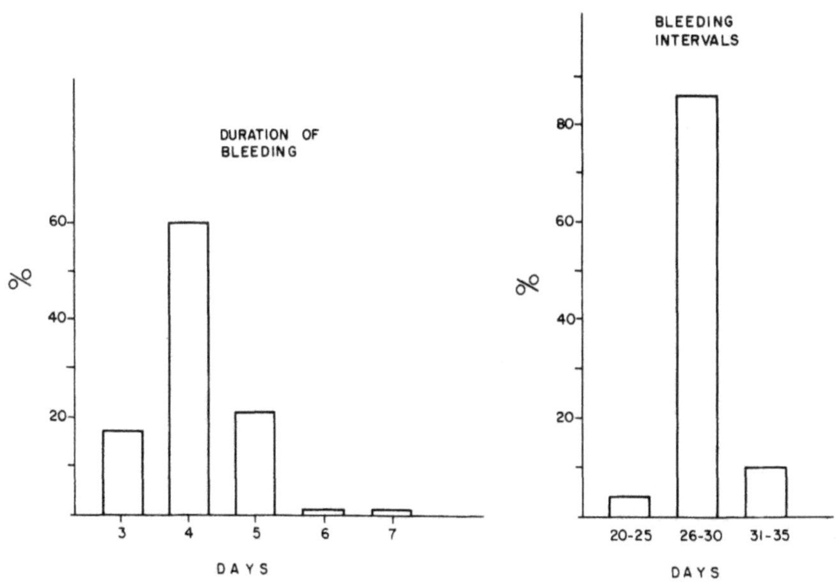

Fig. 7.4. The bleeding interval and duration of withdrawal bleeding in all subjects.

Table 7.1. Pelvic and vaginal conditions detected during use of the vaginal pill.

	No.	%
Cervicitis	12	9.6
Cervical erosion	6	4.8
Vaginal discharge	26	20.9
Vulvitis	3	2.4
Pelvic pain	4	3.2
Pruritus	5	4.0
Nabothian cysts	2	1.6

insertion either "impractical," "distasteful," "indecorous," "tiresome," or simply "tedious" (7).

No significant changes occurred in blood cell counts and hematocrit. Blood glucose, cholesterol urea nitrogen, tryglycerides, transaminases, alkaline phosphatase, creatinine, inorganic phosphorus, calcium, sodium, and potassium all remained within normal levels throughout the treatment.

Levonorgestrel and Ethinyl Estradiol

The combination *l*-norgestrel 0.25 mg with 0.05 mg ethinyl estradiol has been also investigated as a vaginal pill. The formulation is sugar-coated but this coat dissolves rapidly in contact with the vaginal mucosa. Studies with this combination pill in a small number of subjects have shown that both progesterone and estrogen levels are suppressed during the period of pill use, indicating that ovulation is inhibited (Fig. 7.5).

In a larger study that included 135 women, 1,187 woman-months of observation were recorded with no pregnancies. Tolerance of this combination pill is as good as that recorded for *dl*-norgestrel–ethinyl estradiol combination.

Cyproterone Acetate and Ethinyl Estradiol

The combination of cyproterone acetate 2 mg and ethinyl estradiol 0.05 mg has been used as both an oral contraceptive and an antiandrogen in the treatment of hirsutism, acne, and other manifestations of androgenicity. Toleration of this combination pill is not as good as that of other oral contraceptives, probably because the estrogenic effects are reinforced by the antiandrogen. The combination has been used as a vaginal pill by 41 patients for 339 woman-months. No pregnancies occurred. Tolerance of this pill by vaginal route was greatly improved as compared with the oral route. Almost no complaints of nausea, which is common in oral-route users, were made by vaginal-route users.

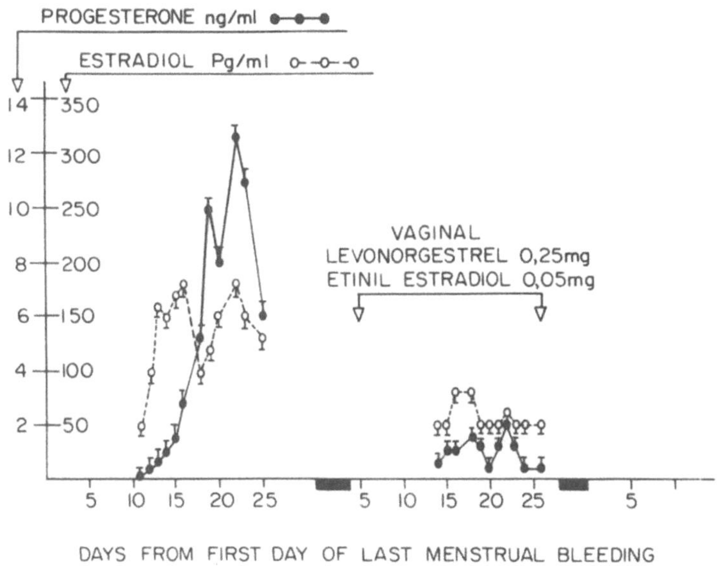

Fig. 7.5. Progesterone and estradiol blood levels in patients using levonorgestrel-ethinyl estradiol pills by vaginal route. Note suppression of both hormones during pill use.

Evaluation of ovulation inhibition in a group of patients revealed that cyproterone acetate–ethinyl estradiol pill inhibited ovulation in eight out of ten patients. Despite the favorable results of the clinical trial, these findings suggest that this pill is probably less safe than the levonorgestrel-ethinyl estradiol pill (Fig. 7.6).

Norethindrone and Mestranol

Norethindrone (NET) was the first 19-norsteroid to be used in oral contraception and its record of safety is excellent. The combination of NET-mestranol was tested as a vaginal contraceptive in a group of 20 women in order to assess the ability of this combination to inhibit ovulation (9). Each of two groups of ten women was instructed to insert either one or two pills daily for 21 days in the vagina, always at the same time of the day. The pill used contained norethindrone 1 mg and mestranol 50 μg. A total of 25 treatment cycles was available for analysis. Of the ten women inserting one pill daily, 13 treatment cycles were available. In six of these cycles, both progesterone and estradiol levels remained within the normal ovulatory pattern, indicating that ovulation occurred. In the remaining seven cycles, no preovulatory estrogen peak or postovulatory progesterone rise occurred, indicating that no ovulation took place (Fig. 7.7).

Of the ten women inserting two pills daily, 12 cycles were available for analysis. In all 12 cycles, no rise in progesterone levels occurred, indicating that ovulation was suppressed. In some cycles estradiol levels rose above 300 pg/mL but

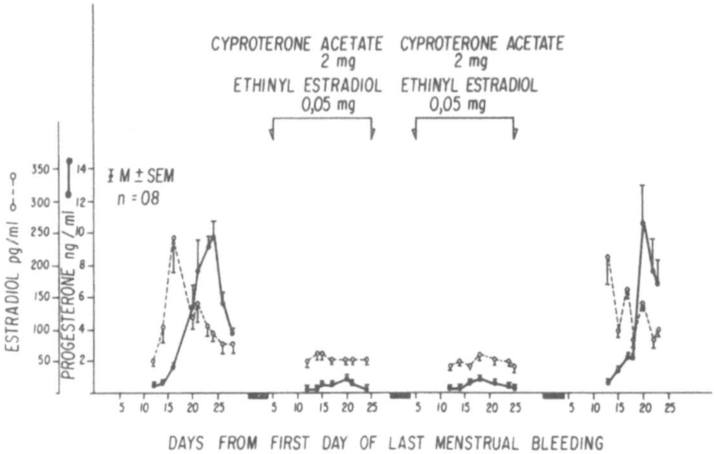

Fig. 7.6. The effect of cyproterone acetate-ethinyl estradiol pills administered by vaginal route on blood levels of progesterone (P) and estradiol (E). Note suppression of both P and E during pill use.

progesterone levels remained below 2 ng/mL. Except for one patient who complained of vaginal irritation with pruritus, no side effects associated with vaginal pill insertion occurred.

Desogestrel and Ethinyl Estradiol

Desogestrel in combination with ethinyl estradiol, which has recently been introduced as a new alternative oral contraceptive, offers several advantages over other widely used combinations because of its favorable effects on blood lipids (12). Desogestrel is a gonane 19-norsteroid derivative. It was synthesized in 1971, and the first tests showed it to be in animal bioassays a potent progestogen with no estrogenic and very weak androgenic properties. In combination with ethinyl estradiol in ovariectomized women, it was found to be about ten times as potent as lynestrenol in its effects on cytology, cervical mucus, and endometrium. Peripheral effects of 0.030 mg levonorgestrel and 0.030 mg desogestrel are comparable. Ovulation-inhibiting dose in women is somewhat below 0.060 mg. It is rapidly metabolized before being effective to 3-ketodesogestrel and is thus a prodrug. Desogestrel and 3-ketodesogestrel have been used synonymously. It has apparently very little or no androgenic effect, as judged from its lack of negative effect on the ethinyl estradiol-induced increase in steroid hormone binding globulin (12).

The ability of a combination containing 0.15 mg of desogestrel and 0.03 mg of ethinyl estradiol to inhibit ovulation has been tested as a vaginal pill in a group of ten women. In all ten women, progesterone levels remained low throughout

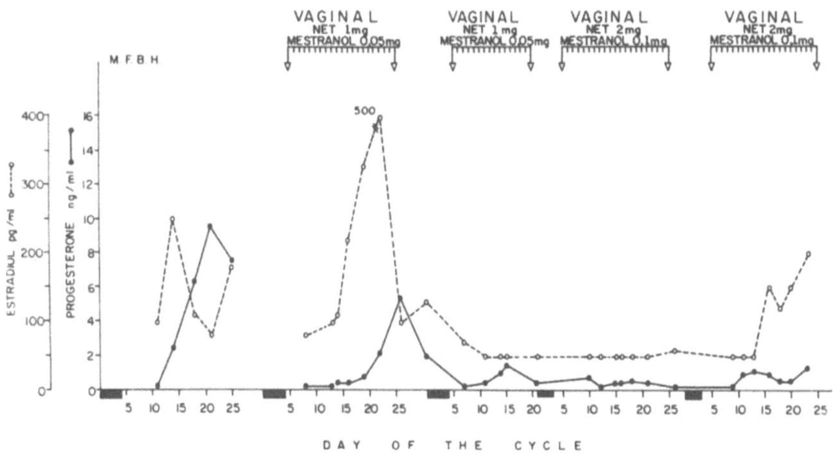

Fig. 7.7. Endocrine profile of a patient using vaginal pill containing mestranol and norethindrone (NET) daily. Note that on the first treated cycle with 1 mg NET and 0.05 mg mestranol, progesterone rises to 5 ng, indicating that ovulation may have occurred. Note also that on the following cycle with the same dose and the next two cycles with double dose, both estradiol and progesterone levels remain low, indicating that oculation was suppressed.

the treatment cycle, indicating that ovulation was inhibited. In another group of four women, blood levels of desogestrel were measured by radioimmunoassay following either vaginal or oral administration. Figure 7.8 shows that whereas a peak follows oral ingestion, no such peak follows vaginal insertion. Blood concentration of desogestrel reaches approximately the same level at the end of six hours and remains so for the remaining 24 hours. This study suggests that the

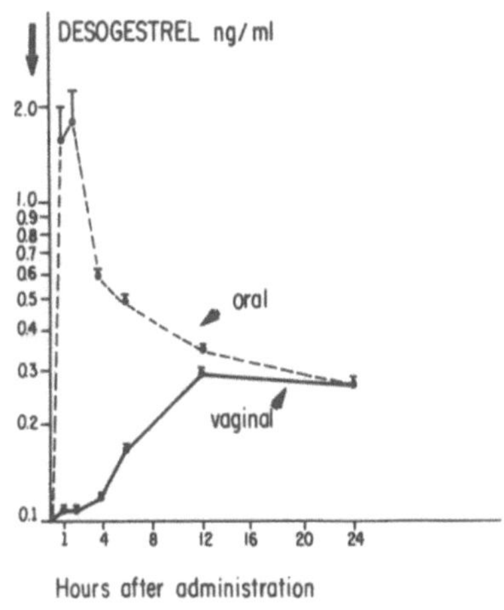

Fig. 7.8. Blood levels of 3-ketodesogestrel in women taking a combination pill containing desogestrel and ethinyl estradiol either by vaginal or oral route. Note the peak that follows oral administration. Curves were drawn on semilogarithmic scale in order to emphasize differences in absorption during the first three hours.

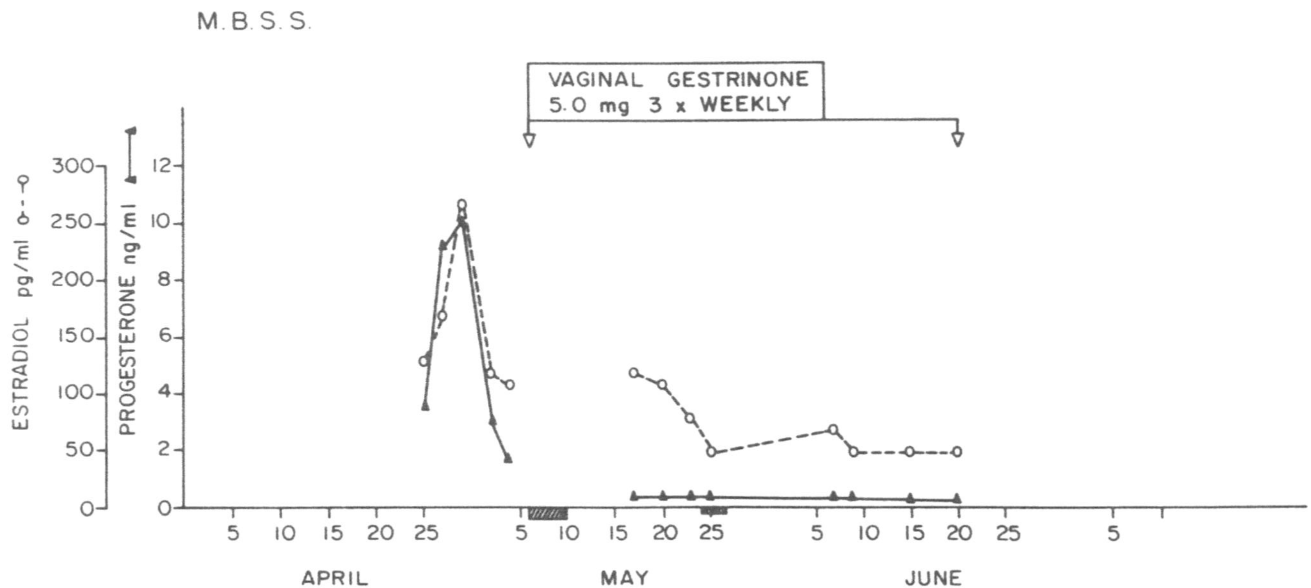

FIG. 7.9. Endocrine profiles and gestrinone blood levels in a patient using gestrinone 5.0 mg three times weekly by vaginal route. Note that on the pretreatment cycle progesterone rises to 10 ng but on the following cycles both estradiol and progesterone levels remain low, indicating ovulation suppression.

peak following oral administration is unnecessary and may, in fact, contribute to provoke side effects.

Gestrinone

Gestrinone is a potent unsaturated 19-trienic norsteroid (synthesized by Roussel-Uclaf, Paris) with antiestrogenic, antiprogesterone properties. The drug has been used experimentally in a variety of conditions, being remarkably effective as a long-acting contraceptive in both women (4,5,15) and men (3).

The antiestrogenic properties of gestrinone (16) provided the basis for its use in the treatment of estrogen-dependent pathologic conditions, such as uterine fibromyoma and benign breast disease (6,11). Its use as a long-acting contraceptive agent is based on its antigonadotropic action but its effect as a midcycle contraceptive (17) probably involves an antiprogesterone action (16). The remarkable effect of this versatile compound on endometriosis results not only from its antigonadotropic action but also from its antiestrogenic effect, which contributes to a rapid and long-lasting atrophy of ectopic endometrium (8).

Gestrinone was also tried as a contraceptive in vaginal rings.[1] Administration of gestrinone as a vaginal pill was carried out either on alternate days (2.5 mg every other day) or twice weekly (5.0 mg), and in both these schedules ovulation suppression occurred. No withdrawal bleeding takes place after the discontinuation of this drug; amenorrhea lasts at least two weeks after the placement of the last pill. The studies with gestrinone extended over two years and involved 80 women. Figure 7.9 shows the ovulation-suppressing effect of vaginally administered gestrinone. Vaginal gestrinone is ideal for women who develop estrogen-dependent pathologies such as benign breast disease, fibromyomas, and endometriosis. It is also to be preferred in cases of breast hypertrophy because, while it acts as a contraceptive, it also treats the condition.

Final Comments

Clinical trials with the various pills reveal a significantly lower occurrence of vaginal problems requiring discontinuation in vaginal pill users than those reported by vaginal ring users. Ring expulsion, interference with coitus, the user's dislike of or difficulty with insertion or removal of the ring, unpleasant ring odor, difficulty with storage or ring loss, which account for the majority of vaginal ring users' discontinuation, are absent among vaginal pill users. Few vaginal pill users discontinue because of vaginal problems. Vaginal discharge itself did not appear to be a serious problem. Its incidence in vaginal pill users (20%) is not higher than that occurring in oral contraceptive or intrauterine device users, suggesting that its occurrence may be unrelated to vaginal pill use. Pathogenic agents identified in the vaginal discharge of women using the vaginal pills were the same agents frequently found in vaginal discharge of women using other contraceptive methods. Furthermore, patients may be successfully treated for vaginitis without necessarily interrupting vaginal pill insertion, even in cases in which vaginal creams are used. This is possible because the vaginal pill remains a relatively short time in the vagina, dissolving quickly and being completely absorbed within one or two hours.

Despite the low plasma levels, cycle control is as good or better than that reported for low-level combined oral contraceptives. The incidence of intermenstrual bleeding and spotting is minimal. The continuation rate of the vaginal pill at one year (64%) also compares favorably with continuation rates recorded in most family planning clinics for ring users (48%) and oral contraceptive users (47%). In addition, it should be emphasized that reasons for discontinuation of the vaginal pill are in many cases trivial, and most patients opting for continuation have previous experience with other methods, including oral contraceptives, whereas many of those who discontinue want to experiment with other methods of which they have heard but with which they have had no personal experience.

Vaginal pills should be indicated for women both at the beginning (adolescents) and at the end of their reproductive life span because they are better tolerated and provoke almost no side effects. The adolescent is afraid of alterations in her body and menstrual function. They are also more sensitive to estrogen-induced nausea. The vaginal pill is almost free of such undesirable side effects. Oral contraceptives are also contraindicated for women after 35 years of age. However, many women refuse to discontinue because they have a sense of well-being while on the pill, which no other method can match or simply because it is the only method they trust. For these women, the vaginal pill offers the advantage of a virtually 100% effective method with much less risk of developing the undesirable changes in lipoproteins, which may lead to cardiovascular problems. This is particularly true with the desogestrel-ethinyl estradiol combination, which seems to be a natural vaginal pill. Vaginal pills should also be tried by women for whom intrauterine devices are contraindicated or those who develop gastrointestinal side effects while on oral contraceptives. The vaginal route may also be used for the administration of postcoital pills or luteolytic agents that require high dosage. Studies now in progress show that the incidence of nausea is significantly reduced when high-dose estrogens or progestogens are administered by the vaginal route as postcoital contraceptives.

References

1. Akinla O, Lakteenmaki P, Jackanicz TM: Intravaginal contraception with the synthetic progestin, R-2323. *Contraception* 1976;14:671.
2. Alvarez F, Faundes A, Johansson E, et al: Blood levels of levo-norgestrel in women following vaginal placement of contraceptive pills. *Fertil Steril* 1983;40:120–123.
3. Coutinho EM, Melo JF: Successful inhibition of spermatogenesis in man without loss of libido. A new method of male contraception. *Contraception* 1973;8:207.
4. Coutinho EM, Da Silva AR, Carreira CM, et al: Contraceptive effectiveness of silastic implants containing an anti-progesterone steroid (R-2323). *Contraception* 1975; 11:625–635.

5. Coutinho EM: Clinical experience with implant contraception. *Contraception* 1978;18:411–427.
6. Coutinho EM: Conservative treatment of uterine leiomyom with the anti-estrogen, anti-progesterone, R-2323. *Int J Gynecol Obstet* 1981;19:357–360.
7. Coutinho EM, Coutinho EJ: Rebouças Gonçalves MT et al: Ovulation suppression in women following vaginal administration of oral contraceptive tablets. *Fertil Steril* 1982;38:380–381.
8. Coutinho EM: Treatment of endometriosis with gestrinone (R-2323), a synthetic anti-estrogen, anti-progesterone. *Am J Obstet Gynecol* 1982;144:895–898.
9. Coutinho EM, Silva AR, Carreira C, et al: Ovulation inhibition following vaginal administration of pills containing norethindrone and mestranol. *Contraception* 1984;29:197–199.
10. Coutinho EM, Silva AR, Carreira C, et al: Conception control by vaginal administration of pills containing ethinyl-estradiol and dl-norgestrel. *Fertil Steril* 1984;42: 478–480.
11. Coutinho EM, Azadian-Boulanger G: Treatment of fibrocystic disease of the breast with Gestrinone, a new tri-enic synthetic steroid with anti-estrogen, anti-progesterone properties. *Int J Gynecol Obstet* 1984;22:363–366.
12. Cullberg G: Pharmacodynamic studies on desogestrel administered alone and in combination with ethinyl-estradiol. Acta Obstet Gynecol Scand [Suppl] 1985;133.
13. Mishell DR Jr, Lumkin M, Stone S: Inhibition of ovulation with cyclic use of progestogen impregnated intravaginal devices. *Am J Obstet Gynecol* 1972;113: 927–932.
14. Mishell DR Jr, Moore DE, Roy S: Clinical performances and endocrine profiles with contraceptive vaginal rings containing a combination of estradiol and d-norgestrel. *Am J Obstet Gynecol* 1978;130:55.
15. Mora G, Faundes A, Patore V: Clinical evaluation of an oral progestin contraceptive, R-2323, 5mg administered at weekly intervals. *Contraception* 1974;10:145–157.
16. Sakiz E, Azadian-Boulanger G, Raynaud JP: Antiestrogens, anti-progesterones. *Excerpta Medica Int Congr Ser* 1972;273:988–994.
17. Sakiz E, Azadian-Boulanger G, Laraguе F, et al: A new approach to estrogen-free contraception based on progesterone receptor blockade by mid-cycle administration of ethyl-norgestrienone (R-2323). *Contraception* 1974;10:467–474.
18. Segal SJ: Acceptance speech. The Axel Munthe Award. Capri, 1985.
19. Sivin I, Mishell D Jr, Victor A, et al: A multi-center study of levonorgestrel-estradiol contraceptive vaginal rings. II Subjective and objective measures of effects. An international comparative trial. *Contraception* 1981;24:359–376.
20. Sivin I, Mishell DR Jr, Victor A, et al: A multi-center study of the levonorgestrel-estradiol contraceptive vaginal rings. I. Use effectiveness. An international comparative trial. *Contraception* 1981;24:4:341–348.
21. Vessey MP, McPherson K, Johansson B: Mortality among women participating in the Oxford Family Planning Association Contraceptive Study. *Lancet* 1977;ii:731–733.
22. Victor A, Jackanicz TM, Johansson EDB: Vaginal progesterone for contraception. *Fertil Steril* 1978;30:631–635.

Part III
Neuroendocrine Regulation

CHAPTER 8

Neuroendocrine Regulation of Fertility

Béla Flerkó

Gonadotropin-Releasing Hormone (GnRH)-Associated Peptide

The decapeptide releasing follicle-stimulating hormone (FSH) and luteinizing hormone (LH) is the luteinizing hormone-releasing hormone (LHRH), called also gonadotropin-releasing hormone (GnRH). According to the findings of Seeburg et al (33), GnRH is biosynthesized in various tissues of human, rat, and mouse origin as part of a 69 amino acid long precursor molecule. The decapeptide sequence in this precursor is preceded by a signal sequence and followed by a site for proteolytic cleavage and C-terminal amidation as well as by a 56 amino acid residue long peptide that is now termed GAP for GnRH-associated peptide. GAP is itself a highly active hypophysiotropic neuropeptide that can be localized by immunocytochemical methods in all brain regions known to contain GnRH and that is secreted into portal blood via the median eminence (ME). Purified GAP produced by recombinant DNA technology was found in vitro to stimulate LH and FSH secretion and to be a potent inhibitor of prolactin secretion by pituitary lactotrophs. These findings provide an explanation for the observation that circulating levels of gonadotropins and prolactin are often inversely related.

LHRH Neuron System

According to the hypothesis of Knobil (20) gonadal function in both sexes is controlled by a cascade of neuroendocrine events beginning in the hypothalamus as a neuronal signal that generates the release of a bolus of the decapeptide LHRH into the pituitary portal circulation approximately once every hour. Each one of these signals causes the secretion by the pituitary gland of a pulse of the gonadotropin hormone into the peripheral circulation.

LHRH Neuron System of the Rat

After having described a system of delicate nerve fibers containing immunoreactive LHRH in the hypothalamus and ME of the rat (14,34,35), we visualized better the same neuron system with the aid of a new immunohistological technique developed in the Pécs Anatomy Department and called the silver intensified peroxidase-antiperoxidase method (16).

In the rat the LHRH synthesizing nerve cells sending their axons to the ME are found in the olfactory bulb and tubercle, diagonal band of Broca, medial septum, medial preoptic and suprachiasmatic area, anterior and lateral hypothalamus, and different regions of the hippocampus. The majority of axons originating from these LHRH synthesizing nerve cells enter the ME (12,15,25,35). Since we found the majority of the LHRH cells sending their axons to the capillary loops of the ME and pituitary stalk in the septal, suprachiasmatic, and especially in the preoptic area, we termed this pathway the preoptico-infundibular LHRH tract (13). The axons of this tract are found mainly in the septal, preoptic, suprachiasmatic, and lateral hypothalamic area as well as in the ME close to the ventral surface of the brain (Fig. 8.1). Many fibers of the same tract, however, pass through the periventricular gray matter on both sides of the third ventricle ("periventricular subependymal fascicle" in Fig. 8.1). Part of the latter fibers descend and pass through the arcuate nuclei and also reach the ME (Fig. 8.1) (31). The superficially running fibers of the preoptico-infundibular LHRH tract form three fascicles (Fig. 8.2). According to our findings (13,26), the median fascicle [1] is unpaired and forms a loop around the optic chiasm. The medial fascicles [2] at first run in the caudal direction, and then – sweeping around the lateral border of the suprachiasmatic nuclei – medially to join the median fascicle in the retrochiasmatic region. Also the lateral fascicles [3] run at first in caudal direction and lie against the dorsomedial margin of the optic tract. Toward the

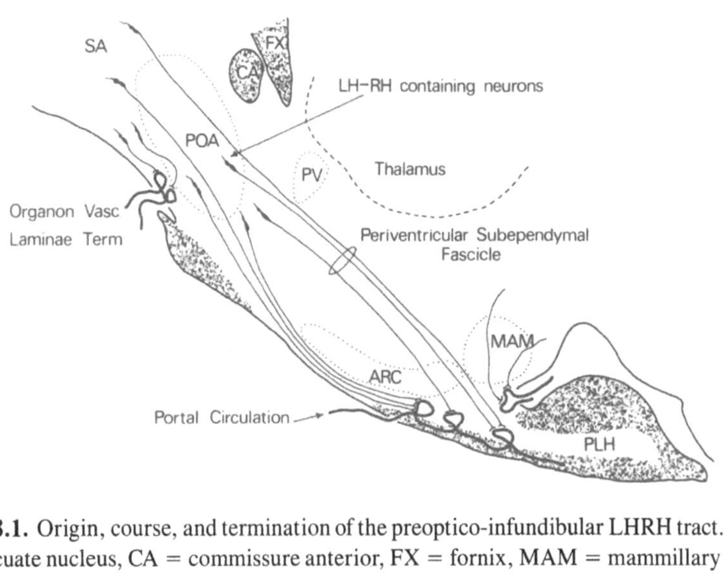

Fig. 8.1. Origin, course, and termination of the preoptico-infundibular LHRH tract. ARC = arcuate nucleus, CA = commissure anterior, FX = fornix, MAM = mammillary body, PLH = posterior lobe of the hypophysis, POA = preoptic area, PV = paraventricular nucleus, SA = septal area.

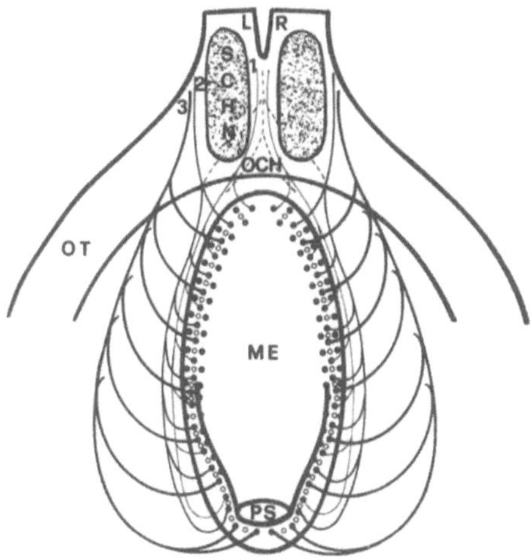

Fig. 8.2. Preoptico-infundibular LHRH tract of the rat compressed into one plane and superimposed on the schematic outline of the base of the brain, as it could be deduced from our deafferentation study (26). Arabic numbers 1, 2, and 3 indicate the median, medial, and lateral fascicles of the preoptico-infundibular LHRH tract, respectively (see also text). Fibers of all these fascicles form axon terminals along the entire extent of the ME. Filled and open circles stand for axon terminals of ipsi- or contralateral origin, respectively. R = right, L = left side; ME = median eminence; OCH = optic chiasm; OT = optic tract; PS = pituitary stalk; SCHN = suprachiasmatic nucleus. From Merchenthaler et al (26) by courtesy of Elsevier/North-Holland Biomedical Press, Amsterdam.

caudal direction their fibers gradually turn into medial direction. All three fascicles form axon terminals along the entire length of the ME and in the oral portion of the pituitary stalk. Similarly, LHRH axons of contralateral origin end along the entire length of the ME. Crossing LHRH fibers serve the anatomical explanation for the earlier finding of Tejasen and Everett (36) that unilateral preoptic stimulation could induce LH release even if the same side of the medial basal hypothalamus (MBH) was injured behind the optic chiasm.

In addition to the regions already mentioned, LHRH-containing axons of varying density could be found in the organon subfornicale and subcommissurale, in the stria medullaris and terminalis, in the temporal lobes around the amygdala, in the pyriform cortex, and in the mesencephalon of the rat. Termination and functional role of these LHRH fibers remain to be elucidated.

LHRH Neurons in LHRH-Deficient Hypogonadal Mouse Bearing Medial Preoptic Area Graft in the Third Ventricle

The basic role of the LHRH neuron system in the maintenance of the male and female fertility has also recently been demonstrated by the transplantation experiments of Charlton (7). In male genetically LHRH-deficient hypogonadal mice fetal medial preoptic area (mPOA), tissue grafted into the third ventricle resulted in an increase in pituitary LHRH receptors, in the synthesis and release of LH and FSH, and in the testicular LH receptors, with stimulation of full spermatogenesis within 60 days. Immunohistochemical analysis revealed a luxuriant outgrowth of LHRH-containing axons from many of the grafts, with more than 90% of these axons reaching the portal vessels of the ME, which is exactly where they should have grown in a normal animal.

In female hypogonadal mice fetal mPOA grafts have also increased pituitary LH and FSH synthesis and release with stimulation of the ovaries and subsequent uterine growth and vaginal opening. However, although waves of follicles developed in the ovaries, no animals have ovulated spontaneously. It appears that the mPOA graft is capable of producing a tonic stimulus to the pituitary, but the cyclic release of LHRH, enough to stimulate a surge of LH and ovulation does not occur. From previous studies it appeared likely that these hypogonadal females with mPOA grafts would display normal female mating behavior, and, indeed, many of them mated within two months of operation. A most unexpected finding, however, was that many of these females became pregnant. In other words, the evidence suggests that the grafting of normal fetal mPOA had converted the hypogonadal female into a reflex ovulator in a period of two months.

LHRH Neuron System of the Rhesus Monkey

Septal, preoptic, and lateral hypothalamic localization of the LHRH-synthesizing nerve cells sending their axons to the ME is characteristic, however, only of the rat and guinea pig. In the dog, about 40% of the LHRH cells are found in the arcuate-ventromedial and premammillary region (4), whereas in primates (3,17,45) and humans (2,6,5) the majority of the LHRH-containing cells mentioned above are found in these regions. This is a remarkable species difference between primates and laboratory rodents.

In the rhesus monkey the locus of the LHRH pulse generator and neuron system appears to be the arcuate nucleus region of the MBH, as determined by Knobil and associates with the aid of classical lesioning and deafferentation techniques (22). It has recently been described electrophysiologically by Wilson et al (40) in female rhesus monkeys by the recording of multiunit electrical activity from that region of the brain either directly or with the use of telemetry. This activity is characterized by a striking, precipitous increase in action potential frequency that, after an overshoot, descends to an elevated plateau lasting up to 20 minutes and then abruptly returns to baseline. These volleys of hypothalamic electrical activity immediately precede, or are coincident with, the beginning of each gonadotropic hormone pulse. The frequency of these events is reduced or arrested altogether by barbiturate anesthesia, by α-adrenergic and dopaminergic blockers as well as by opioids. The last mediates the effects of progesterone, which normally reduces the frequency of the pulse generator (18,21).

Changes in the frequency of LHRH pulses have a profound effect on the response of the pituitary to this stimulus. In the rhesus monkey, supraphysiologic frequencies induce a refractoriness of the pituitary to the releasing hormone, with a resultant reduction or total cessation of gonadotropin secretion. Reductions in the frequency of LHRH signals result in the alteration of normal LH to FSH ratios, with consequent disturbances in follicular maturation and ovulation. In contrast, normal ovulatory menstrual cycles can be reestablished in monkeys in which the source of endogenous LHRH has been interrupted, consequent to hypothalamic lesions or pituitary stalk section, by the administration of exogenous LHRH in the unvarying physiological frequency of one pulse per hour. The same regimen rapidly induces normal ovarian cycles in infantile females, suggesting that the functioning of the hypothalamic pulse generator is limiting in the initiation of puberty (20,21).

Knobil's hypothesis is well supported by the clinical use of pulsatile, long-term administration of LHRH by means of small portable computerized infusion pumps. Although for many years the continuous infusion of LHRH and the use of superagonists of LHRH did not prove helpful where a lack of endogenous LHRH existed, the pulsatile mode of application has shown its efficacy in many different situations. Today, pulsatile LHRH therapy is the treatment of choice for hypothalamic hypogonadism in the male and female (39).

The mechanisms by which the LHRH neurons are linked to fire in synchrony, the organization of the interval generator and its mechanism of modulation by steroids, and other factors present some of the most challenging issues in the field of neuroendocrinology.

Some Influences Modulating the Activity of the LHRH Neuron System

Since brain structures and peptides, putative neurotransmitters, and hormones modulating the activity of the LHRH neuron system have also lately been summarized (11,23,24,27,28,44), only a few recent experimental data bearing on this subject will be mentioned.

Estrogen Feedback and Acetylcholine

Kelly et al. (19) recently investigated the direct membrane actions of estrogen and the effect of estrogen on the response of MBH LHRH neurons to norepinephrine (NE), methoxamine (MX, an α_1-agonist), and acetylcholine (ACh) in the female guinea pig hypothalamic slice preparation using in vitro intracellular recording techniques, immunocytochemistry and radioimmunoassay of LHRH. Estrogen directly inhibits LHRH and other MBH neurons via a slow hyperpolarization and an increased membrane conductance. Moreover, these neurons are excited by NE via an α_1-receptor mechanism, and this response is increased by exogenous estrogen treatment. In many instances NE and MX will cause depolarizations and bursting, an electrophysiological criterion for peptide neurosecretion. Each affect is predominantly excitatory and appears to be unchanged by estrogen treatment. Preliminary studies on monkey brain slices indicate that primate MBH neurons respond similarly to estrogen, NE, and ACh. Therefore, estrogen has direct membrane actions and modulates the α_1 excitatory input to MBH neurons that may be involved in the negative and positive feedback, respectively, of estrogen on pulsatile gonadotropin secretion.

Norepinephrine

Neonatal exposure of female rats to testosterone reduces the ability of the adult hypothalamus to generate surges of prolactin (PRL) and LH in response to estrogen. Since NE and endogenous opioid peptides (EOPs) are intimately concerned with both surges, Dyer and co-workers (10) recently investigated whether neonatal testosterone propionate (TP) influences these systems.

First they established that "surge" amounts of LH and PRL can be released in TP-treated adult rats by electrical stimulation of the mPOA. After stimulation, plasma LH and PRL concentrations rose values higher than those obtained in oil-treated control rats (9). These results indicate that mPOA neurones are able to generate PRL and LH surges but are normally prevented from doing so. To investigate whether the prevention involves reduced NE drive to the LHRH neurons of the mPOA, they stimulated the noradrenergic projection to the hypothalamus and found that released amounts of LH and PRL were similar in TP-treated and control rats. Since direct mPOA stimulation released significantly more hormone in TP-treated rats, Dyer et al. (10) suggest that the NE input to the mPOA may indeed be weakened in these animals. This is in good agreement with our earlier finding that ovulation could be induced in neonatally TP-treated, persistent estrous rats with the intraventricular infusion of NE (37).

It is well-known that opiates inhibit NE release in mPOA (1). Dyer et al. (10) investigated whether this inhibition was greater in TP-treated females than in controls, but found morphine to be equally effective in both groups of animals. However, naloxone potentiated electrically stimulated NE release only in TP-treated rats, which suggests that in these rats there is increased activity of opioid input to NE nerve terminals. These findings suggest that neonatal TP modifies the EOP and ME circuitry so that treated rats are unable to generate LH and PRL surges spontaneously. However, the TP treatment does not alter the mPOA projection to the ME (sufficiently) to prevent electrically stimulated surges of the two hormones.

Although many experimental findings, especially those of Barraclough and co-workers (29,30,41–43), indicate that NE initiates the preovulatory surge of LH by evoking the release of newly accumulated LHRH from ME axon terminals, the results of Coen and co-workers (8) suggest that the contribution of NE to the preovulatory LH surge may be limited to its role in the biosynthetic pathway for epinephrine. It is possible that the process by which intraventricular NE stimulated the release of LH involves an initial uptake of NE into neurons and conversion of epinephrine. The latter amine has indeed been found to be more effective than the former in stimulating ovulation or LH release when administered intraventricularly, as it was demonstrated by Rubinstein and Sawyer (32) as well as by Vijayan and McCann (38). According to more recent findings of Coen and co-workers (8), the anatomical site at which epinephrine contributes to the ovulatory LH surge in the rat is suggested by the coincidence of that event with an increased turnover rate of epinephrine in the mPOA, the main site of the LHRH-containing perikarya which project to the ME. Since the pulsatile release of LH is also blocked by phenylethanolamine *N*-methyltransferase (PNMT) inhibitors and α_1 receptor antagonists, it is suggested by Coen that the increase in the turnover rate of preoptic area epinephrine at the time of the surge is associated with the increase in the frequency and amplitude of the pulses that constitute the surge and that this change in turnover is governed by a circadian oscillator probably located in the suprachiasmatic nucleus. Recent immunohistochemical studies indicate a marked PNMT-containing innervation of the region of the LHRH perikarya.

Conclusions

Although the entire menstrual and probably also the estrous cycle can deveolve in the presence of an unvarying circhoral hypophysiotrophic influence acting in a permissive manner, there can be no doubt that in a physiologic context, the neural component of the control system is profoundly influenced by higher centers as well as by gonadal and other hormones. These influences impinge on the LHRH neuron system by changing its frequency and modulating its activity.

References

1. Augood S, Diez-Guerra FJ, Dyer RG, et al: Evidence for opioid inhibition of noradrenaline (NA) release in the medial preoptic area (MPOA) of female rats. *J Physiol* 1986;371:213.
2. Barry J: Characterization and topography of LH-RH neurons in the human brain. *Neurosci Lett* 1976;3:287–291.
3. Barry J, Carette B: Immunofluorescence study of LRF neurons in primates. *Cell Tissue Res* 1975;164:163–178.
4. Barry J, Dubois MP: Immunofluorescence study of LRF producing neurons in the cat and the dog. *Neuroendocrinology* 1975;18:290–298.
5. Bloch B: Les neurones producteurs de LHRH chez l'homme au cours de la vie foetale. *Ann Sci Univ Franche Comté-Besancon* 1978;5:1–210.
6. Bugnon C, Bloch B, Fellmann D: Mise en évidence cyto-immunologique de neurones a LH-RH chez le foetus humain. *CR Hebd Seanc Acad Sci* 1976;282:1625–1628.
7. Charlton HM: Biological effects of neural transplants containing GnRH producing neurons. *Union Physiol Sci* 1986;XVI:237.
8. Coen C: Hypothalamic biogenic amines and the regulation of luteinizing hormone release in the rat. *Proc Int Union Physiol Sci* 1986;XVI:237.
9. Dyer RG, Mansfield S: Neonatal exposure to testosterone potentiates electrically stimulated gonadotrophin secretion in adult female rats. *J Physiol* 1985;364:53.

10. Dyer RG, Mansfield S, Bicknell RJ, et al: Effect of neonatal testosterone upon hypothalamic mechanisms controlling LH and prolactin secretion in female rats. *Proc Int Union Physiol Sci* 1986;16:312.
11. Flerkó B: Hypothalamic mediation of neuroendocrine regulation of hypophysial gonadotrophic functions, in Greep RO (ed): *MTP International Review of Science, Physiology Series One, vol VIII, Reproductive Physiology.* London, Butterworths, 1974, pp 1–32.
12. Flerkó B: The LH-RH neuron system of the rat and rabbit and a new insight into the hypophysiotrophic area of the hypothalamus. *Folia Endocrinol Jpn* 1981;57:292–297.
13. Flerkó B: LH-RH neuron system, the control of pituitary gonadotrophin secretion, in Endröczi E, Angelucci L, Scapagnini W, DeWied D (eds): *Neuropeptides, Neurotransmitters and Regulation of Endocrine Processes.* Budapest, Akadémiai Kiadó 1983, pp 1–10.
14. Flerkó B, Sètàlò G, Vigh S, et al: The luteinizing hormone-releasing hormone (LH-RH) neuron system in the rat and rabbit, in Scott DE, Kozlowski GP, Weindl A (eds): *Brain-Endocrine Interaction III. Neural Hormones and Reproduction.* Basel, Karger, 1978, pp 108–116.
15. Flerkó B, Sètàlò Gy, Vigh S, et al: Recent immunohistological findings on the LH-RH neuron system of the rat. *Mater Med Pol* 1980;12:119–123.
16. Gallyas F, Görcs T, Merchenthaler I: High-grade intensification of the end-product of the diaminobenzidine reaction for peroxidase histochemistry. *J Histochem Cytochem* 1982;30:183–184.
17. Hoffman GE: Immunocytochemical localization of luteinizing hormone-releasing hormone (LH-RH) in murine and primate brain. *Anat Rec* 1976;184:429–430.
18. Kaufman J-M, Kesner JS, Wilson RC, et al: Electrophysiological manifestation of luteinizing hormone-releasing hormone pulse generator activity in the rhesus monkey: influence of α-adrenergic and dopaminergic blocking agents. *Endocrinology* 1985;116:1327–1333.
19. Kelly MJ, Condon TP, Ronnenkleiv OK: The endogenous electrical activity of hypothalamic parvocellular neurosecretory neurons: direct modulation by estrogen. *Proc Int Union Physiol Sci* 1986;XVI:433.
20. Knobil E: The neuroendocrine control of the menstrual cycle. Rec. *Prog Horm Res* 1980;36:53–88.
21. Knobil E: The hypothalamic control of pulsatile gonadotropic hormone secretion. *Proc Int Union Physiol Sci* 1986;XVI:432.
22. Krey LC, Butler WR, Knobil E: Surgical disconnection of the medial basal hypothalamus and pituitary function in the rhesus monkey. I. Gonadotropin secretion. *Endocrinology* 1975;96:1073–1087.
23. McCann SM: The role of brain peptides in the control of anterior pituitary hormone secretion, in Müller EE, MacLeod RM (eds): *Neuroendocrine Perspectives vol I.* Amsterdam, Elsevier, 1982, pp 1–22.
24. McCann SM, Krulich I, Ojeda SR, et al: Neurotransmitters in the control of anterior pituitary function, in Fuxe K, Hökfelt T, Luft R (eds): *Central Regulation of the Endocrine System.* New York, Plenum Press, 1979, pp 329–347.
25. Merchenthaler I, Görcs T, Sètàlò Gy, et al: Gonadotropin-releasing hormone (GnRH) neurons and pathways in the rat brain. *Cell Tissue Res* 1984;237:15–29.
26. Merchenthaler I, Kovàcs G, Lovàsz G, et al: The preoptico-infundibular LH-RH-tract of the rat. *Brain Res* 1980;198:63–74.
27. Ojeda SR, McCann SM: Control of LH and FSH release by LHRH: influence of putative neurotransmitters. *Clin Obstet Gynaecol* 1978;5:283–303.

28. Ojeda SR, McCann SM: The possible role of prostaglandins in the control of pituitary hormone secretion, in Lederis K, Veale WL (eds): *Current Studies of Hypothalamic Function, vol I.* Basel, Karger, 1978, pp 106–121.
29. Rance N, Barraclough CA: Effects of phenobarabital on hypothalamic LHRH and catecholamine turnover rates in proestrous rats. *Proc Soc Exp Biol Med* 1981;166: 425–431.
30. Rance N, Wise PM, Selmanoff M, et al: Catecholamine turnover rates in discrete hypothalamic areas and associated changes in median eminence luteinizing hormone-releasing hormone and serum gonadotropins on proestrus and diestrous day 1. *Endocrinology* 1981;108:1795–1802.
31. Rèthelyi M, Vigh S, Sètàlò G, et al: The luteinizing hormone releasing hormone-containing pathways and their co-termination with tanycyte processes in and around the median eminence and in the pituitary stalk of the rat. *Acta Morphol Acad Sci Hung* 1981;29:259–283.
32. Rubinstein L, Sawyer CH: Role of catecholamines in stimulating the release of pituitary ovulating hormone(s) in rats. *Endocrinology* 1970;86:988–995.
33. Seeburg PH, Nikolics K, Ramachandran J: Structure and biological properties of the GnRH precursor. *Proc Int Union Physiol Sci* 1986;XVI:237.
34. Sètàlò G, Vigh S, Schally AV, et al: LH-RH-containing neural elements in the rat hypothalamus. *Endocrinology* 1975;96:135–142.
35. Sètàlò Gy, Vigh S, Schally AV, et al: Immunohistological study of the origin of LH-RH-containing nerve fibers of the rat hypothalamus. *Brain Res* 1976;103:597–602.
36. Tejasen T, Everett JM: Surgical analysis of the preoptico-tuberal pathway controlling ovulatory release of gonadotropins in the rat. *Endocrinology* 1967;81:1387–1396.
37. Tima L, Flerkó B: Ovulation induced by norepinephrine in rats made anovulatory by various experimental procedures. *Neuroendocrinology* 1974;15:346–354.
38. Vijayan E, McCann SM: Re evaluation of the role of catecholamines in control of gonadotropin and prolactin release. *Neuroendocrinology* 1978;25:150–165.
39. Wagner TOF: Clinical relevance of episodic hormone secretion. *Proc Int Union Physiol Sci* 1986;XVI:433.
40. Wilson RC, Kesner JS, Kaufman J-M, et al: Central electrophysiologic correlates of pulsatile luteinizing hormone secretion in the rhesus monkey. *Neuroendocrinology* 1984;39:256–260.
41. Wise PM, Camp-Grossman P, Barraclough CA: Effects of estradiol and progesterone on plasma gonadotropins, prolactin, and LHRH in specific brain areas of ovariectomized rats. *Biol Reprod* 1981a;24:820–830.
42. Wise PM, Rance N, Barraclough CA: Effects of estradiol and progesterone on catecholamine turnover rates in discrete hypothalamic regions in ovariectomized rat. *Endocrinology* 1981b;108:2186–2193.
43. Wise PM, Rance N, Selmanoff M, et al: Changes in radioimmunoassayable luteinizing hormone releasing hormone in discrete brain areas of the rat at various times on proestrus, diestrus day 1, and after phenobarbital administration. *Endocrinology* 1981c;108:2179–2185.
44. Zigmond RE: Binding, metabolism and action of steroid hormones in the central nervous system, in Iversen LL, Iversen LD, Snyder SH (eds): *Handbook of Psychopharmacology vol. V.* New York, Plenum Press, 1975, pp 239–328.
45. Zimmerman EA, Antunes JL: Organization of the hypothalamic pituitary system: current concepts from immunohistochemical studies. *J Histochem Cytochem* 1976;24:807–815.

CHAPTER 9

GnRH Analogs in Contraception

Regine Sitruk-Ware

Introduction

We are fast approaching an era in which "zero population growth" will be a vital necessity rather than merely a desirable alternative. Reducing the millions of unplanned and unwanted pregnancies worldwide would solve many economic and health problems and would enable developing countries, particularly, to improve their standard of living (29).

Currently available methods of contraception are either inadequate, inconvenient, or unacceptable to some groups and individuals (35). Safer, simpler, and more convenient methods than the ones now available will help to control population growth and should bring about maximal benefits to health. Among several contraceptive approaches being considered and developed is one based on analogs of the hypothalamic hormone controlling secretion of both luteinizing hormone (LH) and follicle-stimulating hormone (FSH) from the anterior pituitary gland, called luteinizing hormone-releasing hormone (LHRH) or gonadotropin-releasing hormone (GnRH).

General Considerations

Since 1971, date of the first isolation and determination of structure and synthesis of the hormone by Schally et al (30–32), more than 1,000 analogs have been generated. A number of highly potent peptidic derivatives have been described and are in various stages of clinical evaluation (24).

The initial enthusiasm and success in synthesis of potent agonistic analogs of GnRH was related to their expected profertility potential. Early findings of gonadal suppressive effects with these agents led to the discovery of pituitary desensitization and down-regulation caused by continuous presence of GnRH and its analogs (10,23). It was then discovered that physiologically GnRH must be secreted in a pulsatile manner to exert activation of the pituitary-gonadal axis leading to gonadal steroidogenesis and gametogenesis (19).

Therefore the development of potent long-lived analogs have received attention as inhibitors of ovulation. A nonsteroidal alternative in hormonal contraception, free of side effects, has then attracted organizations developing methods of family planning. Schally et al (31) initially forecast that replacement and deletion of some amino acids in GnRH might result in an analog possessing the structural features requisite for effective binding with the receptor but lacking those necessary to elicit the physiological stimulation of LH and FSH (31). These analogs have been developed as inhibitors of GnRH and antagonists. On the other hand it has been established by the same group that maintenance of the first three amino acids was necessary to maintain biological activity and elicit LH and FSH responses. Substitutions in position 6,10 both gave more potent analogs with a more prolonged activity than GnRH itself, leading to potent agonistic analogs 50 to 100 times more potent than the native peptide (13,14).

At the moment different agonistic analogs have reached the stage of clinical development and some of them are marketed. Table 9.1 summarizes the main compounds developed for contraceptive purpose.

Mechanisms of Action of GnRH Analogs

The GnRH analogs exert their major actions through interaction with the receptors on the pituitary gonadotropes. The mechanism by which GnRH agonists exert their paradoxical inhibitory effects on reproductive function is complex. Several possible mechanisms of action have been proposed: desensitization of the pituitary gonadotropes, down-regulation of gonadal gonadotropin receptors with reduction of steroidogenic response of the ovary, and direct inhibitory actions of GnRH agonists on the reproductive organs. It has been shown by autoradiographic analysis in rat pituitary cells that GnRH agonists interact with a binding site on the plasma membrane and are internalized by receptor-mediated endocytosis. Then the peptide is delivered by vesicular transport to Golgi elements and to lysosomes (16). It has also been demonstrated that the specific receptors for GnRH in cultured rat pituitary cells were increased by subnanomolar concentrations of these peptides (23). Therefore, physiological concentrations of GnRH binding to GnRH receptors induces sensitization and achieves hormone secretion whereas exposure to high ligand concentrations causes net loss of GnRH receptors and consequent desensitization of gonadotropin release. GnRH antagonists also bind to the same receptors as agonists (11,18); however, their binding characteristics are different from those of agonists. They do not alter GnRH binding capacity and block the increase in sites induced by GnRH (10,23). They are unable to induce receptor regulation and refractoriness. To summarize, GnRH agonists may cause down-regulation or densensitization of pituitary receptors after a simple exposure, but antagonists must be continuously present to compete with the endogenous peptide for receptor occupancy and do not induce down-regulation of receptors (23).

Table 9.1. GnRH agonists in clinical development for contraception.

Compound	Name	Level of study	Dosage form
[D-Trp^6]GnRH	Tryptorelin	advanced clinical trials	injection, injectable polymer
[D-Trp^6,NMeLeu^7,Pro^{9}NEt]GnRH	Lutrelin	clinical trials	injection
[D-Ser(tBu)^6,Pro^{9}NEt]GnRH	Buserelin	advanced clinical trials	nasal spray
[D-His(Bzl)^6,Pro^{9}NEt]GnRH	Histrelin	clinical trials	nasal

Although GnRH receptors have been demonstrated in human anterior pituitary tissue, they have not been found in human gonadal tissue, either human corpus luteum or testis (12). Therefore a direct gonadal inhibitory action of GnRH analogs is unlikely, and their effect is mediated through pituitary inhibition.

Thus, in humans, desensitization of the pituitary processes responsible for gonadotropin secretion is believed to be the main action of the chronic GnRH agonist treatment. Repeated administration of D-Ser(tBu)6,Pro9-NHEt-GnRH (Buserelin) to either normally ovulating or postmenopausal women reduces the pituitary's ability to release gonadotropins (1,2). Furthermore, failure of positive feedback of estradiol has been demonstrated in chronic GnRH agonist-treated women (5). The normal pulsatile GnRH and LH release from the hypothalamus and pituitary, respectively, may be abolished by the continuous administration of GnRH or by daily administration of potent, long-acting GnRH agonists (37).

Although extrapituitary effects of GnRH and its agonists have been clearly demonstrated in experimental animals, treatment with high doses of GnRH or its agonists did not alter the steroidogenic effect of FSH on cultured human granulosa cells (8). Furthermore, women respond to exogenous human gonadotropins with a similar estrogen increase both during and after chronic GnRH agonist treatment (26). These data, together with the fact that receptors for GnRH have so far not been detected in human gonads (12), indicate that GnRH agonists probably do not directly inhibit the human ovary.

Contraception in Women

Ovulation Inhibition

Several approaches to contraception have been tried since GnRH analogs became available in clinical studies. Disruption of the normal cycle by administration of agonists in the early follicular phase or in the luteal phase has been successful but disappointing as far as pregnancy interruption is concerned in humans (7,36). An immunological approach by using an antiserum to GnRH and also use of GnRH antagonists have not yet proved useful for contraception. However daily intranasal administration of an LHRH agonist has been the basis of a new successful approach to contraception.

Nillius et al (26) conducted long-term studies using daily intranasal spray administration of D-Ser(tBu)6,Pro9-NHEt-LHRH, (Buserelin) with a high rate of effectiveness and acceptability. They showed in 71 women regularly menstruating that a once daily nasal dose of Buserelin was effective in 100% of the cases with no pregnancy occurrence. However, on 564 treatment months only 561 were anovulatory. The bleeding pattern was variable from regular intervals to amenorrhea, with short periods of spotting in three women at the beginning of therapy. Heavy or prolonged bleeding never occurred. Estradiol(E_2) levels were decreased during therapy but remained at levels of 60 to 80 pg/mL, values observed in the normal early follicular phase. The endometrial aspects were either inactive, atrophic, or weakly proliferative, according to the E_2 levels at the time of biopsy (4). After cessation of prolonged therapy a normal reproductive function was rapidly restored. The first normal cycle was 33 days on average. However, prolonged use of GnRH agonists could be questionable. Indeed, in some amenorrheic women with inactive endometrium, very low levels of E_2 could be, on a long-term basis, responsible for hypoestrogenic symptoms and possible osteoporosis. This has not been the case in the Nillius et al study (26). On the other hand, in women whose E_2 levels remain high and unopposed by progesterone, one could question the possibility of endometrial hyperstimulation. In no case of long-term users did Nillius et al (26) report hyperplasia. However, one of the future issues is to envisage GnRH analog treatment associated to a ten-day course of progesterone on a monthly basis as an adjuvant. This alternative has been tested by Lemay et al (20) and proved to be useful as far in bleeding control.

Luteolysis

Previous trials have shown that successful luteolysis was achieved when D-Trp6, Pro9-NHEt GnRH was administered between five and eight days after the LH peak (7,34). Lemay et al (21) made the same observation with high doses of intranasal buserelin. However, it was demonstrated that the luteolytic effect of GnRH agonist was completely prevented by the presence of human chorionic gonadotropin (HCG) (3,9). Therefore, the use of GnRH agonists as postovulatory luteolytic agents is doubtful. Although pregnancy termination in baboons was reported after GnRH agonist administration (15), it failed to induce early abortion in humans (36).

Contraception in Men

The male reponse to daily treatment with GnRH agonist is suppression of circulating levels of testosterone into the castrate range (22). Testicular biopsies performed in men while under therapy revealed the absence of spermatogenesis

(27). These findings have been shown in elderly patients treated with GnRH agonists for prostate cancer. In contrast, in young, healthy men the fall in sperm count is more difficult to achieve. Different trials (6,22,25) have indicated that continuous administration of GnRH agonist enhanced desensitization, and testosterone suppression could be obtained but not azoospermia.

Impotence and hot flashes occur under therapy (22) in relation to the drop in testosterone levels. Therefore, androgen supplementation is needed to maintain normal sexual life. In most trials androgen replacement therapy can partially maintain spermatogenesis (25), and until recently it was unlikely that GnRH agonists could be useful as male contraception. However, recently more promising results have been obtained. D-Trp6-GnRH analog was administered monthly intramuscularly in biodegradable microspheres of poly(D,L lactide-CO-glycolide) (6). The microspheres released 100 μg/d. Androgen replacement therapy was administered after the first month intramuscularly. Although testosterone was replaced, azoospermia was obtained after 28 weeks of treatment (6).

At the moment, the combined use of GnRH agonists and testosterone for contraception in men is still questionable and require extensive investigations.

Antagonists

Several recently synthesized antagonists are already potent enough to be practical for clinical tests (28,40). In women it has been shown that GnRH antagonists can suppress endogenous or GnRH-induced gonadotropin release (40) and inhibit ovulation (41).

The use of antagonistic analogs of GnRH as luteolytic agents proved to be successful when the peptide was administered during midluteal phase (17). However, as well as observed with the agonists, antagonists action is overruled by endogenous HCG and fails to interrupt early pregnancies in rhesus monkeys (39). In male primates GnRH antagonists effectively induce complete azoospermia and total inhibition of fertility (38) even in the face of exogenous testosterone supplementation (37). However, antagonistic analogs are active in both men and women at considerably higher doses than agonists. Therefore, their use will result in a higher daily cost of treatment than with agonists. Also, evaluation of antagonists actions in chronic administration in humans will have to await the completion of long-term toxicity studies.

Modes of Administration of LHRH Analogs

Frequent parenteral administration of GnRH agonists would be inconvenient and hence impractical. Oral administration requires doses one thousand times larger than those effective by the parenteral route (29). Intranasal application has been found to be a convenient and practical way of administering GnRH agonists (26). Injectable microcapsules with GnRH agonists prepared from biodegradable biocompatible polymer were designed for controlled release of the agonists over a 30-day period (33). This once-a-month intramuscular administration would avoid patient compliance problems and failures to follow daily regimens properly.

Conclusion

To date, long-term clinical trials have demonstrated that continuous intranasal GnRH agonist treatment for inhibition of ovulation can provide safe, effective, and reversible contraception in women with no serious side effects. The only drawbacks are related to light bleeding irregularities. No luteolytic effect nor abortifacient effect of superactive GnRH agonists have been observed in humans.

In men, peptide contraception is still at its very early beginning and its effectiveness on a long-term basis remains to be proven. At the moment, ovulation inhibition in women by continuous GnRH agonist treatment appears to be the most promising lead to peptide contraception.

References

1. Bergquist C, Nillius SJ, Wide L: Inhibition of ovulation in women by intranasal treatment with a luteinizing hormone-releasing hormone agonist. *Contraception* 1979; 19:497.
2. Bergquist C, Nillius SJ, Wide L: Reduced gonadotropin secretion in post menopausal women during treatment with a stimulatory LHRH analogue. *J Clin Endocrinol Metab* 1979;49:472.
3. Bergquist C, Nillius SJ, Wide L: Luteolysis induced by a luteinizing hormone releasing hormone agonist is prevented by human chronic gonadotropin. *Contraception* 1980;22:341.
4. Bergquist C, Nillius SJ, Wide L, et al: Endometrial patterns in women on chronic LHRH agonist treatment for inhibition of ovulation. *Fertil Steril* 1981;36:339.
5. Bergquist C, Nillius SJ, Wide L: Failure of positive feedback of Oestradiol during chronic intranasal luteinizing hormone releasing hormone agonist treatment. *Clin Endocrinol* 1982;16:147.
6. Bouchard P, Blondet C, Spitz I, et al: Sperm supression by long acting GnRH agonist therapy in men: critical effect of testosterone substitution and duration of therapy. 68th Annual meeting of the Endocrine Society, Anaheim. Abstract 498.
7. Casper RF, Yen SSC: Induction of luteolysis in the human with a long-acting analog of luteinizing hormone-releasing factor. *Science* 1979;205:408.
8. Casper RF, Eriksson GF, Rebar RW, et al: The effect of luteinizing hormone-releasing factor and its agonist on cultured human granulosa cells. *Fertil Steril* 1982;37:406.
9. Casper RF, Sheehan KL, Yen SSC: Chorionic gonadotropin prevents LRH agonist-induced luteolysis in the human. *Contraception* 1980;21:471.
10. Clayton RN: Gonadotropin-releasing hormone modulation of its own pituitary receptors: Evidence for biphasic regulation. *Endocrinology* 1982;152:161.
11. Clayton RN, Catt KJ: Receptor binding affinity of gonadotropin releasing hormone analogs: analysis by radioligand-receptor assay. *Endocrinology* 1980;106:1154.

12. Clayton RN, Huhtaniemi IT: Absence of gonadotropin-releasing hormone receptors in human gonadal tissue. *Nature (London)* 1982;299:56.
13. Coy DH, Labrie F, Savary M, et al: LH-releasing activity of potent LHRH analogs in vitro. *Biochem Biophys Res Commun* 1975;67:576.
14. Coy DH, Vilchez-Martinez JA, Coy EJ, et al: Analogs of luteinizing hormone-releasing hormone with increased biological activity produced by D. amino acid substitutions in position 6. *J Med Chem* 1976;19:423.
15. Das C, Talwar GP: Pregnancy terminating action of a luteinizing hormone releasing hormone agonist D-Ser(But)6 desGly1O ProEA in baboons. *Fertil Steril* 1983;39: 218.
16. Duello TM, Nett TM, Farquhar MG: Fate of a gonadotropin-releasing hormone agonist internalized by rat pituitary gonadotrophs. *Endocrinology* 1983;112:1.
17. Fraser H, Baird DT, McRae GI, et al: Suppression of luteal progesterone secretion in the stumptailed macaque by an antagonistic analogue of luteinizing hormone-releasing hormone. *J Endocrinol* 1985;104:R1.
18. Hazum E, Keinand D: Gonadotropin releasing-hormone receptors: photo affinity labeling with an antagonist. *Biochem Biophys Res Com* 1983;110:116.
19. Knobil E: The neuro-endocrine control of the menstrual cycle. *Recent Prog Horm Res* 1980;36:53.
20. Lemay A, Faure N, Labrie F, et al: Inhibition of ovulation during discontinuous intranasal luteinizing hormone-releasing hormone agonist dosing in combination with gestagen-induced bleeding. *Fertil Steril* 1985;43:868.
21. Lemay A, Labrie F, Belanger A, et al: Luteolytic effect of intranasal administration of (D-Ser(TBU)6,DES-GLY-NH2 10)- luteinizing hormone- releasing hormone ethylamide in normal women. *Fertil Steril* 1979;32:646.
22. Linde R, Doelle GC, Alexander N, et al: Reversible inhibition of testicular steroidogenesis and spermatogenesis by a potent gonadotropin-releasing hormone agonist in normal men. *N Engl J Med* 1981;305:663.
23. Loumaye E, Catt KJ: Homologous regulation of gonadotropin-releasing hormone receptors in cultured pituitary cells. *Science* 1982;215:983.
24. Nestor JJ Jr: Development of agonist LH RH analogs, in Vickery BH, Nestor JJ Jr, Hafez ESE (eds): *LH RH and Its Analogs: Contraceptive and Therapeutic Applications*. Boston, MTP Press, 1984, p 3.
25. Nieschlag E, Akhtar FB, Weinbauer GF, et al: LHRH analogues for male fertility control: experiments in monkeys and men. *J Steroid Biochem* 1984;20:1374.
26. Nillius SJ, Berguist C, Wide L: Inhibition of ovulation in women by chronic treatment with a stimulatory LHRH analogue. A new approach to birth control? *Contraception* 1978;17:537.
27. Rajfer J, Swerdloff RS, Heber DM: Testicular histology following chronic GnRH agonist treatment. *Fertil Steril* 1984;42:765.
28. Rivier C, Rivier J, Vale N: Antireproductive effects of a potent GnRH antagonist in the female rat. *Endocrinology* 1981;108:1425.
29. Schally AV: Research frontiers in fertility regulation. *PARFR* 1983;2:1.
30. Schally AV, Arimura A, Baba Y, et al: Isolation and properties of the FSH and LH-releasing hormone. *Biochem Biophys Res Commun* 1971;43:393.
31. Schally AV, Arimura A, Kastin AJ, et al: Gonadotropin releasing hormone: one polypeptide regulates secretion of luteinizing and follicle-stimulating hormones. *Science* 1971;173:1036.

32. Schally AV, Nair RMG, Redding TW, et al: Isolation of the luteinizing hormone and follicle-stimulating hormone-releasing hormone from porcine hypothalami. *J Biol Chem* 1971;246:7230.
33. Schally AV, Redding TW: Combination of long-acting micro capsules of the D-tryptophan-6analog of luteinizing hormone-releasing hormone with chemotherapy: investigation in the rat prostate cancer model. *Proc Natl Acad Sci USA* 1985;82:2498.
34. Sheehan KL, Casper RF, Yen SSC: Induction of luteolysis by luteinizing hormone-releasing factor (LRF) agonist: sensitivity, reproducibility and reversibility. *Fertil Steril* 1982;36:664.
35. Sciarra JJ: The continuing need for contraceptive research. *Fertil Steril* 1981;36:697.
36. Tolis G, Comaru-Schally AM, Mehta AE, et al: Failure to interrupt established pregnancy in humans by D-tryptophan-6-luteinizing-hormone-releasing hormone. *Fertil Steril* 1981;36:241.
37. Vickery BH: Comparison of the potential for therapeutic utilities with gonadotropin-releasing hormone Agonists and Antagonists. *Endocrine Rev* 1986;7:115.
38. Weinbauer GF, Surmann FJ, Akhtar FB, et al: Reversible inhibition of testicular function by a gonadotropin hormone-releasing hormone antagonist in monkeys (Macaca fascicularis). *Fertil Steril* 1985;42:906.
39. Wilks JW, Folkers K, Smith CW: Activities of LHRH antagonists in rhesus monkeys during the periovulatory interval of the menstrual cycle and early pregnancy, in Zatuchni GI, Shelton JD, Sciarra JJ (eds): *LHRH Peptides as Female and Male Contraceptives*. Philadelphia, Harper & Row, 1981, p 151.
40. Zarate A, Canales E, Schally AV, et al: The use of LHRH agonists and antagonists as antifertility agents in the human female, in Zatuchni GI, Shelton JD, Sciarra JJ (eds): *LHRH Peptides as Female and Male Contraceptives*. Philadelphia, Harper & Row, 1981, p 227.
41. Zarate A, Canales ES, Sthory I, et al: Anovulatory effect of a LHRH antagonist in women. *Contraception* 1981;24:315–320.

CHAPTER 10

A Novel Hypothalamic Peptide Inhibitor of Prolactin Secretion

J. Ramachandran, K. Nikolics, and P.H. Seeburg

Introduction

The concept of neural regulation of the pituitary gland was developed four decades ago (23). Since then a great deal has been learned about the function and structural organization of the complex and highly integrated system controlling the secretion of trophic hormones from the pituitary. The basal secretion of the pituitary is controlled by the mediobasal hypothalamus whereas most stress-induced and physiologic patterns of pituitary secretion are regulated by both hypothalamic and extrahypothalamic regions of the brain. Neurotransmitters are involved primarily in transferring information from extrahypothalamic sites to hypothalamic peptidergic neurons, which are the major controlling links between the brain and the endocrine system.

As a result of efforts in several laboratories during the past two decades, hypothalamic peptide factors regulating the secretion of all but one pituitary hormone were isolated and characterized. The exception was prolactin (PRL), the secretion of which is under chronic inhibitory control of the hypothalamus. It has been observed that removal of the influence of the hypothalamus by lesions, transection of the pituitary stalk, or transplantation of the pituitary to noncranial sites, results in elevated serum PRL levels. Furthermore, Talwalker et al (46) demonstrated that the increased secretion of PRL by the pituitary in vitro could be inhibited by rat hypothalamic extracts. Although dopamine is known to inhibit PRL secretion (6,44), there is evidence that additional factors are involved in the hypothalamic inhibition of PRL secretion (11,14). Attempts to isolate hypothalamic PRL release inhibiting factor(s) (PRIF) have shown that areas containing dopamine-free PRIF activity were located in the mediobasal hypothalamus and the organum vasculosum of the lamina terminalis (OVLT) (13), regions that are known to have many neurons producing and secreting gonadotropin-releasing hormone (GnRH) (45). Studies with ovine and porcine hypothalamic tissue indicated that the PRIF activity was associated with peptide(s) of molecular weight $<8,000$ (13,14,22).

Since attempts to isolate this peptidic PRIF from hypothalamic tissue were not successful, dopamine came to be considered as the major PRIF during the late seventies and eighties. The elucidation of the structure of a hypothalamic peptide fulfilling the criteria for a PRIF was recently accomplished by the use of recombinant DNA technology. The discovery of this peptide and its biological properties are discussed below.

Biosynthetic Precursor of GnRH

Although mature GnRH is a decapeptide blocked at both the amino and carboxyl termini, high molecular weight immunoreactive forms of GnRH were reported by several laboratories (4,16,33,19). High molecular weight precursors of GnRH

```
                                         -23         -20
                                         MET lys pro ile gln lys leu leu ala gly leu
H   GGGATCTTTTTGGCTCTCTGCCTCTAAACAGA     ATG AAG CCA ATT CAA AAA CTC CTA GCT GGC CTT      65
R   TGGTATCCCTTTGGCTTTCACATCCAAACAGA     ATG GAA ACG ATC CCC AAA CTG ATG GCC GCT GTT
                                         MET glu thr ile pro lys leu met ala ala val

        -10                                     1
    ile leu leu thr trp cys val glu gly cys ser ser gln his trp ser tyr gly leu arg
H   ATT CTA CTG ACT TGG TGC GTG GAA GGC TGC TCC AGC CAG CAC TGG TCC TAT GGA CTG CGC      125
R   GTT CTG TTG ACT GTG TGT TTG GAA GGC TGC TCC AGC CAG CAC TGG TCC TAT GGG TTG CGC
    val leu leu thr val cys leu glu gly cys ser ser gln his trp ser tyr gly leu arg

        10                                      20
    pro gly gly lys arg asp ala glu asn leu ile asp ser phe gln glu ile val lys glu
H   CCT GGA GGA AAG AGA GAT GCC GAA AAT TTG ATT GAT TCT TTC CAA GAG ATA GTC AAA GAG      185
R   CCT GGG GGG AAG AGA AAT ACT GAA CAC TTG GTT GAT TCT TTC CAA GAG ATG GGC AAG GAG
    pro gly gly lys arg asn thr glu his leu val asp ser phe gln glu met gly lys glu

        30                                      40
    val gly gln leu ala glu thr gln arg phe glu cys thr thr his gln pro arg ser pro
H   GTT GGT CAA CTG GCA GAA ACC CAA CGC TTC GAA TGC ACC ACG CAC CAG CCA CGT TCT CCC      245
R   GAG GAT CAA ATG GCA GAA CCC CAG AAC TTC GAA TGC ACT GTC CAC TGG CCC CGT TCA CCT
    glu asp gln met ala glu pro gln asn phe glu cys thr val his trp pro arg ser pro

        50                                      60
    leu arg asp leu lys gly ala leu glu ser leu ile glu glu glu thr gly gln lys lys
H   CTC CGA GAC CTG AAA GGA GCT CTG GAA AGT CTG ATT GAA GAG GAA ACT GGG CAG AAG AAG      305
R   CTT AGG GAT CTG CGA GGA GCT CTG GAA CGT CTG ATT GAA GAG GAA GCT GGG CAG AAG AAG
    leu arg asp leu arg gly ala leu glu arg leu ile glu glu glu ala gly gln lys lys

     69
    ile END
H   ATT TAA ATCCATTGGGCCAGAAGGAATGACCATTACTAACATGACTTAAGTATAATTCTGACATTGAAAATTTATAA      382
R   ATG TAA -TGCACTGGCCCAGAAGG-AT--CCA----CAACAT---CCGAGTGTGACATTGACGCTGAGA-TCCATGA
    met END

H   CCCATTAAATACCTGTAAATGGTATGAATTTCAGAAATCCTTACACCAAGTTGCACATATTCCATAATAAAGTGCTGTG      461
R   CCTGTTACAGCTCTGTAATTG-TGTGGGCTTCCGCATTTCTATACCC-AGCAGCGTAGATTTCATAATAAAGTAATGTG

H   TTGTGAATG-polyA
R   TTGTGGATC-polyA
```

Fig. 10.1. Nucleotide and amino acid sequences of cloned human (H) and rat (R) hypothalamic cDNAs encoding the GnRH-PIF protein precursor. Amino acid numbers appear above the sequences, with negative numbers corresponding to the presumed signal peptides. The number of the last nucleotide in each line is indicated on the right and corresponds to the cloned human cDNA sequence. The 3′ untranslated regions were aligned to maximize homology. The signal for polyadenylation is underlined. Arrows indicate the positions of introns in the corresponding genes (2).

were also detected during cell-free translation of rat and sheep hypothalamic messenger RNA in the presence of labeled amino acids (9,10,26). To elucidate the structure of the biosynthetic precursor of GnRH, Seeburg and Adelman (43) cloned genomic and complementary (c) DNA sequences encoding the precursor form of GnRH. A human genomic clone containing the GnRH gene was identified by screening a human genomic library with an oligonucleotide probe encoding a hypothetical peptide containing the GnRH sequence with a Lys–Arg (paired basic processing site) extension at the amino terminus and an additional glycine residue (amide donor) at the carboxy terminus. A 600 base pair restriction fragment containing the GnRH coding sequence was then used to screen a large human placental cDNA library and isolate cDNA clones encoding the precursor of GnRH (43). Hypothalamic cDNA clones were subsequently isolated from human and rat hypothalamic cDNA libraries and sequenced (2). The structures of the human and rat GnRH precursors are shown in Figure 10.1.

In the peptide structure predicted from the nucleotide sequences of the cDNA, the decapeptide sequence of GnRH is preceded by a 23 amino acid signal sequence and followed by a Gly–Lys–Arg sequence, which is required for enzymatic processing and carboxy terminal amidation of GnRH. This is followed by a peptide sequence of 56 amino acid residues which has been referred to as GnRH-associated peptide (GAP) (37). The amino acid sequence of GAP does not contain any *N*-glycosylation sites, and it bears no resemblance to sequences of other known precursors of neuroendocrine peptides. The carboxy terminal of GAP contains the sequence Lys–Lys–Ile, which may be a cleavage site for further processing.

The human and rat GnRH precursor proteins have identical numbers of amino acid residues and display 75% overall identity of sequence. In the signal sequence there are five conservative and four nonconservative changes whereas the GnRH sequence and the Gly–Lys–Arg sequence following it are completely conserved between the two species. Rat GAP has 17 amino acid substitutions relative to human GAP, of which eight are conservative changes. The GAP sequences of both the human and the rat GnRH precursors are acidic and highly hydrophilic.

Anatomical Distribution of GAP

In order to investigate whether GAP is, indeed, expressed in hypothalamic neurons and to analyze the anatomical distribution of this peptide, antisera were generated against a synthetic peptide corresponding to the sequence of amino acid residues 40–53 of the human proGnRH (GAP 27–40). Radioimmunoassay of rat hypothalamic extracts with this antiserum revealed the presence of an antigen capable of parallel displacement with the synthetic peptide (39). Immunohistochemical studies with this antiserum and GnRH antiserum demonstrated that the distribution of cells displaying GAP immunoreactivity was identical to that observed for GnRH and included the medial septal nucleus, the diagonal band of Broca (Fig. 10.2A), the nucleus triangulairs septum, the medial preoptic area, the anterior hypothalamus, and the paraolfactory region, which extends

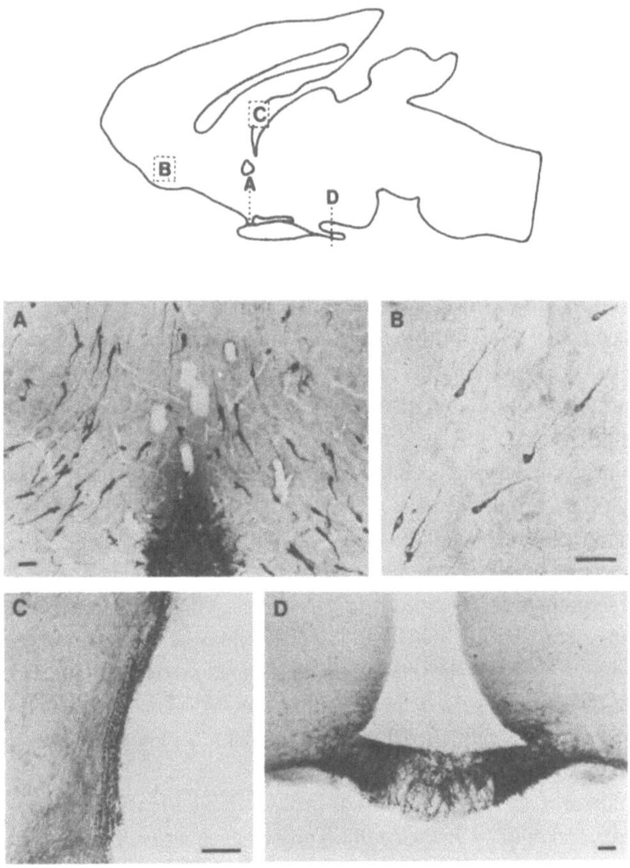

Fig. 10.2. Distribution of proGnRH-40–53 immunoreactivity in rat brain. Schematic diagram of sagittal section through rat brain depicts the location of micrographs A-D. **A**: Coronal section demonstrating numerous immunoreactive perikarya in the diagonal band of Broca and a dense terminal field in the OVLT (arrowheads). **B**: Sagittal section displaying immunoreactive perikarya in the parolfactory region. **C**: Sagittal section depicting immunoreactive fibers along the surface of the third ventricle in the region of the subfornical organ. **D**: Coronal section through median eminence with numerous terminals immunostained for proGnRH-40–53. Scale bars, 50 μm.

from the anterior commissure to the olfactory bulb (Fig. 10.2B). The largest concentration of immunoreactive perikarya was found in the diagonal band of Broca near the OVLT. Immunoreactivity with the synthetic GAP peptide antiserum was observed in axons and terminals as well as in neuronal perikarya. Bundles of immunoreactive axons were observed along the ependymal lining of the third and lateral ventricles (Fig. 10.2C). Dense networks of terminals were also observed in the OVLT (Fig. 10.2A) and median eminence (Fig. 10.2D). In all areas examined, the appearance of projections immunoreactive for proGnRH

40–53 was identical to that observed for GnRH. Perfect correspondence between axons immunoreactive for GAP and GnRH was revealed when tissue from the septal preoptic area and hypothalamic region was examined with dual immunofluorescence staining (39). Immunocytochemical analysis at the ultrastructural level indicated coexistence of GnRH and GAP immunoreactivity within individual vesicles (39).

These studies strongly suggest that GAP or a fragment of GAP is co-secreted with GnRH into the hypothalamo-hypophyseal portal circulation. Recent studies of hypophyseal portal blood of ovariectomized sheep using radioimmunoassays based on antisera generated against GAP peptides showed that GAP is indeed secreted into the hypophyseal portal blood in the same characteristic pulsatile manner as observed for GnRH (38). The finding that GAP is co-secreted into the portal circulation with GnRH suggests that it may influence the secretion of pituitary hormones. In order to test this, GAP was produced by recombinant DNA techniques.

Preparation of GAP by Expression in *Escherichia coli*

The coding sequences of the human placental cDNA encoding the GnRH precursor were used to construct a bacterial expression vector for the production of GAP (37). Since the predicted sequence of human GAP contains no methionine residues, this peptide was expressed as the carboxy-terminal part of a fusion with the *E coli* protein trpΔLE1413 (35) after introducing a methionine residue between the bacterial and human coding sequences. The presence of a methionine residue at this position enabled the release of GAP by cyanogen bromide (CNBr) cleavage of the fusion protein produced by *E coli* (24).

A DNA restriction fragment coding for residues 8–56 of the GAP sequence was isolated from the plasmid containing the cloned GnRH precursor cDNA by *Hin* f I digestion and ligated to a synthetic DNA, encoding the first six amino residues of GAP preceded by a methionine codon and an *Eco* RI cohesive end. This ligation step also reconstituted GAP codon 7. An in-phase termination codon was created by cleaving the semisynthetic DNA fragment with endonuclease *Aha* III and inserting the resulting 177-base pair DNA into *Eco* RI/*Hin* c II cleaved M13mp18 replication form DNA. The newly constructed GAP gene was isolated following primed DNA synthesis from recombinant M13 DNA as an *Eco* RI/*Hin* d III DNA fragment and joined in-frame to the coding sequences of *E coli* trpΔLE1413 gene. *E coli* cells were transformed with the resulting plasmid ptrpLE-GAP and grown under conditions of tryptophan depletion. Analysis of the cell extracts by sodium dodecyl sulfate-polyacrylamide gel electrophoresis (SDS-PAGE) and Western blots (40) using a GAP-specific antibody showed that the bacterial cells produced high levels of a GAP-immunoreactive protein of relative molecular weight, M_r ~27,600, which is the expected size for the fusion protein composed of 190 amino acid residues of the trpΔLE1413 protein and the met-GAP peptide of 57 amino acid residues. *E coli* cells containing the plasmid ptrpLE-GAP were grown by high-density fermentation which yielded the fusion protein in the form of insoluble refractile bodies which facilitated purification (28).

The fusion protein was first purified by reversed phase high-pressure liquid chromatography (HPLC), and GAP was released from the fusion protein by treatment with CNBr. The mixture of peptides released from the fusion protein was separated by a combination of gel exclusion chromatography and reversed phase chromatography. GAP was monitored by its immunoreactivity with GAP peptide antisera and obtained in a homogeneous state (37). Amino acid analysis and amino terminal sequence analysis confirmed the identity of GAP produced by bacterial expression of GAP cDNA.

Biological Properties of GAP

The biological activities of GAP were investigated first using primary cultures of rat anterior pituitary cells, have been highly useful in studying the actions of several hypothalamic-releasing hormones, including corticotropin-releasing factor (CRF) and growth hormone-releasing factor (GRF), as well as the inhibitory factor somatostatin. At a concentration of 10^{-8} mol/L (which for most hypophysiotropic hormones is sufficient for eliciting maximal response in anterior pituitary cell or organ cultures) GAP had no effect on basal or stimulated secretion of ACTH or growth hormone (GH) (37). On the other hand, the secretion of the gonadotropins luteinizing hormone (LH) and follicle-stimulating hormone (FSH), as well as PRL was profoundly influenced by GAP.

Effects of GAP on Gonadotropin Release

GAP stimulated the release of both LH and FSH from rat anterior pituitary cells (Fig. 10.3). GAP was less potent (ED_{50} 2 nM) than GnRH (ED_{50} 0.8 nM) in releasing LH but was equipotent with GnRH (ED_{50} 0.5 nM) in releasing FSH. The maximal rate of release of the gonadotropins induced by GAP was only ~70% of the maximal rate caused by GnRH. There was no significant additive effect when cells were incubated with both GAP and GnRH.

The higher potency of GAP in stimulating FSH secretion compared with LH secretion has new implications for the selective regulation of LH and FSH release. Whereas GnRH is active only in the mature processed form, GAP may be active even in the unprocessed prohormone form. This raises the question whether selective release of FSH may be accomplished by the secretion of the GnRH-GAP peptide in the unprocessed state.

Effects of GAP on PRL Release

The release of PRL from pituitary cell cultures was investigated as a function of time and concentration of GAP (Fig. 10.4). Maximal inhibition of the basal secretion of PRL was induced by 0.1 nM GAP after three to four hours of incuba-

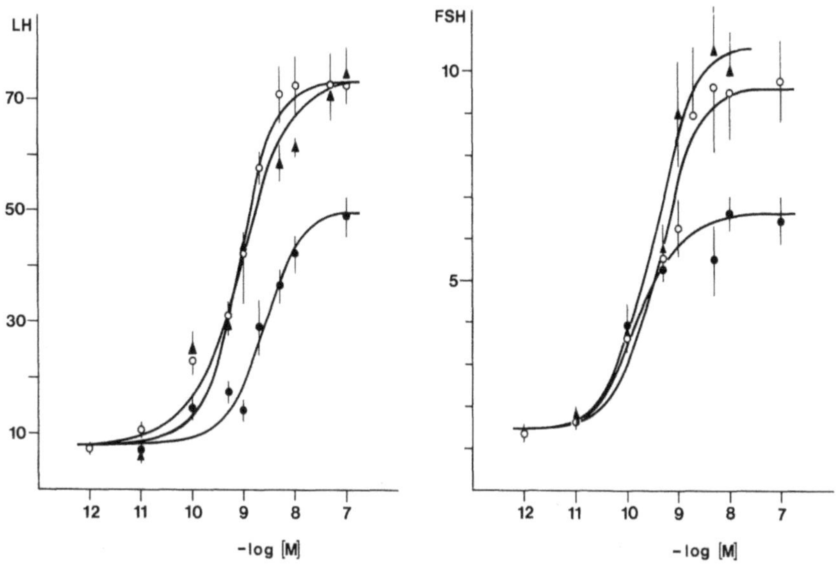

Fig. 10.3. Gonadotropin-releasing activity of GAP. LH and FSH secretion from rat anterior pituitary cell culture in the presence of increasing concentrations of GAP (●), GnRH (○), and an equimolar ratio of the two peptides (Δ). The indicated concentrations of GAP, GnRH, and their mixture were incubated with 5 × 10^5 cells in 1 mL culture medium for four hours. Media were assayed by RIA, and values are expressed in terms of rLH-RP-2 and rFSH-RP-2, respectively in nanograms.

tion. The gradual decrease in inhibition at longer incubation times may be due to degradation of the peptide or the result of desensitization. Half maximal inhibition of PRL release was observed at a GAP concentration of 25 pM. The maximal degree of inhibition produced by GAP was comparable to that caused by 100 nM dopamine. There was no significant synergistic effect when GAP and dopamine were tested together. GAP was able to inhibit the secretion of PRL stimulated by thyrotropin-releasing hormone (TRH) (unpublished results).

The concentration range of GAP eliciting inhibition of PRL release was found to be an order of magnitude lower than the concentration range over which GAP stimulated gonadotropin secretion. The concentration of GAP found in the hypothalamo-hypophyseal portal circulation is much lower than that required for gonadotropin release but is compatible with its action as an inhibitor of PRL release. In view of this, inhibition of PRL release appears to be the primary physiological role of GAP, and the peptide qualifies as a potent PRIF. The degree of inhibition induced by GAP is comparable to that observed in earlier attempts to isolate a putative PRIF from hypothalamic extracts (13,14).

Further evidence of the role of GAP as a PRIF was obtained from studies in vivo. The peptide inhibited suckling-induced PRL secretion in rats (42). When hypogonadal mice that lacked hypothalamic GnRH and GAP (8,32) were administered multiple daily injections of GAP for two weeks, serum PRL levels decreased with a concomitant increase in pituitary PRL content (unpublished

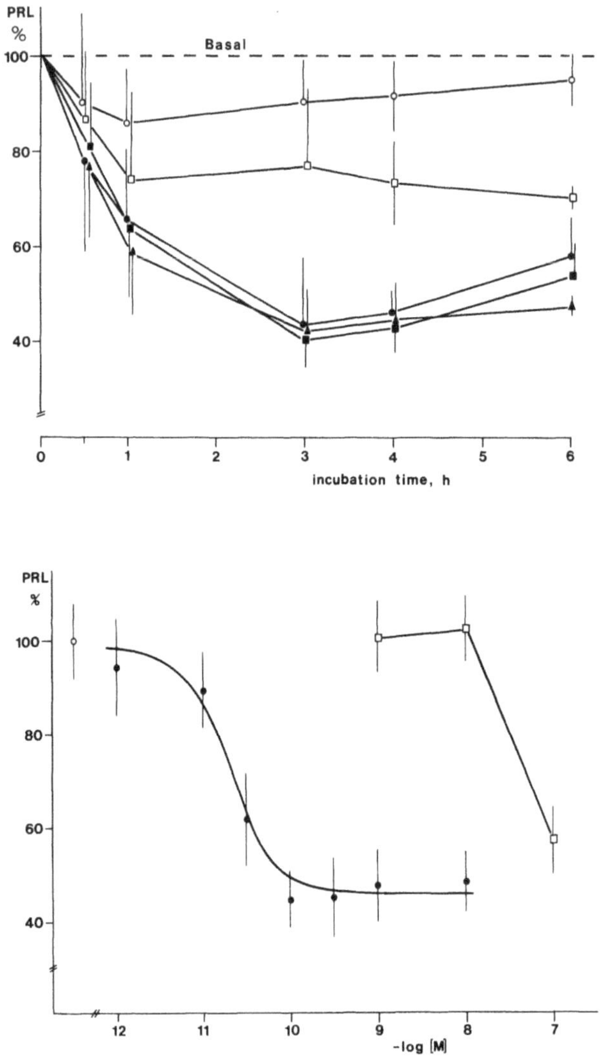

Fig. 10.4. Prolactin release-inhibiting effect of GAP. **Top**: PRL secretion in the presence of 10^{-11} mol/L (○), 3×10^{-11} mol/L, 10^{-10} mol/L (●), 10^{-9} mol/L (□), and 10^{-8} mol/L (△) concentrations of GAP as a function of incubation time. PRL secretion is expressed as percent of uninhibited basal levels. **Bottom**: PRL secretion from rat anterior pituitary cell cultures in the presence of increasing concentration of GAP (●) and dopamine (○). Control value without additions is shown by (○). The indicated concentrations of GAP or dopamine were incubated with 5×10^5 cells in 1 mL medium for six hours and assayed by RIA. Values are expressed in terms of rPRL-RP-3.

results). Hyperprolactinemic rats were found to have reduced hypothalamic content of GnRH and GAP (36).

The physiological importance of GAP in regulating PRL secretion was also assessed by monitoring serum levels of PRL in rabbits immunized with two synthetic peptide antigens corresponding to the amino terminal and middle regions of GAP. Whereas no significant changes were observed with control animals immunized with unrelated peptide antigens, PRL levels were elevated dramatically in all animals immunized with GAP antigens (Fig. 10.5). The increase in PRL levels correlated with the ability of the corresponding antibodies to recognize GAP itself as measured by radioimmunoassay (RIA) and immunocytochemistry. GH levels were unaffected in all rabbits. The elevation of serum PRL levels in response to neutralization of endogenous GAP by active immunization strongly suggests that GAP is an important PRIF. The failure to inhibit PRL secretion in normal rats by the administration of GAP (42) needs to be investigated further.

Coupled Regulation of PRL and Gonadotropin Secretion

The fact that a prolactin-inhibiting peptide shares a precursor with GnRH suggests a new model for the hypothalamic control of gonadotropin and prolactin secretion. According to this model, the secretion of these pituitary hormones is coupled through the influence of the two hypothalamic peptides, GnRH and PRIF, produced from a single precursor molecule expressed from a unique gene in a population of hypothalamic neurosecretory cells. Because of the differential actions of these peptides, our model predicts that gonadotropin and prolactin levels are inversely related. Thus, when hypothalamic neurones are triggered to secrete GnRH and PRIF in a regular episodic pattern, circulating gonadotropin

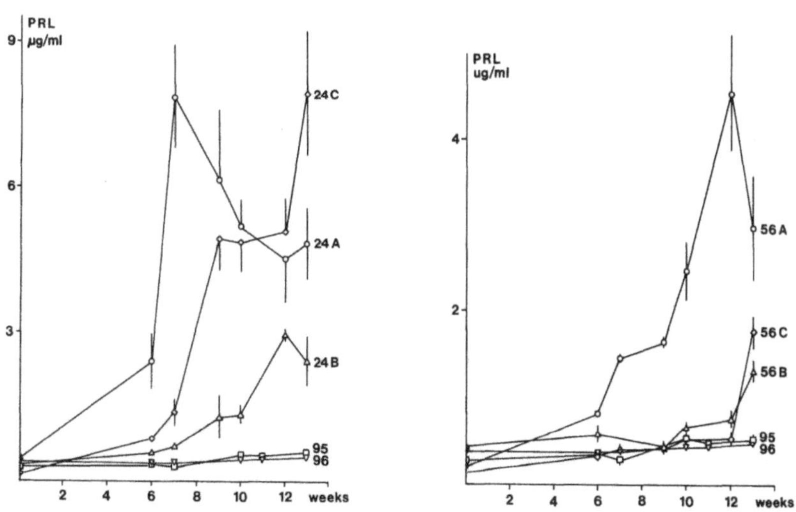

Fig. 10.5. Effect of immunization with proGnRH-14–24 (24A,B,C), proGnRH-33–56 (56A,B,C), and human insulin receptor-1–12 (95,96) on serum prolactin levels in rabbits. Female rabbits were injected intramuscularly with the soybean trypsin inhibitor (STI) conjugates of the peptides at 0, 3, 5, 7, 10, and 13 weeks. Values are expressed in μg rPRL-RP-3/mL serum.

levels will be elevated and prolactin levels repressed. Conversely, irregular firing of the neurosecretory cells and low output of GnRH and PRIF result in low gonadotropin and high prolactin release from the anterior pituitary. Whereas the former situation is predicted to occur during reproductive competence, the latter should govern lactation after parturition.

It is well understood that the control of PRL secretion is more complex than is suggested by the action of GAP owing to the influence of additional hypothalamic and extrahypothalamic factors on lactotrophs. Thus, for example, the co-surge of prolactin and LH found in certain physiological states could result from the overriding effect of prolactin-releasing factor(s) (1) and/or from the stimulation of lactotrophs by factor(s) from GnRH-primed gonadotrophs (12). However, a wealth of physiological and biochemical evidence supports our model of a central concerted and inverse regulation of LH/FSH and prolactin by the concerted action of the two peptides derived from the GnRH precursor. Indeed, the early recognition of concomitant inverse changes in gonadotropin and prolactin levels led to the belief that stimulation of gonadotropin release and inhibition of prolactin secretion might be two activities of the same molecule (13,15). Thus, suckling stimulates prolactin release but inhibits gonadotropin secretion, both effects being mediated by the hypothalamus (31,34). Reproductive disorders such as hypogonadism and amenorrhea are often associated with hyperprolactinemia when pulsatile GnRH secretion is disturbed and low while prolactin is under insufficient inhibition (29,41,48). The feedback action of estradiol on the hypothalamus acutely increases prolactin and decreases LH secretion (3,25). The ontogeny of the hypothalamo-pituitary-gonadal system in humans and sheep is critically timed and correlates with prolactin secretion (27). The first step is the development of the hypothalamo-hypophyseal GnRH-LH axis followed by the development of the negative feedback circuit via the hypothalamus from the gonads. The first appearance of the negative feedback effect of estradiol on LH and FSH secretion on the 105th day in the ovine fetus has been found to be concomitant with an increase in prolactin secretion (20), evidence that is in good agreement with our model.

Opiates inhibit the secretion of GnRH (17), and the opiate antagonist naloxone increases LH and FSH levels with a concomitant reduction of prolactin secretion (7,21). A transplantable prolactin-secreting tumor was found to develop and act through the hypothalamus by inhibiting GnRH release paralleled by high prolactin output (47). In this study, low levels of LH and FSH were associated with very high prolactin levels despite increased portal blood levels of dopamine. Furthermore, γ-aminobutyric acid (GABA) displays PRIF activity, and an antagonist of GABA was found to suppress LH and FSH secretion through blocking GnRH output from the hypothalamus (18). Such GABAergic neurones are estrogen-receptive and reside in the medial preoptic/anterior hypothalamic area also controlling LH and FSH secretion (30).

In conclusion, a tightly coupled regulation of pituitary gonadotropin and prolactin secretion occurs, manifested through the hypothalamic production of a prohormone generating the two hypophysiotropic hormones GnRH and PRIF.

The fact that the same prohormone is synthesized in extrahypothalamic brain areas and in gonadal tissue suggests that GAP could have additional central and paracrine functions involved in the regulation of reproduction.

References

1. Abe H, Engler D, Molitch ME, et al: Vasoactive intestinal peptide is a physiological mediator of prolactin release in the rat. *Endocrinology* 1985;116:1383–1390.
2. Adelman JP, Mason AJ, Hayflick JS, et al: Isolation of the gene and hypothalamic cDNA for the common precursor of human and rat. *Proc Natl Acad Sci USA* 1986;83:179.
3. Ajika K, Krulich L, Fawcett CP: Effects of estrogen on plasma and pituitary gonadotropins and prolactin, and on hypothalamic releasing and inhibiting factors. *Neuroendocrinology* 1972;9:304–315.
4. Barnea A, Porter JC: Demonstration of a macromolecule cross-reacting with antibodies to luteinizing hormone releasing hormone and its tissue distribution. *Biochem Biophys Res Commun* 1975;67:1346.
5. Ben-Jonathan N, Oliver C, Weiner HJ, et al: Dopamine in hypophyseal portal plasma of the rat during the estrus cycle and throughout pregnancy. *Endocrinology* 1972;100:452–458.
6. Birge CA, Jacobs LS, Hammer CT, et al: Catecholamine inhibition of prolactin secretion by isolated rat adenohypophyses. *Endocrinology* 1970;86:120–130.
7. Carter DA, Cooper JS, Inkster SE, et al: Evidence for an increased opioid inhibition of LH secretion in hyperprolactinaemic ovariectomized rats. *J Endocrinol* 1984;101: 57–61.
8. Cattanach BM, Iddon CA, Charlton HM, et al: Gonadotrophin releasing hormone deficiency in a mutant mouse with hypogonadism. *Nature* 1977;269:338.
9. Curtis A, Fink G: A high molecular weight precursor of luteinizing hormone releasing hormone from rat hypothalamus. *Endocrinology* 1983;112:390.
10. Curtis A, Lyons V, Fink G: The human hypothalamic LHRH precursor is the same size as that in rat and mouse hypothalamus. *Biochem Biophys Res Commun* 1983; 117:872.
11. Demarest KT, Riegle GD, Moore KE: Adenohypophyseal dopamine content during physiological changes in prolactin secretion. *Endocrinology* 1984;115:2091–2097.
12. Denef C, Andries M: Evidence for paracrine interaction between gonadotrophs and lactotrophs in pituitary cell aggregates. *Endocrinology* 1983;112:813–822.
13. Dhariwal APS, Grosvenor CE, Antunes-Rodrigues I, et al: Studies on the purification of ovine prolactin-inhibiting factor. *Endocrinology* 1968;82:1236–1241.
14. Enjalbert A, Moor F, Carbonell L, et al: Prolactin inhibiting activity of dopamine-free subcellular fractions from rat mediobasal hypothalamus. *Neuroendocrinology* 1977;24:147–161.
15. Everett JW: Regulation of prolactin secretion. *Physiol Rev* 1964;44:373–431.
16. Fawcett CP, Beezley EA, Wheaton JE: Chromatographic evidence for the existence of another species of luteinizing hormone-releasing factor (LRF). *Endocrinology* 1975;96:1311.
17. Fink G, Stanley HF, Watts AG: Central nervous system control of sex and gonadotropin release: peptide and nonpeptide transmitter interactions, in Krieger DT, Brownstein MJ, Martin JB (eds): *Brain Peptides*. New York, Wiley, 1983, pp 413–435.

18. Fuchs E, Mansky T, Stock KW, et al: Involvement of catecholamines and gluatmate in GABAergic mechanism regulatory to luteinizing hormone and prolactin secretion. *Neuroendocrinology* 1984;38:484–489.
19. Gautron JP, Pattou E, Kordon C: Occurrence of higher molecular forms of LHRH in fractionated extracts from rat hypothalamus, cortex and placenta. *Mol Cell Endocrinol* 1981;24:1.
20. Gluckman PD, Marti-Hennenberg C, Kaplan S, et al: The effect of 17 β-estradiol infusion on fetal plasma gonadotropin and prolactin and the maturation of sex steroid-dependent negative feedback. *Endocrinology* 1983;111:1619–1623.
21. Grosselin RE, Blankstein J, Dent DW, et al: Effects of naloxone and an enkephalin analog on serum prolactin, cortisol, gondaotropin in the chimpanzee. *Endocrinology* 1983;112:2168–2173.
22. Greibrokk T, Currie B, Johanson KNG, et al: Purification of a prolactin inhibiting hormone and the revealing of hormone D-GHIH which inhibits the release of growth hormone. *Biochem Biophys Res Commun* 1974;59:704–709.
23. Harris GW: Neural control of the pituitary gland. *Physiol Rev* 1948;28:139–170.
24. Itakura K, Hirose T, Crea R, et al: Expression in Escherichia coli of a chemically synthesized gene for the hormone somatostatin. *Science* 1977;198:1056–1063.
25. Johnson MD, Crowley WR: Acute effects of estradiol on circulatory LH and PRL concentrations and on serontonin turnover in individual brain nuclei. *Endocrinology* 1983;113:1935–1941.
26. Jutisz M, Counis R, Corbani M: Biosynthesis of gonadotropin releasing hormone (GnRH): present status. *Psychoneuroendocrinology* 1983;8:251.
27. Kaplan SL, Grumbach MM, Aubert ML: The ontogenesis of pituitary hormones and hypothalamic factors in the human fetus: maturation of central nervous system regulation of anterior pituitary function. Recent Prog Horm Res 1976;32:161–243.
28. Kleid DG, Yansura D, Small B, et al: Cloned viral protein vaccine for foot-and-mouth disease: responses in cattle and swine. *Science* 1981;214:1125–1129.
29. Klibanski A, Beitins IZ, Merriam GR, et al: Gonadotropin and prolactin pulsations in hyperprolactinemic women before and during bromocriptive therapy. *J Clin Endocrinol Metab* 1984;58:1141–1147.
30. Lamberts R, Vijayan E, Graf M, et al: Involvement of preoptic-anterior hypothalamic GAGA neurons in the regulation of pituitary LH and prolactin release. *Exp Brain Res* 1983;52:356–362.
31. Lu KH, Chen HT, Huang HH, et al: Relation between prolactin and gonadotropin secretion in post-partum lactating rats. *J Endocrinol* 1976;68:241–250.
32. Mason AJ, Hayflick JS, Zoeller RT, et al: A deletion truncating the GnRH gene is responsible for hypogonadism in the hpg mouse. *Science* 1986;234:1366–1371.
33. Millar RP, Aehnelt C, Rossier G: Higher molecular weight immunoreactive species of luteinizing hormone releasing hormone: possible precursors of the hormone. *Biochem Biophys Res Commun* 1977;74:720.
34. Minaguchi H, Meites J: Effects of suckling on hypothalamic LH-releasing factor and prolactin inhibiting factor, and on pituitary LH and prolactin. *Endocrinology* 1967;80:603–607.
35. Miozzari GF, Yanofsky C: Translation of the leader region of the *Escherichia coli* tryptophan operon. *J Bacteriol* 1979;133:1457–1466.
36. Negro-Vilar A, Culler MD, Johnston CA, et al: Orchidectomy and hyperprolactinemia induce marked changes in hypothalamic and preoptic LHRH precursor levels. 68th Ann Meeting Endocrine Soc Anaheim. 1986, Abstract #476.

37. Nikolics K, Mason AJ, Szonyi E, et al: A prolactin-inhibiting factor within the precursor for human gonadotropin-releasing hormone. *Nature* 1985;316:511.
38. Nikolics K, Seeburg PH, Clarke IJ: Secretion of gonadotropin releasing hormone (GnRH) and GnRH-associated peptide (GAP) into the hypothalamo-pituitary portal circulation in sheep. 1986, 68th Ann Meeting Endocrine Soc, Anaheim, Abstract #33.
39. Phillips HS, Nikolics K, Branton D, et al: Immunocytochemical localization in rat brain of a prolactin release-inhibiting sequence of gonadotropin-releasing hormone prohormone. *Nature* 1985;316:542.
40. Renaut J, Sandoval IV: Western Blots. *Method Enzymol* 1984;104:455–460.
41. Sauder SE, Frager M, Case GD, et al: Abnormal patterns of pulsatile LH secretion in women with hypoprolactinemia and Amenorrhea: Responses to bromoceirprine. *J Clin Endocrinol Metab* 1984;59:941–948.
42. Schalley AV, Olsen DB, Gulyas J, et al: 1986, *In vitro* and *In vivo* studies with GnRH-associated peptide (GAP), a proposed PRL release inhibiting peptide. 68th Ann Meeting Endocrine Soc, Anaheim, Abstract #8.
43. Seeburg PH, Adelman JP: Characterization of cDNA for precursor of human luteinizing hormone releasing hormone. *Nature* 1984;311:666.
44. Shaar CJ, Clemens JA: The role of catecholamines in the release of anterior pituitary prolactin *in vitro*. *Endocrinology* 1974;95:1202–1212.
45. Shivers BD, Harlan RE, Morrell JI, et al: Immunocytochemical localization of luteinizing hormone-releasing hormone in male and female rat brains: Quantitative studies on the effect of gonadal steroids. *Neuroendocrinology* 1983;36:1.
46. Talwalker PK, Ratner A, Meites J: *In vitro* inhibition of pituitary prolactin synthesis and release by hypothalamic extract. *Ann J Physiol* 1963;205:213–218.
47. Weber RFA, Degreef WJ, Dekoning J, et al: LH-RH and dopamine levels in hypophysial stalk plasma and their relationship to plasma gonadotropins and prolactin levels in male rats bearing a prolactin- and adrenocorticotrophin-secreting pituitary tumor. *Neuroendocrinology* 1983;36:205–210.
48. Winters SJ, Troen P: Altered pulsatile secretion of luteinizing hormone in hypogonadal men with hyperprolactinaemia. *Clin Endocrinol* 1984;21:257–263.

CHAPTER 11

Inhibin: Chemistry, Measurement, Physiology, and Its Potential as a Fertility-Regulating Agent

Anil R. Sheth, Sudhir B. Moodbidri, and Nandita J. Arbatti

Introduction

Follicle-stimulating hormone (FSH) has a vital role to play in various reproductive processes. In the male, FSH is necessary for the quantitative maintenance of normal spermatogenesis. The neutralization of endogenous FSH by the administration of FSH antibodies to nonhuman primates severely affects spermatogenesis (68,83). In the female, FSH is involved in the growth and maturation of follicles as well as in the selection of the dominant follicle (27). Therefore, the selective suppression of FSH synthesis, release, and action by any agent will offer new approaches to the regulation of fertility in both men and women. Inhibin is one such substance that fits the bill. Inhibin is a putative gonadal hormone that, in recent years, has established itself as a factor responsible for the control of FSH. Besides, certain dimers of the β-subunit of inhibin are known to activate the release of FSH (44,77). It is tempting to speculate, therefore, that the native inhibin and the β-dimer may account for the differential control of FSH and LH, which could not be explained by the prevailing concept of the regulation of gonadotropins by a single decapeptide–gonadotropin-releasing hormone (GnRH) (Fig. 11.1).

The purpose of this chapter is to review the current status of our knowledge on inhibin and its potential as a contraceptive.

Chemistry

Inhibin From Male Sources

A plethora of reports on the isolation and purification of inhibin has appeared since the mid-1970s. Inhibin has been isolated from a variety of sources such as testicular extracts, testicular lymph, rete testis fluid, and seminal plasma. This has been reviewed (13,23,33,66) from time to time by several investigators and will not be detailed here. A quick review of the literature reveals that by and large, inhibin from male sources has remained elusive, so much so that some of

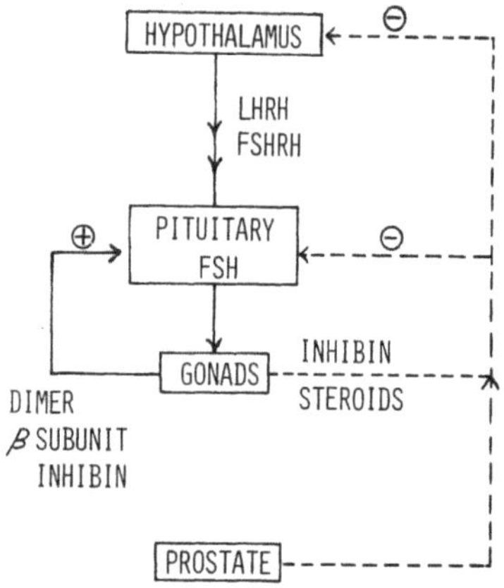

Fig. 11.1. Feedback regulation of FSH.

the early proponents have left the field in disbelief. The main cause of concern has been the multiplicity of forms of inhibin isolated from different sources. Further, two of the most extensively characterized inhibin preparations isolated from seminal plasma—a source used more often than others—are now reported to originate from the prostate and seminal vesicles and not the testes.

One of these was isolated in our laboratory from human seminal plasma and further characterized in collaboration with Dr. Hans Jornvall of the Karolinska Institute in Sweden and Dr. Michael Chretien of the Clinical Research Institute of Montreal. The work carried out at these two centers independently demonstrated that this protein is made up of 94 amino acid residues (40,64) (Fig. 11.2) and that it has a sequence homology with sperm-coating protein from the prostate (40). The other inhibin preparation was reported by Ramasharma et al (60). This relatively smaller peptide is made up of only 35 amino acid residues bearing a 40% sequence homology with a bacterial enolase and is suspected to originate from the seminal vesicles (43).

In view of these observations, it would be interesting to know the primary structure of true testicular inhibin. Using a purification protocol similar to that used by us for the isolation of human seminal inhibin, we have successfully isolated human testicular inhibin in a fairly purified form (Moodbidri and Sheth (1986) unpublished data). The preparation, which appears to be different from the one isolated from human semen, is being analyzed for its primary structure.

It may be mentioned here that we have previously reported the isolation of a low molecular weight species (1.5 kDa) of inhibin from human, sheep, rat, and chicken testes using a simple purification method involving gel filtration (6,53).

Inhibin From Female Sources

Since inhibin was conceived as a male hormone, almost all the early work centered around the isolation of inhibin from male sources. Later, as it became clear that inhibin is not unique to the male and can exist in the female, attempts were

	1				5					10					15
NH2.	Ser	Cys	Tyr	Phe	Ile	Pro	Asn	Glu	Gly	Val	Pro	Gly	Asp	Ser	Thr
					20					25					30
	Arg	Lys	Cys	Met	Asp	Leu	Lys	Gly	Asn	Lys	His	Pro	Ile	Asn	Ser
					35					40					45
	Glu	Trp	Gln	Thr	Asp	Asn	Cys	Glu	Thr	Cys	Thr	Cys	Tyr	Glu	Glu
					50					55					60
	Glu	Ile	Ser	Cys	Cys	Thr	Leu	Val	Ser	Thr	Pro	Val	Gly	Tyr	Asp
					65					70					75
	Lys	Asp	Asn	Cys	Gln	Arg	Ile	Phe	Lys	Lys	Glu	Asp	Cys	Lys	Tyr
					80					85					90
	Ile	Val	Val	Glu	Lys	Lys	Asp	Pro	Lys	Lys	Thr	Cys	Ser	Val	Ser
				94											
	Glu	Trp	Gly	Ile	.COOH										

Fig. 11.2. Amino acid sequence of a inhibin isolated from human seminal plasma.

made to isolate and characterize inhibin from female sources. Although discovered decades later, we can talk about female inhibin in more definitive terms than we can about male inhibin. This, of course, does not mean that it did not pass through a phase of uncertainties.

The works of deJong (21,22), Chari (14), and Williams (84,85) have greatly contributed to our knowledge of the physicochemical properties of follicular fluid inhibin. However, it is the work reported during the last two years that has given us a deeper insight into this molecule.

Robertson et al (61) have reported that inhibin from bovine follicular fluid is a glycoprotein with an apparent molecular weight of 58 kDa, composed of two subunits linked together by disulfide bonds. On the other hand, two 32-kDa forms of porcine follicular fluid inhibin, designated A and B, were reported by Ling et al (44) and Miyamoto et al (52). Each form is composed of two cross-linked subunits, a larger common α-subunit of 18 kDa, and a smaller distinctive β-subunit of 14 kDa. These subunits are designated as α_A- and β_A-subunits for the A form of inhibin and α_B- and β_B-subunits for the B form. The subunits occur as dimers of the β-subunits joined together by disulfide bonds. They are designated as FSH-releasing protein, which is composed of two β_A-subunits (77) and activin, which is composed of one β_A- and one β_B-subunit (44). An additional finding of great interest is the striking homology between the β-dimers and the transforming growth factor-β (31,48). The homology is such that they are considered to be products of the same gene family.

Inhibin From Extragonadal Sources

The existence and partial purification of inhibin from human term placenta was reported for the first time from our laboratory (4,5). Two molecular forms—one of 14 kDa and the other of 1.5 kDa—were isolated. The relationship between these two forms of inhibin is not known.

Semipurified preparations of inhibin have also been obtained from human prostate (42) and the gut (67). These tissues, when affected with certain types of carcinomas, are found to secrete excessive amounts of inhibin, and it is suggested that the quantitation of serum inhibin levels in such cases is of diagnostic importance (72).

Methods of Detection

Bioassays

Earlier workers have used changes in castration cells in the pituitary gland (49) or prevention of ovarian maturation in parabiotic pairs of immature female-mature ovariectomized female rats (47) as an index of inhibin activity. The more recent bioassays depend on a direct measurement of the reduction in FSH secretion by the pituitary in in vivo and in vitro systems.

Since inhibin is of gonadal origin, the suppression of the rise in serum FSH caused by gonadectomy is taken as an index of inhibin activity. There are many variations of the assay system using male and female castrate animals as models.

In vitro systems for the assay of inhibin assesses the reduction in basal and GnRH-induced FSH release into the culture medium of rat pituitary halves or dispersed pituitary cells (3,19).

A reduction in ovarian or uterine weight caused by a decrease in the circulating levels of FSH in human chorionic gonadotropin (HCG)-primed animals has also been taken as an index of inhibin activity. The administration of HCG to immature rats or mice causes an increase in ovarian and uterine weight as a result of the synergism between endogenous FSH and administered HCG. Inhibin sup-

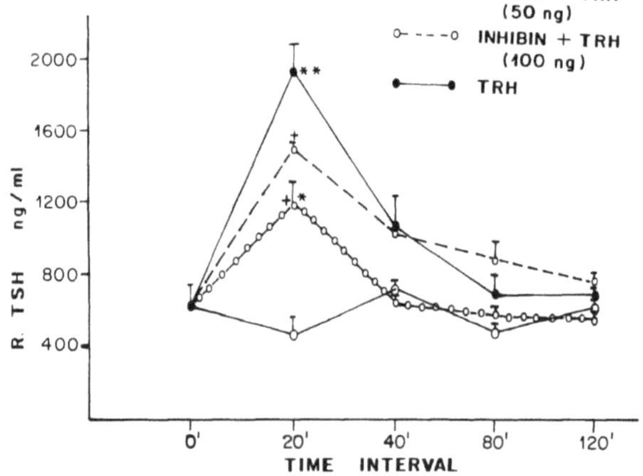

Fig. 11.3. Modulation of the TRH response by inhibin isolated from human seminal plasma in terms of TSH levels in male rats.

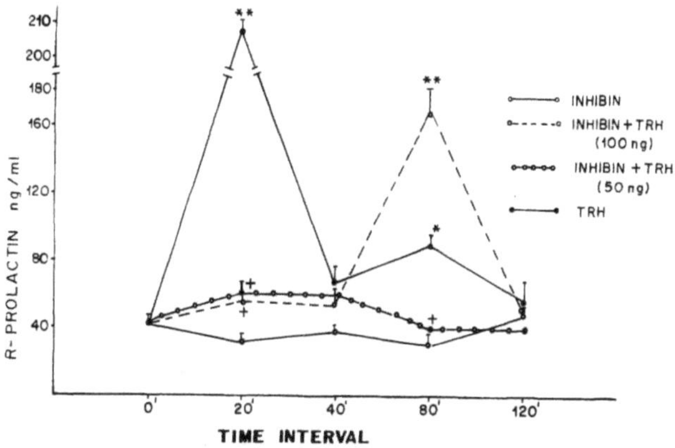

Fig. 11.4. Modulation of the TRH response by inhibin isolated from human seminal plasma in terms of prolactin levels in male rats.

presses endogenous FSH secretion and as a result inhibits the increase in ovarian and uterine weight following HCG treatment (12,59,65). These assays, at best, are detection methods and have produced data that, at times, are confusing. This stems from observations that some of the inhibin preparations suppress both luteinizing hormone (LH) and FSH in animals. The effect on LH secretion is observed only if a frequent sampling protocol is used, and it has been suggested that this effect is rapid and generally short-lived whereas the effect on FSH is slow and long-acting (8). There are also instances wherein inhibin has suppressed the release of both LH and FSH by either pituitary halves or dispersed pituitary cells in response to GnRH, in in vitro studies (29). Further, although the basal prolactin and thyroid-stimulating hormone (TSH) levels were not affected by the administration of inhibin to intact rats, thyrotropin-releasing hormone (TRH)-induced release of these hormones was suppressed (73) (Figs. 11.3 and 11.4).

Most of these observations have been made with relatively crude preparations of inhibin and may need revision when purer preparations are available and when we have a better understanding of the chemistry of inhibin. In this context, it is interesting to note the recent findings of Vale et al (77) and Ling et al (44). These workers demonstrated that follicular fluid inhibin is composed of two subunits, α and β. The β-subunit occurs as dimers in several forms. Two of these β-dimers stimulate FSH release while there are others that suppress the secretion of growth hormone (GH) and ACTH by the pituitary (77). Although the existence of α-dimers has not been demonstrated, such a possibility cannot be ruled out. Also, we do not as yet have such information on testicular inhibin, and if such dimers are found to be present, it would add a new dimension to inhibin research.

Radioimmunoassays

The insufficient sensitivity of bioassays makes them impractical for physiological studies. Moreover, they are influenced by steroidal hormones. In this context, radioimmunoassays (34,50,70,87) with their high capacity, greater sensitivity, and freedom from steroidal influences are of immense value.

Site of Action

Although the mechanism of action of inhibin is not known with certainty, its action may be multifarious, affecting the hypothalamus, pituitary, and gonads.

Inhibin has been reported to bring about a decrease in circulating levels of FSH by acting directly on the pituitary via the hypothalamus. Our studies (78) in rats have shown that this decrease is not likely to be due to an increase in the FSH clearance rate. Moreover, in vitro experiments (16,32) have provided convincing evidence that inhibin acts on the pituitary to reduce the synthesis rather than the release of FSH. Studies on the kinetics of inhibin action on gonadotropin secretion in castrated rams (2,41) and castrated male rabbits (33) suggest its action on FSH synthesis.

At the hypothalamic level, inhibin acts to reduce the synthesis of GnRH (24, 45,55).

In addition to its action on the pituitary and hypothalamus, inhibin appears to act at the gonadal level also. It has been shown to interfere with the binding of FSH to gonadal receptors (18,54,87) which may result in reduced cAMP formation by gonadal tissue in response to FSH (82).

Further, inhibin also appears to modulate some of the enzymes involved in vital gonadal processes. For instance, it has been demonstrated that inhibin inhibits aromatase (15), stimulates ornithine decarboxylase (58), and reduces protein transcription and translation (62).

Inhibin-Gonadal Function Relationship

With the advent of sensitive assays for inhibin, it has been possible to reexamine some of the postulates based on mere circumstantial evidence and obtain more direct evidence in their favor.

Sheth and his co-workers (71) have observed that circulating levels of inhibin were maximal during the late follicular phase and, as expected, were undetectable or very low in peri/postmenopausal women, thereby supporting the hypothesis of Sherman and Korenman (69). This has also been supported by studies carried out in rats (26), hamsters, rhesus monkeys, and women (10,11). These studies have demonstrated that ovarian content and secretion of inhibin is greatest during the late follicular phase immediately prior to the ovulatory surge of LH and FSH. These observations suggest that the concentration of inhibin in maturing follicles may be a primary factor for regulating FSH secretion (76) and reflect the viability of the follicles (51).

Alterations in the ratio of LH/FSH levels are known to occur in the polycystic ovarian syndrome (Stein-Leventhal syndrome). Yen (86) postulated that the polycystic ovary may secrete increasing amounts of inhibin, which may cause depressed FSH levels. Tanabe et al (75) observed that granulosa cells from polycystic ovaries have enhanced inhibin activity and that the secretion per granulosa cell is greater in such ovaries than in normal ovaries. In laboratory animals, experimentally induced polycystic ovaries can be restored to normalcy by the administration of antiserum to inhibin, which neutralizes endogenous inhibin and thereby brings about a selective increase of FSH (38) (Fig. 11.5).

Active immunization of sheep with inhibin produces superovulation in the immunized animals in a fashion similar to that observed after injections of FSH preparations (17).

Several authors have reported that elevated levels of FSH in serum and urine are associated with severe damage to the germinal epithelium as in Klinefelter's syndrome or Sertoli cell-only syndrome, resulting in reduced inhibin production (66). It is not surprising, therefore, that low or undetectable concentrations of inhibin in serum or seminal plasma were observed in these patients with azoospermia or severe oligospermia (1,63,80).

Future of Inhibin as a Contraceptive

Effect of Inhibin on Spermatogenesis

There has been considerable interest in exploring the use of inhibin as a potential male contraceptive. The logic behind this is that if FSH is essential for the initiation and maintenance of spermatogenesis, then selective suppression of FSH secretion by inhibin should render a man infertile without affecting libido since LH and testosterone secretion would not be altered. In this regard, there are some interesting observations that hold promise.

The administration of inhibin to juvenile rats for two weeks resulted in a reduction in testicular weight without affecting the accessory organs and a decrease in the number of intermediate spermatogonia and pachytene spermatocytes (37), probably caused by the inhibitory effect of inhibin on the synthesis of DNA by actively dividing spermatogonia at the beginning of spermatogenesis (33).

In adult male rats, chronic administration of inhibin resulted in a decrease in both testicular and epididymal sperm counts along with a decrease in testicular weight. Serum LH and testosterone in these animals remained unaffected. This effect of inhibin was more pronounced when testosterone was administered simultaneously either in the form of implants or as injections of testosterone propionate (79) (Fig. 11.6).

However, contrary to these observations, Lugaro et al (46) were unable to find any effect of their inhibin preparation on spermatogenesis, although serum FSH

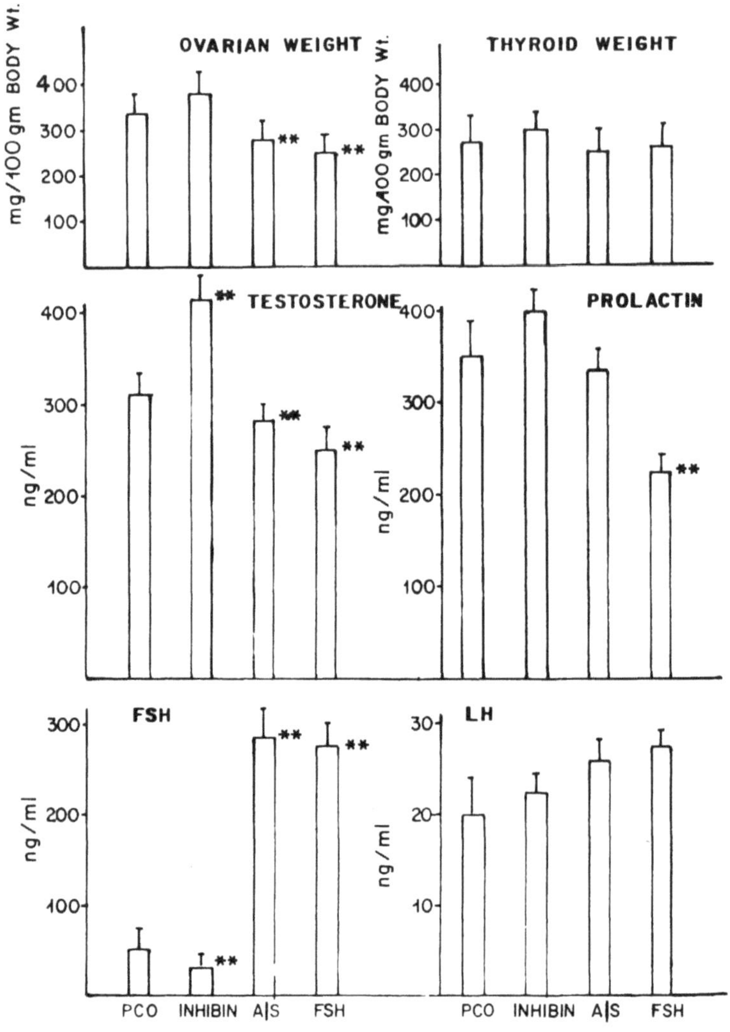

Fig. 11.5. Endocrine effects of antibodies to inhibin isolated from human seminal plasma in rats having polycystic ovaries.

was decreased by 80% and LH by 50% at the end of 25 days of treatment. Similarly, Davies et al (20) did not find any change in the testicular sperm count of adult male rats treated with a crude preparation of inhibin from rete testis fluid for 90 days.

Effect of Rabbit Anti-Inhibin on Ejaculated Sperms

Bandivdekar et al (7) demonstrated in in vitro experiments that rabbit antibodies to inhibin derived from human seminal plasma bring about agglutination of human sperms. Further, perhaps as a consequence of the spermagglutination

Fig. 11.6. Testicular □ and epididymal ■ sperm count in male rats treated with inhibin, testosterone, and inhibin + testosterone for a one-month period.

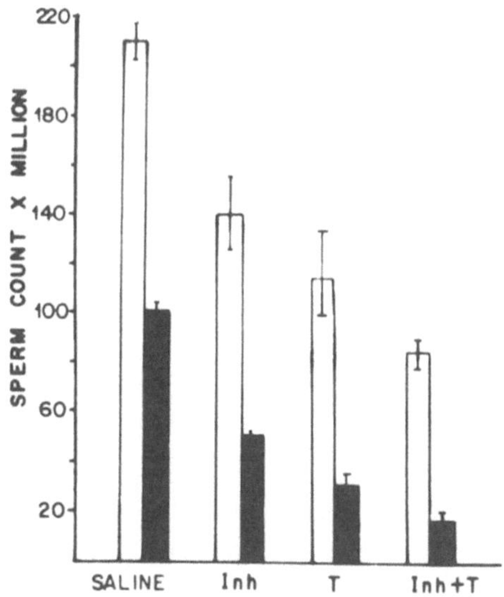

effect, sperm motility, the ability of the sperms to penetrate cervical mucus, and their fertilizing capacity were also impaired.

Effect of Inhibin on Ovulation and Implantation

There are reports indicating that the administration of follicular fluid during the normal ovarian cycle in rats suppresses the FSH surge at proestrus as well as estrus (36) and also affects follicular growth and ovulation (9). In contrast to these findings, DePaolo et al (25,26) and Nandedkar et al (57) observed that inhibin did not affect ovulation in cyclic rats, nor did it alter oocyte maturation or ovarian morphology.

Inhibin has also been reported to interfere with progesterone synthesis by ovarian tissue. Stouffer et al (74) demonstrated that the administration of charcoal-treated follicular fluid containing inhibin during the early follicular phase in rhesus monkeys induced a defective luteal phase with decreased progesterone levels. Jayaraman et al (39) observed that inhibin from sheep ovaries administered in bonnet monkeys during the first three days of the cycle led to a delay in the occurrence of the preovulatory estradiol peak, a reduction in the secretion of estradiol and progesterone, and a consistent reduction in the luteal phase (Figs. 11.7 and 11.8). Inhibition of progesterone synthesis by bovine granulosa cells in culture in the presence of inhibin has also been reported (35).

Implantation has been another point of attack. Nandedkar et al (57) demonstrated that inhibin administered between days 4 and 6 of pregnancy suppressed implantation in mice. This is presumably due to estrogen deprivation resulting from suppression of FSH, which is reported to be very high during the preimplantation period and in early pregnancy (56). Finn and Martin (30) reported that on

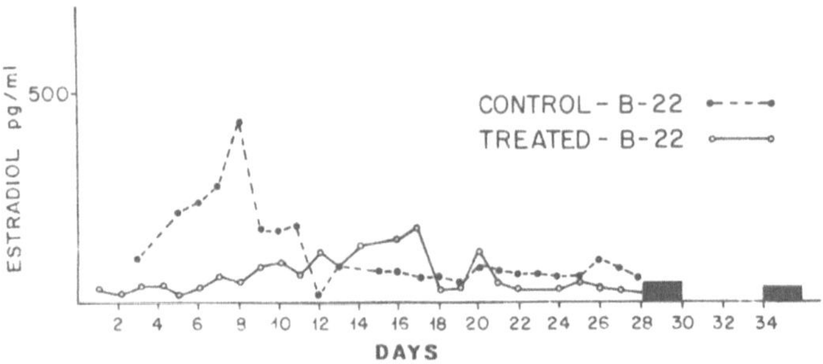

Fig. 11.7. Effect of low molecular weight sheep ovarian inhibin (~1,500 daltons) on serum estrogen levels.

day 3 of pregnancy in the rodent, cell division in the epithelium of the uterine lumen and glands is dependent on estrogen secretion by the ovary, which requires FSH (28).

Implantation in hamsters, unlike in mice, does not depend on estrogens. However, inhibin administered to hamsters during the preimplantation period prevents implantation (79). This indicates the possibility of another mode of

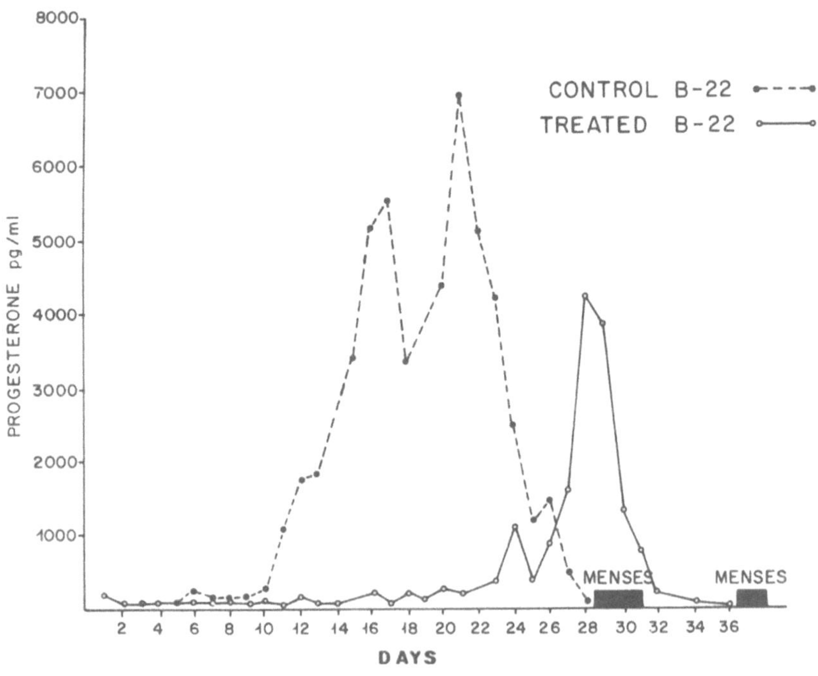

Fig. 11.8. Effect of low molecular weight sheep ovarian inhibin (~1,500 daltons) on serum progesterone levels.

action of inhibin on implantation besides its action via the suppression of FSH. It has been speculated that inhibin may interefere with the binding of progesterone with its endometrial receptors and thus affect implantation, or it may interfere with the endometrial secretions, which help to build the uterine receptive phase. This is based on the observation in hamsters and that this effect of inhibin could be reversed by simultaneous administration of progesterone.

It is clear from the foregoing discussion that, at least in experimental animals, inhibin causes a significant reduction in testicular and epididymal spermatozoa in the male and luteal phase defect in the female. The precise mechanism by which inhibin brings about these effects is not currently known and its elucidation will greatly help in the more effective use of inhibin. The lack of unanimity in the observations in some cases, which is due to differences in source and purity of the preparations used, would be overcome with the availability of synthetic peptides.

References

1. Asch RH, Vaze AY, Thakur AN, et al: Concentrations of immunoreactive inhibin in seminal plasma and serum from normospermic, oligospermic, vasectomized, klinefelter's syndrome and Sertoli-cell only syndrome subjects. *J Androl* 1980;1:252–254.
2. Baker HWG, Bremner WJ, Burger HG, et al: Testicular control of follicle stimulating hormone secretion. *Recent Prog Horm Res* 1976;32:429–465.
3. Baker HWG, Burger HG, deKretser DM, et al: Studies on purification of inhibin from ovine testicular secretions using an in-vitro bioassay. *Int J Androl* [Suppl] 1978;2: 115–123.
4. Bandivdekar AH, Vijayalakshmi S, Jaswaney V, et al: Inhibin in human placenta. *Indian J Exp Biol* 1981;19:744–745.
5. Bandivdekar AH, Moodbidri SB, Sheth AR: Isolation of inhibin like peptides from human placenta. *J Biol Sci* 1985;7:175–190.
6. Bandivdekar AH, Vijayalakshmi S, Moodbidri SB, et al: Low molecular weight inhibin from sheep, human, rat and chicken testes. *J Androl* 1982;3:140–144.
7. Bandivdekar AH, Sheth AR: Studies on antibodies to human seminal plasma inhibin and sperm fertilizing ability, in Sheth AR (ed): *Inhibins: Isolation, Estimation and Physiology.* Florida CRC Press (in press).
8. Blanc MR, Hochereau-de-Reviers MH, Cohareau C, et al: Inhibin: Effects on gonadotropin secretion and testis function in the ram and rat, in Franchimont P, Channing CP (eds): *Intragonadal Regulation of Reproduction.* New York, Academic Press, 1981, pp 3–65.
9. Chappel SC, Selkar F: Relation between the secretion of FSH during the periovulatory period and ovulation during the next cycle. *Biol Reprod* 1979;21:347–352.
10. Chappel SC, Holt JA, Spies HG: Inhibin: Differences in bioactivity within human follicular fluid in the follicular and luteal stages of menstrual cycle. *Proc Soc Exp Biol Med* 1980;163:310–314.
11. Chappel SC, Spies HG: Control of gonadotropin and prolactin secretion in rhesus monkeys and rodents, in McKerns KW (ed): *Reproductive Processes and Contraception.* New York, Plenum Press, 1981, pp 3–25.
12. Chari S, Duraiswami S, Franchimont P: A convenient and rapid bioassay for inhibin. *Horm Res* 1976;7:129–137.

13. Chari S: Chemistry and physiology of inhibin 'Review.' *Endokrinologie* 1977;70: 99–107.
14. Chari S, Hopkinson CRN, Daume E, et al: Purification of 'Inhibin' from human ovarian follicular fluid. *Acta Endocrinol* [Suppl] (Copenh) 1979;90:157–166.
15. Chari S, Daume E, Sturm G, et al: Regulation of steroidal secretion and inhibin activity in human ovarian follicular fluid. *Moll Cell Endocrinol* 1985;41:137–145.
16. Chowdhury MA, Steinberger A, Steinberger E: Inhibition of de novo synthesis of FSH by the Sertoli cell factor (SCF). *Endocrinology* 1978;103:644-647.
17. Cummins LJ, O'Shea I, Al-Obaidi SAR, et al: Increase in ovulation rate after immunization of merino ewes with a fraction of bovine follicular fluid containing inhibin activity. *J Reprod Fertil* 1986;77:365-372.
18. Daume E, Chari S, Hopkinson CRN, et al: Two distinct inhibin-like factors in human and bovine follicular fluid, abstracted. *Acta Endocrinol* [Suppl] (Copenh) 1979;225:145.
19. Davies RV, Mains SJ, Setchell BP: Inhibin: Evidence for its existence, methods of bioassay and nature of the active material. *Int J Androl* [Suppl] 1978;2:102–112.
20. Davies RV, Main SJ, Laurie MS, et al: The effects of long term administration of either a crude inhibin preparation or an antiserum to FSH or serum hormone levels, testicular function and fertility of adult male rats. *J Reprod Fertil* [Suppl] 1979;26:183–191.
21. deJong FH, Sharpe RM: Evidence for inhibin like activity in bovine follicular fluid. *Nature* 1976;263:71–72.
22. deJong FH, Jensen EHJ, Hermans WP, et al: Purification, characterization and physiological significance of inhibin from ovine follicular fluid, in Channing CP, Segal SJ (eds): *Non-Steroidal Regulators in Reproductive Biology and Medicine*. New York, Plenum Press, 1982, pp 73–82.
23. deJong FH, Robertson DM: Inhibin: 1985 Update on action and purification. *Mol Cell Endocrinol* 1985;42:95–103.
24. Demoulin A, Koulischer L, Hustin J, et al: Organ culture of mammalian testis III Inhibin secretion. *Horm Res* 1979;10:177–190.
25. DePaolo LV, Hirshfield AN, Anderson LD, et al: Suppression of pituitary secretion of follicle stimulating hormone by porcine follicular fluid during proestrus and estrus in the rat: Effects on gonadotropin and steroid secretion, follicular development and ovulation during the following cycle. *J Endocrinol* 1979;83:355-368.
26. DePaolo LV, Shander D, Wise PM, et al: Identification of inhibin-like activity in ovarian venous plasma of rats during the estrous cycle. *Endocrinology* 1979;105:647-654.
27. diZerega GS, Turner CK, Stouffer RL, et al: Suppression of FSH-dependent folliculogenesis during the primate ovarian cycle. *J Clin Endocrinol Metab* 1981;52:451–458.
28. Dorrington JH, Moon YS, Armstrong DT: Estradiol-17-biosynthesis in cultured granulosa cells from hypophysectomized immature rats, stimulation by folliculostatin. *Endocrinology* 1975;97:1328–1331.
29. Eddies LW, Baker HWG, Higginson RE, et al: A bioassay for inhibin using pituitary cell cultures. *J Endocrinol* 1979;81:49–60.
30. Finn CA, Martin L: The control of implantation. *J Reprod Fertil* 1974;39:195–206.
31. Forage RG, Ring JM, Brown RW, et al: Cloning and sequence analysis of cDNA species coding for the two subunits of inhibin from bovine follicular fluid. *Proc Natl Acad Sci USA* 1986;83:3091-3095.
32. Franchimont P, Demoulin A, Verstaelen-Proyard J, et al: Nature and mechanisms of action of inhibin: Perspective in regulation of male fertility. *Int J Androl* [Suppl] 1978;2:69–79.

33. Franchimont P, Verstraeten-Proyard J, Hazee-Hagelstein MT, et al: Inhibin: From concept to reality. *Vitam Horm* 1979;37:243–302.
34. Hasegawa Y, Suzuki T, Ui M, et al: Measurement of inhibin production for porcine granulosa cells by specific radioimmunoassay. 68th Annual Endocrine Society meetings, 1986, abstract no 168.
35. Henderson KM, Franchimont P: Regulation of inhibin production by bovine ovarian cells in vitro. *J Reprod Fertil* 1981;63:431–440.
36. Hermans WP, Dehets MHM, deJong FM: Effects of single injection of bovine follicular fluid on gonadotrophin concentrations throughout the estrus cycle of the rat. *J Endocrinol* 1982;92:425–432.
37. Hochereau-De-Reviers MT: Control of spermatogonial multiplication, in McKerns KN (ed): *Reproductive Process and Contraception*. New York, Plenum Press, 1981, pp 307–331.
38. Hurkadli KS, Jayaraman S, Gopalkrishnan K, et al: Role of inhibin in polycystic ovarian syndrome. *Int J Fertil* 1986;30:165–170.
39. Jayaraman S, Krishman KA, Vanage GR, et al: Interference in primate follicular maturation: An approach with inhibin for female contraception. *Adv Contracept* 1986;2:337–342.
40. Johansson J, Sheth AR, Cederlund E, et al: Analysis of an inhibin preparation reveals apparent identity between a peptide with inhibin-like activity and a sperm coating antigen. *FEBS Lett* 1984;176:21–26.
41. Keogh EJ, Lee VWK, Rennie GC, et al: Selective suppression of FSH by testicular extracts. *Endocrinology* 1976;98:997–1004.
42. Krishnan KA, Panse GT, Sheth AR: Comparative study of inhibin from human testis, prostate and seminal plasma. *Andrologia* 1982;14:409–415.
43. Lilja H, Jeppsson JO: Amino acid sequence of the predominant basic protein in human seminal plasma. *FEBS Lett* 1985;182:181–182.
44. Ling N, Ying S-Y, Ueno N, et al: Isolation and partial characterization of a Mr 32000 protein with inhibin activity from porcine follicular fluid. *Proc Natl Acad Sci USA* 1985;82:7217–7221.
45. Lugaro G, Cassellato MM, Mazzola G, et al: Evidence for the existence in spermatozoa of a factor inhibiting hormone synthesis. *Neuroendocrinology* 1974;15: 62–68.
46. Lugaro G, Casellato MM, Manera E, et al: Effect of the inhibin-like factor from bull spermatozoa on sexual maturation of the male rat. *J Reprod Fertil* [Suppl] 1979;26: 193–196.
47. Martins T, Rocha A: Regulation of the hypophysis by testicle and some problems of sexual dynamics. *Endocrinology* 1931;15:421–434.
48. Mason AJ, Hayflick JS, Ling N, et al: Complementary DNA sequence of ovarian follicular fluid inhibin show precursor structure and homology with transforming growth factor-β. *Nature* 1985;318:659–663.
49. McCullagh DR: Dual endocrine activity of testes. *Science* 1932;76:19–20.
50. McLachlan RJ, Robertson DM, Burger HG, et al: The radioimmunoassay of bovine and human follicular fluid and serum inhibin. *Mol Cell Endocrinol* 1986;46:175–185.
51. McLachlan RJ, Robertson DM, Healy DL, et al: Plasma inhibin levels during gonadotropin-induced ovarian hyperstimulation for IVF: A new index of follicular function? *Lancet* 1986;1:1233–1234.
52. Miyamoto K, Hasegawa Y, Fukuda M, et al: Isolation of porcine follicular fluid inhibin of 32 K daltons. *Biochem Biophys Res Commun* 1985;129:396–403.

53. Moodbidri SB, Joshi LR, Sheth AR: Isolation of an inhibin like substance from ram testes. *IRCS Med Sci* 1976;4:217.
54. Moodbidri SB, Joshi LR, Sheth AR: On the mechanism of action of low molecular weight inhibin from ram testes. *Indian J Exp Biol* 1980;18:100–101.
55. Moodbidri SB, Vijayalakshmi S, Bandivdekar AH, et al: Inhibition of hypothalamic GnRH synthesis by inhibin. *Experientia* 1981;37:525–526.
56. Murr SM, Stabenfelt GM, Bradford GE, et al: Plasma progesterone during pregnancy in the mouse. *Endocrinology* 1974;94:1209–1211.
57. Nandedkar TD, Vijayalakshmi S, Sarvamangala BS, et al: Effect of inhibin on ovulation and implantation in mice. *Contraception* 1981;24:695–698.
58. Natraj U, Vanage GR, Didolkar A, et al: Stimulation of ornithine decarboxylase in rat prostate with inhibin. *Int J Androl* 1986;9:218–228.
59. Ramasharma K, Shashidhara Murthy HM, Moudgal NR: A rapid bioassay for measuring inhibin activity. *Biol Reprod* 1979;20:831–835.
60. Ramasharma K, Sairam MR, Seidah MG, et al: Isolation, structure and synthesis of a Human Seminal Plasma Peptide with Inhibin-like activity. *Science* 1984;223: 1199–1202.
61. Robertson DM, Foulds LM, Leversha L, et al: Isolation of Inhibin from bovine follicular fluid. *Biochem Biophys Res Commun* 1985;126:220–226.
62. Sarvamangala BS, Narayani R, Pradhan DS, et al: Effect of inhibin on in-vitro transcription. *Indian J Exp Biol* 1984;22:1–4.
63. Scott RS, Burger HG: Inhibin is absent from azoospermic semen of infertile men. *Nature* 1980;285:246–247.
64. Seidah NG, Arbatti NJ, Rochemont J, et al: Complete amino acid sequence of human seminal plasma β-inhibin. Prediction of post Gln-Arg cleavage as a maturation site. *FEBS Lett* 1984;175:349–355.
65. Setchell BP, Sirinathsinghji DJ: Antigonadotropic activity in rete testis fluid, a possible inhibin. *J Endocrinol* 1972;53:lx–lxi.
66. Setchell BP, Davies RV, Main J: Inhibin, in Johnson A, Gomes WR (eds): *The Testis, vol 4.* New York, Academic Press, 1977, pp 189–238.
67. Shanbaugh SA, Sheth AR, Nanivdekar SA, et al: Immunoreactive inhibin-like material in serum and gastric juice of patients with benign and malignant diseases of the stomach. *Br J Cancer* 1985;51:877–882.
68. Sheela Rani CS, Murty GSRC, Moudgal NR: Effect of chronic neutralization of endogenous FSH on testicular function in the adult male bonnet monkey–assessment using biochemical parameters. *Int J Androl* 1978;1:489–500.
69. Sherman BM, Korenman SG: Hormonal characteristics of the human menstrual cycle throughout reproductive life. *J Clin Invest* 1975;55:699–706.
70. Sheth AR, Vaze AY, Thakur AN: Development of a radioimmunoassay for inhibin. *Indian J Exp Biol* 1979;16:1025–1026.
71. Sheth AR, Vaze AY, Thakur AN, et al: Inhibin levels in women during the menstrual cycles. *Indian J Med Res* 1981;74:848–851.
72. Sheth NA, Shanbaugh SA, Sheth AR: Inhibin in gastric cancer, in Sheth AR (ed): Inhibins: Isolation, Estimation and Physiology. Florida CRC Press, in press.
73. Sheth AR, Sheth PR, Roy R: Interaction of TRH with inhibin in male rats. *J Endocrinol* 1983;98:1–6.
74. Stouffer RL, Coensgen JL, diZerega GS, et al: Induction of defective corpus luteum function by administration of follicular fluid to monkey during the follicular phase of the menstrual cycle, in Schwartz NB, Hunzicker-Dunn M (eds): *Dynamics of Ovarian Function*. New York, Raven Press, 1981, pp 185–190.

75. Tanabe K, Shiraishi S, Kowaguchi U, et al: Physiology and pathophysiology of inhibin in women, in Sheth AR (ed): *Inhibins: Isolation, Estimation and Physiology.* USA CRC Press, in press.
76. Tsukamoto I, Taya K, Watanabe G, et al: Inhibin activity and secretion of gonadotropin during the period of follicular maturation. *Life Sci* 1986;39:119–125.
77. Vale W, Rivier J, Vaughan J, et al: Purification and characterization of an FSH releasing protein from porcine ovarian follicular fluid. *Nature* 1986;321:776–779.
78. Vanage GR, Thakur AN, Kadam MS, et al: Metabolic clearance rate of inhibin in mature and immature male rats. *Biol Reprod* 1980;23:606–610.
79. Vanage GR, Sheth AR: Potential use of inhibin for regulation of fertility, in Sheth AR (ed): *Inhibins: Isolation, Estimation and Physiology.* Florida CRC Press, in press.
80. Vaze AY, Thakur AN, Sheth AR: Levels of inhibin in human semen and accessory reproductive organs. *Andrologia* 1980;12:66–71.
81. Vijayalakshmi S, Moodbidri SB, Bandivdekar AH, et al: Modulation of FSH action by inhibin. *Arch Androl* 1980;5:231–235.
82. Vijayalakshmi S, Sheth PR, Dandekar SP, et al: Biological studies with low and high molecular weight inhibin preparations. *Int J Androl* 1981;4:691–702.
83. Wickings EJ, Nieschlag E: Suppression of spermatogenesis over two years in rhesus monkeys actively immunized with follicle stimulating hormone. *Fertil Steril* 1980; 34:269–274.
84. Williams AJ, Rush ME, Lipner H: Isolation and preliminary characterization of inhibin-F, in Channing CP, Morgh J, Sadler WA (eds): *Ovarian Follicular and Corpus Luteum Function (Advances in Experimental Medicine and Biology, Vol 112).* New York, Plenum Press, 1979, pp 429–435.
85. Williams AJ, Lipner H: Partial purification of porcine follicular fluid gonadostatin, in Channing CP, Segal SJ (eds): *Intragonadal Control Mechanisms.* New York, Plenum Press, 1982, pp 99–116.
86. Yen SSC: The polycystic ovary syndrome. *Clin Endocrinol* 1980;12:177–207.
87. Ying S-Y, Ling N, Ueno N, et al: Immunoneutralization and radioimmunoassay of inhibin isolated from porcine follicular fluid. 68th Annual Endocrine Society meetings. 1986, Abstract No 766.

Part IV
Immunobiology of Pregnancy

CHAPTER 12

Immunobiology of Human Trophoblast

Y.W. Loke

Trophoblast Populations

The placenta is interposed between fetus and mother. It, therefore, plays a central role in the exchange of immunological information between the two. The placental cells that form the ultimate fetal-maternal interface are the trophoblasts. They comprise a heterogeneous population of cells that come into direct contact with a variety of maternal tissues at different locations throughout pregnancy (17). This is represented diagrammatically by Figure 12.1:

1. Enveloping each chorionic villus are two layers of trophoblasts, an inner *cytotrophoblast* or *Langhans' layer* and an outer *syncytiotrophoblast*. The latter can also be seen to line the surface of the intervillous blood sinusoids and to dip into the openings of some uterine veins.
2. At the tips of some villi, cytotrophoblast cells proliferate through the overlying syncytiotrophoblast to become *cytotrophoblast columns*, which anchor the villi to the uterus.
3. From these columns, groups of irregularly shaped cells spread outward to form the *cytotrophoblast shell*, which comes to intercede between the chorionic villi and the underlying decidua.
4. Cytotrophoblast cells, either individually or in small groups, continue to migrate into the decidua and even to reach the myometrium. These are the *interstitial cytotrophoblast*.
5. Most of these interstitial trophoblasts remain as mononuclear cells, but some are transformed into multinuclear *placental bed giant cells*.
6. The walls of the uterine spiral arteries are frequently infiltrated by cytotrophoblasts (*mural trophoblast*), which eventually replace parts of the endothelial lining of the vessel (*endovascular trophoblast*) and may even reach the lumen of the arteries (*intravascular trophoblast*).

From this it can be seen that placental trophoblast encroaches onto maternal territory over an extensive area, coming into close association either with maternal arterial and venous blood or with a variety of uterine tissues. Since trophoblast is of fetal origin, the central question in reproductive immunology is how

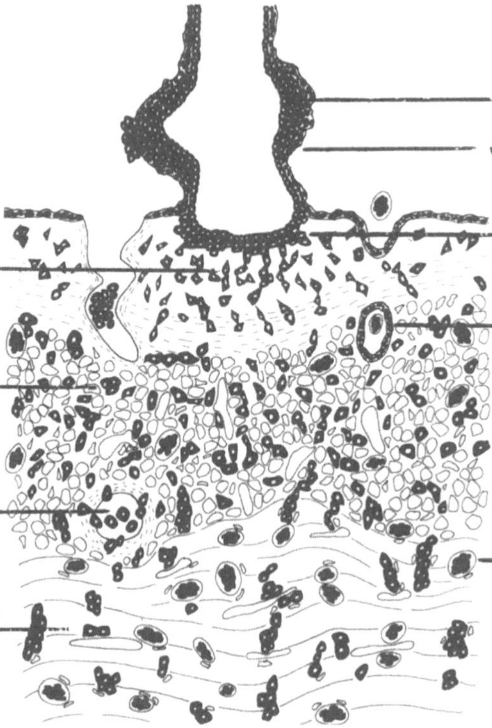

Fig. 12.1. Diagram of trophoblast populations enveloping a chorionic villus and infiltrating the placental implantation site. (From Loke YW, Butterworth BH: *Immunoregulation & Fetal Survival*. New York, Oxford University Press, 1986, with permission.)

these cells manage to evade maternal rejection. This chapter describes those phenotypic and functional characteristics of trophoblasts that may contribute to their survival.

Expression of Major Histocompatibility Complex (MHC) and Non-MHC Antigens by Trophoblast

As mentioned in the preceding section, syncytiotrophoblast forms the outermost covering of all the chorionic villi and lines the venous sinusoids at the placental implantation site. This tissue, therefore, is the major fetal component of the hemochorial interface. It is generally agreed that this layer is devoid of both class I and class II MHC antigens (Fig. 12.2), which could be a valid explanation for the lack of maternal allogeneic recognition [see review by Faulk and Hsi (10)]. This also applies to the underlying layer of villous cytotrophoblast. Although the absence of antigenic expression is accepted, there is still some debate as to the

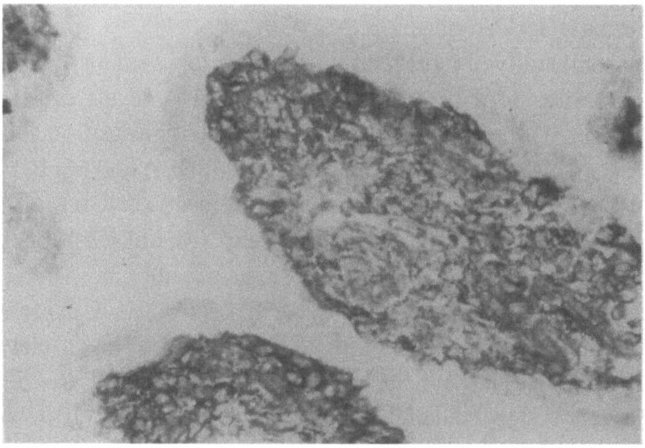

Fig. 12.2. Frozen section of first trimester chorionic villi stained with anti-HLA monoclonal antibody W6/32. Note that only the cells of the villous mesenchyme are reactive while both the cytotrophoblast and syncytiotrophoblast layers remain unstained.

exact mechanism involved. There are several hypotheses: 1) The surface antigens are masked by some immunorefractory substance; 2) the antigens are constantly being swept off the membrane by maternal antibody; 3) the antigens fail to be inserted into the membrane; or 4) the antigens are not synthesized.

It was Kirby and his colleagues (14) who originated the idea of antigen masking when they claimed that mouse trophoblast was covered by a layer of immunologically inert sialo-mucin coating. However, the experimental evidence on which this conclusion is based is not entirely satisfactory so that some degree of uncertainty still remains. With human syncytiotrophoblast, this hypothesis has not attracted much of a following because a similar sialomucoprotein layer has never been convincingly demonstrated. Nevertheless, the human syncytiotrophoblast cell surface is highly sialylated, and it is possible that the carboxyl groups of the sialic acid residues may, in turn, be responsible for the high surface negative charge observed on these cells [see review by Whyte (33)]. It is likely that these membrane features of syncytiotrophoblast are primarily related to their role as an antithrombotic lining to prevent clotting of the slow-moving blood in the venous sinusoids, but a possible immunological function could be that the negatively charged trophoblast and similarly charged maternal lymphocytes would result in mutual electrostatic repulsion and thus interfere with maternal effector lymphocyte cytotoxicity. Such an evasive mechanism has been suggested for tumor cells.

Some investigators believe that MHC antigens are not detected on syncytiotrophoblast because these molecules are constantly being removed by maternal antibodies (31). It is claimed that the shedding ability of syncytiotrophoblast microvilli is such that when maternal antibodies bridge the paternal polymorphic determinants, the whole membrane portion is extruded. This would explain why all human lymphocyte antigen (HLA) molecules are modulated and not just paternally derived ones.

The recent availability of molecular probes have focussed on the last two of the hypotheses. Kawata et al (13) looked at the mRNA levels of HLA and B_2-microglobulin (B_2M) levels in trophoblast by cDNA hybridisation and concluded that the absence of HLA antigens is due to a lack of HLA and not B_2M transcription. The mechanism, therefore, is not failure of insertion like the DAUDI lymphoblastoid cell line, but there appears to be a selective transcriptional repression of HLA-A,B,C molecules by cells of the trophoblast lineage.

Although human villous trophoblasts are devoid of HLA expression, other types of alloantigenic molecules appear to be present. Of most relevance to the fetal-maternal relationship are the trophoblast-lymphocyte cross-reactive antigens (TLX) described by Faulk [see review by Faulk and Hsi (10)] which exhibit a degree of polymorphism. It has been proposed that these antigens belong to a distinct trophoblast histocompatibility system coded by genes situated near the MHC (22). The antigenic expression by extravillous trophoblast is also highly complex. Surprisingly, these cells appear to bear HLA class I antigens but not class II (27). These include the endovascular and intravascular trophoblast that have invaded the uterine spiral arteries (17). The hemochorial interface of the human placenta, therefore, is less straightforward than originally envisaged, with maternal blood coming into contact with HLA positive cytotrophoblast on the arterial side and HLA negative syncytiotrophoblast lining the intervillous sinusoids on the venous side. What this means in immunological terms remains to be explored. The transition from HLA negative villous trophoblast to HLA positive extravillous trophoblast is seen to be at the cytotrophoblast columns where staining by anti-class I antibody becomes progressively more intense as the trophoblast cells fan out toward maternal decidua (3). However, these HLA antigens expressed by extravillous trophoblast may not be exactly analogous to that found on other somatic cells. Redman and his colleagues (25) observed that although extravillous trophoblast reacted with a monomorphic anti-HLA antibody like W6/32, they failed to bind to four antisera directed at HLA-A and HLA-B polymorphic determinants. Similarly, Wells et al (32) observed that these trophoblasts are recognized by W6/32 and an antibody to B_2M, but not by another 61D2, which is also directed at HLA common determinants. These observations may be interpreted either as due to some structural modifications of the HLA molecule when expressed by extravillous trophoblast or that these are not classical MHC antigens but represent some kind of HLA-related molecules. So here again we encounter a similar situation with extravillous trophoblast as that observed with villous trophoblast TLX antigens. It is interesting to note that the rat trophoblast has also been reported to bear a broadly reactive antigen with public specificities, which is encoded by a locus in the MHC region (23). Minor antigenic systems are not too well documented in man. Perhaps the trophoblast would be a fruitful source for the study of such antigens.

The part played by these nonclassical MHC antigens in trophoblast survival is unclear, but it has been proposed that they may actually stimulate the mother to produce some protective factor (e.g., blocking antibodies). If true, then it means that the survival of trophoblast is not entirely a passive phenomenon due to an absence of MHC antigens, but may be an active process mediated by maternal recognition of some trophoblast allotypic system. This hypothesis has a certain elegance because reproduction can thus contribute to the generation of antigenic diversity in the population.

Besides the antigens already mentioned, a variety of other surface determinants are readily demonstrated on trophoblast cells by polyclonal and monoclonal antibodies (Figs. 12.3 and 12.4). This is not too surprising since the trophoblast plasma membranes can be seen to contain a large number of proteins (30). A great deal of research is now involved in characterizing these proteins not only to understand the membrane biology of these cells, but also to define those molecules that are sufficiently trophoblast-specific as to be used as potential immunogen for a vaccine in the control of fertility. A surface molecule that is frequently recognized by antitrophoblast antibodies is the isoenzyme placental alkaline phosphatase (PLAP) (1,26,29). Its role in pregnancy is not understood. Sialic acid is a prominent component of the PLAP molecule and the main saccharide is *N*-acetyl-glucosamine, so its presence may well contribute to the high degree of surface sialylation of syncytiotrophoblast and the ability of these cells to bind the lectin wheat germ agglutinin. Another antigen on trophoblast is the transferrin receptor whose primary function must be in the transport of iron from mother to fetus; however, Faulk and Hsi (10) have suggested that the binding of

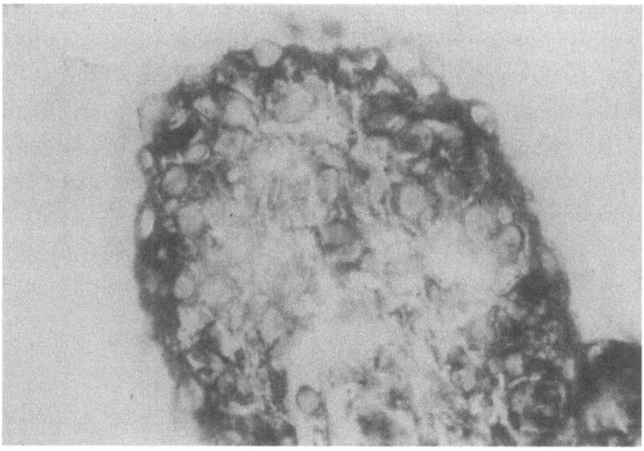

Fig. 12.3. Frozen section of first trimester chorionic villus double-stained with anti-HLA monoclonal antibody W6/32 (blue) and with antitrophoblast monoclonal antibody 18B/A5 (red). Note that only the cytotrophoblast are reactive with 18B/A5 and not the syncytiotrophoblast. (From Butterworth BH, Loke YW: Immunocytochemical identification of cytotrophoblast from other mononuclear cell populations isolated from first-trimester human chorionic villi. *J Cell Sci* 1985;76:189–197, with permission.)

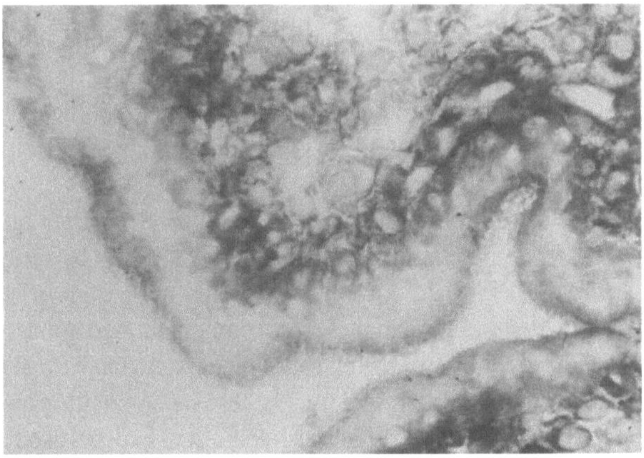

Fig. 12.4. Frozen section of first trimester chorionic villus double-stained with anti-HLA monoclonal antibody W6/32 (blue) and with antitrophoblast monoclonal antibody 18A/C4 (red). Note that both cytotrophoblast and syncytiotrophoblast layers are stained with 18A/C4. (From Butterworth BH, Loke YW: Immunocytochemical identification of cytotrophoblast from other mononuclear cell populations isolated from first-trimester human chorionic villi. *J Cell Sci* 1985;76:189–197, with permission.)

maternal serum proteins on these surface receptors might serve as a disguise against immunological recognition, a situation analogous to that postulated to explain the survival of schistosomiasis in an immunocompetent host.

The nature of the antigens recognized by the vast majority of polyclonal and monoclonal antibodies generated against trophoblast plasma membranes has not been determined, but some of these antisera appear to have rather interesting cross-reactivities. Some are found to bind to cell lines of different lineages (34), and many others react with human cancers (12,18,21). We have also reported on the cytotoxicity of an antitrophoblast antibody on retroviral infected cells (28). It is interesting that cancer and trophoblast cells should share such a similar membrane antigenic profile.

Trophoblast Products

Villous syncytiotrophoblasts are found to produce substances that may play an immunoprotective role during pregnancy, but they have not been well characterized. Some appear to be of low molecular weight and nondialyzable (6). Others could be one of the placental hormones that are synthesized by villous syncytiotrophoblast throughout pregnancy (7). Of particular interest to reproductive immunologists is human chorionic gonadotropin (HCG) because this hormone has been found to have immunosuppressive properties by a variety of in vitro tests (16). Besides being a secreted product, HCG is also very closely bound to the syncytiotrophoblast plasma membrane, which may be a reflection on the way the hormone is glycosylated and excreted. Like PLAP, HCG is also a sialoglycoprotein and may therefore similarly contribute to the high sialic content of syncytiotrophoblast membrane. There is, thus, two forms of trophoblastic HCG: 1) a secreted form involved in endocrinological and possibly immunological functions at the hemochorial interface, and 2) a membrane-bound form that may play a local immunoprotective role on the surface of the cell. HCG and PLAP are two trophoblast products that are most frequently reexpressed by cancer cells, even by those that are not of trophoblast derivation. These onco-placental proteins, therefore, provide further evidence for the similarity between trophoblasts and cancer cells.

In contrast to villous syncytiotrophoblasts, extravillous trophoblasts do not synthesize HCG or PLAP. Indeed, apart from a few intermediate cells that contain human placental lactogen (HPL) (15), extravillous trophoblasts are relatively unproductive. Even the multinuclear placental bed giant cells appear to be inert (17). The immunoprotective effects of HCG, therefore, cannot apply to this trophoblast population. However, there is evidence, at least in the pregnant mouse, that some extravillous trophoblast products could trigger the generation of uterine suppressor cells which provide protection for these class I-bearing trophoblast among the decidua (5).

Susceptibility of Trophoblast to Immune Damage

A great deal of data have accumulated in the literature as to whether or not trophoblast is susceptible to immune damage, but a general consensus has still not been reached. Some of the more recent studies have shown that human trophoblast cells are not susceptible to NK cell killing (24) nor are they capable of stimulating maternal leukocytes in a one-way mixed lymphocyte reaction (11). Similar findings are reported for murine trophoblast, which are also resistant to cytotoxic cell-mediated lysis (4). These observations seem to support the immunological neutrality of trophoblast. In contrast, Zuckermann and Head (35) demonstrated that antipaternal strain alloantisera are cytotoxic to Fl trophoblast, which point to the presence of MHC antigenic targets that are susceptible to complement-mediated lysis. The authors concluded that in vivo the repair capacity of the placenta is such that it continues to survive in spite of maternal allogeneic attack. These findings of Zuckermann and Head are more in accord with older experiments where maternal lymphocytes were shown to be cytotoxic to trophoblast (8,9) and where trophoblast cells were lysed by anti-HLA tissue typing antisera (19).

A major problem in all such in vitro studies is the difficulty in obtaining sufficiently pure trophoblast without contamination by other cell types. It would be immediately apparent that here is a further use for antitrophoblast monoclonal antibodies which will aid the identification and separation of trophoblast cells for subsequent biological investigations (2,20). When this is done, the question on the susceptibility of trophoblast to immune lysis in vitro may then be convincingly answered.

Conclusion

As can be seen from the preceding discussion, the mechanisms by which the allogeneic trophoblast evade maternal immunological rejection remain unresolved. The problem is complicated by the fact that trophoblasts comprise a heterogeneous group of cells, each of which is in contact with a different kind of maternal tissue. The absence of HLA antigens combined with the production of some trophoblast immunoprotective substances would seem to be at least one valid explanation for trophoblast survival. However, there is now the additional interest in the expression of non-MHC antigens by trophoblast and their role in the fetal-maternal relationship. The possibility that reproductive performance may be improved by maternal recognition of these antigens is an intriguing concept.

References

1. Brown PJ, Molloy CM, Johnson PM: Immunochemical identification of human trophoblast membrane antigens using monoclonal antibodies. *J Reprod Immunol* 1983;5:351–361.
2. Butterworth BH, Loke YW: Immunocytochemical identification of cytotrophoblast from other mononuclear cell populations isolated from first-trimester human chorionic villi. *J Cell Sci* 1985;76:189–197.
3. Butterworth BH, Khong TY, Loke YW, et al: Human cytotrophoblast populations studied by monoclonal antibodies using single and double biotin-avidin-peroxidase immunocytochemistry. *J Histochem Cytochem* 1985;33:977–983.
4. Chaouat G, Kolb J-P: Immunoactive products of placenta. IV. Impairment by placental cells and their products of CTL function at effector stage. *J Immunol* 1985;135:215–222.
5. Clark DA, Slapsys R, Chaput A, et al: Immunoregulatory molecules of trophoblast and decidual suppressor cell origin at the maternofetal interface. *Am J Reprod Immunol Microbiol* 1986;10:100–104.
6. Colouscou J-M, Remacle-Bonnet MM, Pommier G, et al: Immunosuppressive properties of human placenta: study of supernatants from short-term syncytiotrophoblast cultures. *J Reprod Immunol* 1986;9:33–47.
7. Contractor SF: Metabolic and enzymatic activity of human trophoblast, in Loke YW, Whyte A (eds): *Biology of Trophoblast*. Amsterdam, Elsevier/North Holland Biomedical Press, 1983, pp 236–281.
8. Currie GA, Bagshawe KD: The masking of antigens on trophoblast and cancer cells. *Lancet* 1967;1:708–710.
9. Douthwaite RM, Urbach GI: In vitro antigenicity of trophoblast. *Am J Obstet Gynecol* 1971;109:1023–1029.
10. Faulk WP, Hsi B-L: Immunobiology of human trophoblast membrane antigens, in

Loke YW, Whyte A (eds): *Biology of Trophoblast*. Amsterdam, Elsevier/North Holland Biomedical Press, 1983, pp 535–570.

11. Hunt JS, King CR, Wood GW: Evaluation of human chorionic trophoblast cells and placental macrophages as stimulators of maternal lymphocyte proliferation in vitro. *J Reprod Immunol* 1984;6:377–391.
12. Johnson PM, Cheng HM, Molloy CM, et al: Human trophoblast-specific surface antigens identified using monoclonal antibodies. *Am J Reprod Immunol* 1981;1:246–254.
13. Kawata M, Parnes JR, Herzenberg LA: Transcriptional control of HLA-A,B,C antigen in human placental cytotrophoblast isolated using trophoblast- and HLA-specific monoclonal antibodies and the fluorescence-activated cell sorter. *J Exp Med* 1984;160:633–651.
14. Kirby DRS, Billington WD, Bradbury S, et al: Antigen barrier of the mouse placenta. *Nature (London)* 1964;204:548–549.
15. Kurman RJ, Main CS, Chen H-C: Intermediate trophoblast: a distinctive form of trophoblast with specific morphological, biochemical and functional features. Placenta 1984;5:349–369.
16. Loke YW: *Immunology and Immunopathology of the Human Foetal-Maternal Interaction*. Amsterdam, Elservier/North Holland Biomedical-Press, 1978.
17. Loke YW, Butterworth BH: Heterogeneity of human trophoblast populations, in Gill, Wegmann (eds): *Immunoregulation and Fetal Survival*. New York, Oxford University Press, 1987, pp 197–209.
18. Loke YW, Whyte A, Davies SP: Differential expression of trophoblast-specific membrane antigens by normal and abnormal human placentae and by neoplasms of trophoblastic and non-trophoblastic origin. *Int J Cancer* 1980;25:459–461.
19. Loke YW, Joysey VC, Borland R: HL-A antigens on human trophoblast cells. *Nature (London)* 1971;232:403–405.
20. Loke YW, Butterworth BH, Margetts JJ, Burland K: Identification of cytotrophoblast colonies in cultures of human placental cells using monoclonal antibodies. *Placenta* 1986;7:221–231.
21. McIntyre JA, Faulk WP: Cross-reactions between cell surface membrane antigens of human trophoblast and cancer cells. *Placenta* 1980;1:197–207.
22. McIntyre JA, Faulk WP, Verhulst SJ, et al: Human trophoblast-lymphocyte cross-reactive (TLX) antigens define a new alloantigen system. *Science* 1983;222:1135–1137.
23. Macpherson TA, Ho HN, Kunz HW, et al: Localization of the Pa antigen on the placenta of the rat. *Transplantation* 1986;41:392–394.
24. Pross H, Mitchell H, Werkmeister J: The sensitivity of placental trophoblast cells to intra-placental and allogeneic cytotoxic lymphocytes. *Am J Reprod Immunol Microbiol* 1985;8:1–9.
25. Redman CWG, McMichael AJ, Stirrat GM, et al: Class-1 major histocompatibility complex antigens on human extra-villous trophoblast. Immunology 1984;52:457–468.
26. Rettig WJ, Cordon-Cardo C, Koulos JP, et al: Cell surface antigens of human trophoblast and choriocarinoma defined by monoclonal antibodies. *Int J Cancer* 1985;35:469–475.
27. Sunderland CA, Redman CWG, Stirrat GM: HLA A,B,C antigens are expressed on nonvillous trophoblast of the early human placenta. *J Immunol* 1981;127:2614–2615.

28. Thiry L, Loke YW, Whyte A, et al: Heterologous antiserum to human syncytiotrophoblast membrane is cytotoxic to retrovirus-producing cells and to some cancer cell lines. *Am J Reprod Immunol* 1981;1:240–245.
29. Travers P, Bodmer W: Preparation and characterization of monoclonal antibodies against placental alkaline phosphatase and other human trophoblast-associated determinants. *Int J Cancer* 1984;33:633–641.
30. Truman P, Ford HC: Proteins of human placental microvilli: II. Identification and topology of the plasma membrane proteins. *Placenta* 1986;7:111–120.
31. Underwood JL, Ruszkiewicz M, Barnden KL, et al: Does antigenic modulation cause the absence of major histocompatibility complex antigens on the syncytiotrophoblast. *Transplant Proc* 1985;17:921–924.
32. Wells M, Hsi B-L, Faulk WP: Class I antigens of the major histocompatibility complex on cytotrophoblast of the human placental basal plate. *Am J Reprod Immunol* 1984;6:167–174.
33. Whyte A: Biochemistry of the human syncytiotrophoblast plasma membrane, in Loke YW, Whyte A (eds): *Biology of Trophoblast*. Amsterdam, Elsevier/North Holland Biomedical Press, 1983, pp 513–533.
34. Whyte A, Ragge N, Loke YW, et al: Human syncytiotrophoblast membrane proteins defined using heterologous antiserum. *Clin Exp Immunol* 1985;59:227.
35. Zuckermann F, Head JR: Susceptibility of mouse trophoblast to antibody and complement-mediated damage. *Transplant Proc* 1985;17:925–927.

CHAPTER 13

Maternal-Placental Interactions in Early Human Pregnancy

Peter M. Johnson

Introduction

The normal and remarkable ability of the semiallogeneic conceptus to evade maternal immune rejection responses throughout the gestation of human pregnancy highlights a particular immunological paradox. A central role has clearly emerged over the past decade for trophoblastic cells, the separate populations of which form the continuous sites in extraembryonic tissues of the fetal cellular contact with the maternal host (9,10,13). Both passive and active mechanisms are now known to contribute to this unique and successful example of transplantation required for mammalian viviparity (14). This article aims to identify these mechanisms effective in early pregnancy and to comment on abnormal situations that may lead to pregnancy interruption following natural or induced disturbance to normal maternal-fetal immunological interplay.

Antigen Expression by Human Fetal Trophoblast

Major Histocompatibility Complex (MHC) Antigens

A considerable amount of data accumulated by different experimental approaches demonstrate that throughout gestation, villous trophoblast of the human chorionic placenta fails to express any known histocompatibility antigens (human leukocyte antigen, HLA) which are recognized to be of considerable immunological importance because of their pivotal role as the focus of both cytotoxic and regulatory immune responses (4,14,23). This is of major significance in the passive protection of essential trophoblastic tissue from maternal cytotoxic attack at the maternal-fetal interface (10). The possibility that occasional examples of spontaneous early pregnancy loss may be resultant from abnormal HLA expression by villous trophoblast remains largely unexplored. However, our understanding of the molecular genetics of how villous trophoblast normally fails to transcribe HLA molecules is very much in its infancy and may require the development of uncontaminated homogeneous villous cytotrophoblast cell cultures from early placental tissues before being able to test mechanistic models. This could be of future significance for a novel approach to birth control by very early pregnancy interruption (possibly even prior to implantation) using biologically active molecules to up-regulate the cell surface expression of HLA in the trophoblastic shell.

Recent immunohistological and biochemical studies have shown that certain human extravillous cytotrophoblast populations in the placental bed and chorion laeve do express a novel 40 kDa class I MHC antigen that is associated with β_2-microglobulin (4,23,25). These appear most likely to be related to Qa-like antigens, and it has been suggested that they may play an active role in early pregnancy for trophoblastic induction of maternal humoral or cellular immunoregulatory responses that assist fetal evasion of maternal immune rejection (14).

Trophoblast-Leukocyte Common (TLX) Antigen

A murine monoclonal antibody, H316, raised against isolated term placental syncytiotrophoblast microvilli, recognizes a cell surface epitope expressed also on many other cell types, including all peripheral blood leukocytes, certain epithelia, testicular germ cells, and capacitated sperm (2,4,20). It is thought that the molecules reactive with H316 are analogous to those described previously using rabbit xenoantisera raised against placental syncytiotrophoblast microvilli and termed the TLX molecular system (7). These xenoantisera exhibit HLA-independent inhibition of mixed lymphocyte culture (MLC) reactions (7), although the H316 mAb does not impair MLC reactions (Barbati and Johnson, unpublished, 1986). The H316-defined TLX epitope is carried on two nonassociated cell surface glycoproteins of approximately 65 and 55 kDa, which exhibit size heterogeneity between individuals (26). The current data suggest that these major components of the surface membrane of all trophoblast populations throughout gestation have a role in recognition events influencing maternal-fetal cellular interactions (14).

Oncotrophoblast and Other Cell Surface Antigens

The syncytiotrophoblast plasma membrane expresses numerous and complex biochemical components including enzymes, growth factor and hormone receptors (33,34). Much less biochemical detail is known about the surface membrane of extravillous cytotrophoblast populations. However, the development and application of monoclonal antibodies (mAbs) has identified numerous cellular antigens that, taken together, constitute the characteristic cell-surface phenotypes of different anatomically defined human placental trophoblast populations (4,17). Several mAbs that recognize cell-surface antigens with a relatively restricted tissue distribution have been described. Most of these antigens are also expressed on a variety of human tumor cells and hence may be described under a generic term of oncotrophoblast antigens (16,20,21,24,27,30,35). One such cell-surface oncotrophoblast antigen – the heat-stable isoenzyme, placental alkaline phosphatase (PLAP) – has been extensively investigated and proven to be of some clinical value as a tumor marker (18,34). PLAP is expressed by placental syncytiotrophoblast only after the first-trimester of pregnancy and, hence, is not of relevance as a candidate target for a birth control vaccine since it is not expressed on preimplantation embryos.

This approach may identify a specific cell-surface epitope, not represented in any other normal adult or fetal tissues, that could form the basis for future development of a birth control vaccine (and, coincidentally, could influence the growth of certain malignancies). Such a target trophoblast membrane epitope would be expressed at a defined anatomic site, initially in small amounts and only following conception. This approach would be analogous to that developed for very early interruption of pregnancy by vaccination against antigenic determinants of the β-chain of human chorionic gonadotropin, a secreted protein hormone of trophoblast (1). However, a recent World Health Organization-supported workshop, involving prior worldwide distribution of 45 rodent mAbs directed against trophoblast-associated epitopes for assessment using a wide variety of laboratory techniques, has confirmed the elusive nature of an entirely "trophoblast-specific" membrane antigen (2a). This workshop confirmed that PLAP and the transferrin receptor are strong immunogens in rodent immunization with human trophoblast membranes, but did also identify five mAbs that may justify further studies in the context of birth control vaccines.

It is likely that there will be further developments along these lines on the specificity of immunological reagents generated against oncotrophoblast antigens. For example, it is known from murine studies that a monoclonal antibody derived from syngeneic pregnancy and reactive with teratocarcinoma cells can alter in vitro fertilization and development (8); similarly, preimmunization of mice with a particular immunogenic fibrosarcoma causes pregnancy loss (28). In addition, Wahlstrom et al (32) described a murine mAb recognizing a component of normal placental syncytiotrophoblast that is immunologically related to mammalian retrovirus structural proteins.

Maternal Responses to Trophoblast

Antibody Responses

It is clear that the mother can respond immunologically to the conceptus in pregnancy and alloantibody activity has been demonstrated in many mammalian species (9). Anti-HLA antibody induced during or following human pregnancy is likely to result from leakage of nontrophoblastic fetal cells, since all trophoblast populations lack the expression of regular HLA antigens. However, precise HLA alloantigen (tissue type)-specific antibodies are not common in human pregnancy, although it is recognized that pregnancy sera often contain immunoglobulins that are capable of directly inhibiting most laboratory assays of lymphocyte proliferation and function. A recent important study by Konoeda et al (15) examined 50,000 human placental extracts for lymphocyte-reactive antibodies and concluded that most specificities were directed against public (framework) epitopes rather than private alloantigen specificities. This is in accord with data from experimental animals associating the specificity of pregnancy-induced anti-MHC antibodies with broadly shared antigenic determinants (9). There is also additional serological data showing that pregnancy-induced antibodies may recognize novel class I MHC antigens (31). These are usually interpreted as an active maternal response to extraembryonic gestational tissue that may be part of the maternal immunoregulatory mechanisms normally underpinning the maintenance and growth of the fetal graft. However, other pregnancy-induced antibody activities may contribute additionally to this effect, including auto-anti-idiotypic antibodies reactive with maternal T-cell receptors for paternal alloantigens. The fine specificities of lymphocyte-reactive antibody activities induced in human pregnancy and their significance, if any, to responses "protecting" the fetoplacental graft remain exciting avenues of exploration.

Maternal antibody activity directed against oncotrophoblast antigens has been similarly elusive to define. We have consistently found it difficult to detect any maternal antibody directly reactive with components of isolated human syncytiotrophoblast microvilli by various immunoassay procedures using pregnancy and postpartum sera or placental eluates (8a,11). A notable feature of PLAP is its extensive genetic polymorphism and it may be relevant that maternal antibodies reactive with paternally inherited fetal PLAP alleles, or indeed with the H316-defined TLX antigens, have not been readily described in normal human pregnancy. There are various possible explanations, such as absorption by excess antigen in pregnancy. Possibly, and more attractively, the continuous deportation of syncytiotrophoblastic elements into maternal blood may be involved in induction of a form of maternal systemic immunological tolerance to fetal syncytiotrophoblast membrane components: this would not be operative for fetal antigens expressed only on other cell types or trophoblast populations. Disruption of maternal tolerance to oncotrophoblast antigens may provide a novel approach to early pregnancy interruption.

Cellular Responses

It is now recognized that local cellular changes in maternal endometrial tissues in pregnancy also involve the early appearance of extensive numbers of leukocytes, including numerous T-lineage lymphoid cells (5). There is now much research aimed toward the phenotypic and functional dissection of different lymphocyte populations in uteroplacental tissues throughout gestation, including suppressor cells, NK cells, and cytotoxic T-cell precursors. However, a currently attractive hypothesis centers on the interleukin-2 (IL2) requirement for proliferation of effector cytotoxic lymphocyte populations that could be detrimental to gestational trophoblastic tissue (14). Locally acting factors that may regulate IL2 activity in uteroplacental tissue would be expected to provide a powerful milieu of immunosuppression for deleterious antifetal responses. An attactive area of future research would appear to be the definition of such biologically active factors and the possibility of generating agonists that might negate their effect, thus allowing IL2-activated cytotoxic attack on embryonic tissue. Indeed, a brief report has claimed that bolus administration of IL2 to pregnant mice causes pregnancy loss (29).

Recurrent Spontaneous Abortions

Laboratory and clinical evidence suggest an absence of normal maternal immunoregulatory cellular and humoral responses to fetoplacental antigens in the majority of patients with unexplained recurrent miscarriages (3,19). It is also unlikely that there will be a single immunological basis to unexplained recurrent miscarriage and, indeed, subgroups of these women are being identified by laboratory criteria (12). Aberrant maternal responses overtly reactive with fetal trophoblast components are not usually found in recurrent spontaneous abortion (8a,12). However, a small group with high-titer serum antipaternal lymphocytotoxins has been identified (19), and preliminary collaborative work has confirmed that these women do have serum antibodies that are also clearly reactive with isolated placental syncytiotrophoblast microvilli by immunoassay procedures (8a). Interestingly, it has been shown by two groups that such sera contain an IgG fraction that can inhibit the attachment or disrupt growth of murine peri-implantation blastocysts (6,22). Such sera, together also with mAbs directed against trophoblast-associated epitopes, may prove to be valuable reagents for the future definition of functional molecules essential for peri-implantation events and against which an immunological response would result in early pregnancy loss.

References

1. Ada GL, Basten A, Jones WR: Prospects for developing vaccines to control fertility. *Nature* 1985;317:288.
2. Anderson DJ, Johnson PM: The monoclonal antibody 316 which recognises polymorphic antigens expressed on human trophoblast also reacts with human testicular germ cells and capacitated sperm. *J Reprod Immunol* [Suppl] 1986;31.

2a. Anderson DJ, Johnson PM, Griffin PD, et al: Monoclonal antibodies to human trophoblast and sperm antigen: report of two WHO-sponsored workshops. *J Reprod Immunol* 1987;10:231.

3. Beer AE, Semprini AE, Xiaoyn Z, et al: Pregnancy outcome in human couples with recurrent spontaneous abortion: HLA antigens profiles, HLA antigen sharing, female serum MLR blocking factors and paternal leukocyte immunization. *Exp Clin Immunogenet* 1985;2:137.
4. Bulmer JN, Johnson PM: Antigen expression by trophoblast populations in the human placenta and their possible immunobiological significance. *Placenta* 1985;6:127.
5. Bulmer JN, Johnson PM: The T lymphocyte population in first-trimester human decidua does not express the interleukin-2 receptor. *Immunology* 1986;58:685.

6. Chavez DJ, McIntyre JA: Sera from women with histories of repeated pregnancy losses cause abnormalities in mouse peri-implantation blastocysts. *J Reprod Immunol* 1984;6:273.
7. Faulk WP, McIntyre JA: Immunological studies of human trophoblast: markers, subsets and functions. *Immunol Rev* 1983;75:139.
8. Hamilton MS, Vernon RB, Eddy EM: A monoclonal antibody, EC-1, derived from a syngeneically multiparous mouse alters in vitro fertilization and development. *J Reprod Immunol* 1985;8:45.

8a. Hole N, Cheng HM, Johnson PM: Antibody reactivity against human trophoblast membrane antigens in the context of normal pregnancy and unexplained recurrent miscarriage, in Chaouat G (ed): *INSERM Colloque on Materno-Fetal Relations*. Paris, INSERM publication, in press.

9. Hunziker RD, Wegmann TG: Placental immunoregulation. *CRC Crit Rev Immunol* 1986;6:245.
10. Johnson PM: Immunobiology of human trophoblast, in Crighton DB (ed): *Immunological Aspects of Reproduction in Mammals*. London, Butterworths Press, 1984, p 109.
11. Johnson PM, Cheng HM, Stevens VC, et al: Antibody reactivity against trophoblast and trophoblast products. *J Reprod Immunol* 1985;8:347.
12. Johnson PM, Chia KV, Risk JM: Immunological question marks in recurrent spontaneous abortion, in Clark DA, Croy BA (eds): *Reproductive Immunology 1986*. Amsterdam, Elsevier Biomedical Press, 1986, p 239.
13. Johnson PM, Risk JM, Bulmer JN, et al: Antigen expression at human materno-fetal interfaces, in Gill TJ III, Wegmann TG (eds): *Immunoregulation and Fetal Survival*. Oxford University Press, 1987, p 181.
14. Johnson PM, Stern PL: Antigen expression at human materno-fetal interfaces, in Cinader B, Miller RG (eds): *Progress in Immunology VI*. Orlando, Academic Press, in press.
15. Konoeda Y, Terasaki PI, Wakisama A, et al: Public determinants of HLA indicated by pregnancy antibodies. *Transplantation* 1986;41:253.
16. Lipinski M, Parks DR, Rouse RV, et al: Human trophoblast cell-surface antigens defined by monoclonal antibodies. *Proc Natl Acad Sci USA* 1981;78:5147.
17. Loke YW, Butterworth BH, Margetts JJ, et al: Identification of cytotrophoblast colonies in cultures of human placental cells using monoclonal antibodies. *Placenta* 1986;7:221.
18. McDicken IW, McLaughlin PJ, Tromans PM, et al: Detection of placental-type alkaline phosphatase in ovarian cancer. *Br J Cancer* 1985;52:59.
19. McIntyre JA, Faulk WP, Nichols-Johnson VR, et al: Immunologic testing and immunotherapy in recurrent spontaneous abortion. *Obstet Gynec* 1986;67:169.
20. McLaughlin PJ, Cheng HM, Slade MB, et al: Expression on cultured human tumor cells of placental trophoblast membrane antigens and placental alkaline phosphatase defined by monoclonal antibodies. *Int J Cancer* 1982;30:21.
21. Mueller UW, Hawes CS, Jones WR: Cell surface antigens of human trophoblast: definition of an apparently unique system with a monoclonal antibody. *Immunology* 1986;59:135.
22. Oksenberg JR, Brautbar C: In vitro suppression of murine blastocyst growth by sera from women with reproductive disorders. *Am J Reprod Immunol Microbiol* 1986;11:118.

23. Redman CWG, McMichael AJ, Stirrat GM, et al: Class I MHC complex antigens on human extravillous trophoblast. *Immunology* 1984;52:45.
24. Rettig WJ, Cordon-Cardo C, Koulos JP, et al: Cell surface antigens of human trophoblast and choriocarcinoma defined by monoclonal antibodies. *Int J Cancer* 1985;35:469.
25. Stern PL, Morris AC, Beresford N, et al: Molecular characterization of human teratocarcinoma-trophoblast cell surface antigens, in Clark DA, Croy BA (eds): *Reproductive Immunology 1986*. Amsterdam, Elsevier Biomedical Press, 1986, p 67.
26. Stern PL, Thompson S, Beresford N, et al: Characterization of the human trophoblast-leukocyte common antigen molecules defined by a monoclonal antibody. *J Immunol* 1986;137:1604.
27. Sunderland CA, Bulmer JN, Luscombe M, et al: Immunohistological and biochemical evidence for a role for hyaluronic acid in the growth and development of the placenta. *J Reprod Immunol* 1985;8:197.
28. Tartakovsky B: Apparent immune disruption of gestation: abortion induced by injection of fetal or tumor cells, in Gill TJ III, Wegmann TG (eds): *Immunoregulation and Fetal Survival*. Oxford University Press, 1987, p 233.
29. Tezabwala BU, Johnstone AP: Effects of administration of interleukin-2 in pregnancy. *J Reprod Immunol* [Suppl] 1986;147.
30. Travers P, Bodmer WF: Preparation and characterization of monoclonal antibodies against placental alkaline phosphatase and other human trophoblast-associated determinants. *Int J Cancer* 1984;33:633.
31. Van Leeuwen A, Giphart MJ, de Groot G, et al: Two different T-cell systems in humans, one of which is probably equivalent to Qa or Tla in mice. *Hum Immunol* 1985;12:335.
32. Wahlstrom T, Nieminen P, Narvanen A, et al: Monoclonal antibody defining a human syncytiotrophoblastic polypeptide immunologically related to mammalian retrovirus structural protein p30. *Placenta* 1984;5:465.
33. Webb PD, Evans PW, Molloy CM, et al: Biochemical studies of human placental microvillus plasma membrane proteins. *Am J Reprod Immunol Microbiol* 1985;8:113.
34. Webb PD, Hole N, McLaughlin PJ, et al: Biochemical and immunological aspects of the human trophoblast cell surface: a review. *Trophoblast Res* 1987; in press.
35. Yamashita K, Nakamura T, Shimizu T, et al: Monoclonal antibodies to human choriocarinoma. *Am J Reprod Immunol Microbiol* 1986;11:130.

CHAPTER 14

Maternal-Blastocyst Interaction at Implantation

Michael J.K. Harper and Marjorie A. Jones

Introduction

During the past six years we have been using the rabbit as a model system for studying the interaction of the developing zygote and the maternal environment, specifically the endometrium, up to and during the process of implantation, which has occurred by day 7, i.e., seven days after mating or insemination and injection of human chorionic gonadotropin (HCG). In 1975 the first report of the presence of prostaglandins (PGs) in day 6 rabbit blastocysts appeared (7). Two more reports confirming and expanding these observations appeared in 1980 by Dey et al (6) and Sharma (38). In the same and subsequent years we reported (11,33) an extensive series of experiments, also in rabbits, that demonstrated that large increases in prostaglandins E (PGE) and F (PGF) levels occurred with increasing developmental age of the blastocysts from day 4 onward, but that although uterine tissue contained PGs, the levels did not change greatly during the same time period. These experiments confirmed that blastocysts not only contained PGs but also released them into the environment during culture in vitro. The amount released also increased with increasing developmental age. When the quantity of PGs released was expressed in terms of ng/blastocyst or ng/mg of blastocyst protein, the increase over the three-day period from day 4 to day 7 of pregnancy was several hundredfold. Similarly, endometrial release of PGs was significantly greater on day 7 than on day 4 of pregnancy, but the magnitude of the change was only twofold.

Thus, the data that were published in full in 1983 (11) pointed to the blastocyst as the key source of the PGs, which had been implicated in the implantation process in the rabbit (12,13) and in other species, e.g., hamsters, rats, and mice (3,9,14,21–23,25–27,31,34,37). However, in these other species it is not certain that the PGs are derived from the blastocysts, since the presence of PGE and PGF in such blastocysts could not be demonstrated in mice (32) or in rats (24), even though one early report had claimed the synthesis of PGs by mouse blastocysts (4).

In large animals, such as cattle (28), sheep (16), and pigs (5), PG synthesis by blastocysts and endometrium has been convincingly demonstrated. Thus, the synthesis of PGs by blastocysts in the rabbit (6,11,32) is not unique to this species and may indicate the suitability of this species as a model for all except the small rodents. No information on synthesis of PGs by primate blastocysts is yet available.

The large amounts of PGs found in rabbit blastocysts and also released during in vitro incubation suggested, however, that de novo synthesis could not be the only source of blastocyst PGs.

Transport of PGs Into Blastocysts

Our initial experiments (17) were done in vitro. Blastocysts at various stages of pregnancy were recovered from superovulated donors and incubated in Tyrode's buffer with [^{3}H]-labeled $PGF_{2\alpha}$ and PGE_2. Accumulation of PGs was assessed by separation of bound and free ligands using filtration. Blastocysts of the same age accumulated approximately the same amounts of both PGs, but showed no saturation over a concentration range of 1 to 1,000 nmol/L. However, blastocysts of different ages accumulated different amounts of PGs, with PGE_2 uptake increasing sevenfold from days 5 to 6 and threefold from days 6 to 6.8. Over the same time periods $PGF_{2\alpha}$ uptake increased 12-fold and threefold. These differences were probably due solely to the increases in blastocyst volume taking place over the same periods, i.e., 23-fold and sixfold, as calculated from the increase in diameter. There was no apparent difference in uptake between PGE_2 and $PGF_{2\alpha}$. Approximately 90% of the PGs are found in the blastocoelic fluid, with only 10% in the blastomeres and/or zona pellucida. This transport of PGs into blastocysts in vitro was inhibited at 0°C and was affected by addition of 10 μg/mL of indomethacin to the incubation medium. Indomethacin is, however, not only a PG synthesis inhibitor (41), but also an inhibitor of PG transport (1,29) and other enzyme systems (39,42). PG accumulation by the blastocysts was not readily inhibited by metabolic inhibitors or an organic anion transport inhibitor, which indicates that this uptake process is not due to active transport or anion transport. The lack of saturation over a large concentration range indicates that the process is not receptor mediated. Since the uptake was both time and temperature dependent, it may be concluded that a facilitated diffusion process is involved.

In vivo we have found that a similar mechanism operates (18). At laparotomy of day 6-pregnant rabbits, both uterine horns were exposed, and 3.5 $\pm$ 0.3 μCi [^{3}H] PGE_2 or 3.7 $\pm$ 0.3 μCi [^{3}H] $PGF_{2\alpha}$ in 20 μL Krebs Ringer bicarbonate buffer (pH 7.4) were injected into one uterine horn via the uterotubal junction, which was then ligated. The contralateral horn acted as the control. At times shown in Table 14.1 the animals were killed and various organs evaluated for radioactivity. The total disintegrations per minute (DPM) recovered were 71.8% of the [^{3}H] PGE_2 and 83.4% of the [^{3}H] $PGF_{2\alpha}$ injected. Little radioactivity was found in the peritoneal cavity, indicating that there was no leakage from the point of ligation or injection.

Table 14.1. Location of radioactivity recovered at various times after intrauterine injection of day 6-pregnant rabbits.[a]

	Hours in vivo	Mean no. blastocysts recovered	DPM per blastocyst in injected horn	DPM per 100 mg injected horn	DPM per 100 mg noninjected horn	Total DPM in urine
PGE_2						
	1	8 ± 4	145985 ± 78160	5797 ± 1083	665 ± 213	$3.3 \times 10^6 \pm 8.9 \times 10^5$
		(4)	(4)	(3)	(3)	(3)
	2	9 ± 2	240477 ± 148825	4221 ± 728	613 ± 120	$3.2 \times 10^6 \pm 6.3 \times 10^5$
		(6)	(6)	(8)	(8)	(8)
	19–21	8 ± 3	31845 ± 27730	1240 ± 460	545 ± 94	$7.0 \times 10^6 \pm 1.3 \times 10^6$
		(3)	(3)	(5)	(5)	(5)
$PGF_{2\alpha}$						
	1	12 ± 4	278674 ± 94925	4838 ± 1778	607 ± 140	$4.1 \times 10^6 \pm 2.7 \times 10^5$
		(5)	(5)	(3)	(3)	(3)
	2	8 ± 2	123503 ± 42924	2995 ± 584	648 ± 125	$5.2 \times 10^6 \pm 9.0 \times 10^5$
		(5)	(5)	(8)	(8)	(6)
	19–21	6 ± 3	24700 ± 7207	1237 ± 195	791 ± 223	$3.5 \times 10^7 \pm 2.9 \times 10^7$
		(4)	(4)	(5)	(5)	(4)

[a] mean ± Sem
( ) = number of rabbits

Uterine tissue and blastocysts from the injected horn had significantly more radioactivity than those from the control horn (Table 14.1). Indeed, blastocysts from the control horn contained no detectable radioactivity. Clearly, the blastocysts from the injected horn preferentially sequestered PGs, containing 57 and 41 times more [^{3}H] PGE_2 and [^{3}H] $PGF_{2\alpha}$, respectively, than uterine tissue at two hours, and still 26 and 20 times more, respectively, at 19 to 21 hours. Radioactivity in the blastocysts was directly related to their distance from the uterotubal junction, i.e., site of radioactive deposition. The injected PGs were rapidly cleared from the uterus. By two hours, 76.9% of the radioactivity was excreted as catabolites in the urine. Despite this accumulation of radioactive PGs, these blastocysts were still viable because implantation of such blastocysts occurred normally. The diameters of implantation sites at day 7.16 (i.e., seven days and four hours after insemination and HCG administration) were similar whether [^{3}H] PGE_2, [^{3}H] $PGF_{2\alpha}$, or buffer had been injected into the uterine horn on day 6.

These data show that rabbit blastocysts accumulate PGs from their environment both in vitro and more importantly in vivo, suggesting that this is the normal physiological process.

Storage of PGs by Blastocysts

In our original in vitro experiments (17) we had found that although uptake of PGs by blastocysts was rapid, release was slow. More than 50% of the [^{3}H] $PGF_{2\alpha}$ taken up by days 5, 6, or 6.8 blastocysts was retained two hours later. For [^{3}H] PGE_2 the results were similar with days 5 and 6 blastocysts, but day 6.8 blastocysts appeared to release this PG more rapidly, perhaps implying an obligate role in the implantation process for PGE_2. Later times could not be satisfactorily studied in vitro.

However, in subsequent experiments we similarly loaded day 6 blastocysts with radioactive PGs in vitro and then transferred them to the uteri of nontreated pseudopregnant rabbits (19). As in the in vitro experiments, approximately 50% of the accumulated radioactivity was lost within one hour, but this percentage had not increased by three hours in vivo, and even after 20 hours in vivo (equivalent to six days, 20 hours of pregnancy) the transferred blastocysts still retained $12.2 \pm 4.1\%$ (mean $\pm$ SEM) of the initial [^{3}H] PGE_2 and $13.0 \pm 6.6\%$ of the initial [^{3}H] $PGF_{2\alpha}$. In contrast, [^{3}H] H_2O exhibited a single first-order rate constant and only $1.0 \pm 0.2\%$ of the label was retained after one hour in vivo.

These results are based solely on monitoring of the radioactivity, and this could be associated with catabolites rather than the original parent compounds. Normally, PGs are not stored intercellularly but rapidly metabolized (36), but in the blastocyst this is not the case. In both in vitro and in vivo experiments, evaluation of the blastocoelic fluid by the high-performance liquid chromatography (HPLC) procedure of Eling et al (8) revealed that little metabolism of the stored PGs occurred (less than 1%) even after the blastocysts loaded with radioactivity in vitro had resided for 20 hours in the uterus of pseudopregnant recipients (19). In addition, in other recent experiments (18) in which the blastocysts accumulated radioactivity after intrauterine instillation of the radiolabeled PGs, even 19 to 21 hours later considerable amounts of the original radioactivity were unmetabolized (Table 14.2). In contrast, less than 2% of the radioactive PGs recovered from urine at two hours eluted with authentic PG standards, indicating almost complete catabolism of the injected PGs. The difference between the results for blastocysts in these two in vivo experiments may reside in the uptake of PGs already metabolized by the uterus in the experiment where uptake occurred in vivo rather in vitro.

Nonetheless, these data suggest that either blastocyst cells (trophoblast and inner cell mass) do not possess the major enzyme responsible for PG metabolism (15-hydroxy protaglandin dehydrogenase) or that the PGs during uptake and release pass between the cells rather than through them. Histochemical studies were unable to demonstrate the presence of this enzyme in hamster blastocysts, even though it was present at earlier stages of development (30).

Blastocyst PGs and Uterine Vascular Permeability Changes

The above experiments demonstrate that in a physiological situation, as near normality as possible, blastocysts sequester PGs from the uterine fluid, store them in an unmetabolized form, and release them in some predetermined pattern during the day preceding the implantation process. These experiments do not, however, prove that it is the blastocyst PGs that are involved in the vascular permeability changes associated with implantation and necessary for subsequent normal embryonic development. To examine this question we conducted the following experiments.

Isolated rabbit uteri were placed in in vitro incubations, and the transport of [^{3}H] PGE_2 and [^{3}H] $PGF_{2\alpha}$ from the uterine lumen to the exterior bathing

Table 14.2. HPLC[a] evaluation of PGs: Percentage of radioactivity as original (unmetabolized) PGs (mean ± SEM).

		PGE_2			$PGF_{2\alpha}$		
Hours in vivo	N	Control[b]	Extracted from blastocysts	Extracted from urine	Control	Extracted from blastocysts	Extracted from urine
2	3	83.9 ± 2.3	73.0 ± 4.2	1.9 ± 0.6	82.4 ± 2.6	63.8 ± 16.5	1.8 ± 0.4
19–21	3	85.4 ± 0.9	34.1 ± 11.7	–	76.9 ± 3.8	53.2 ± 24.2	–

[a]HPLC system of Eling et al (8).

[b]Control = PG not injected in uterine lumen but kept at room temperature for same time before extraction.

N = number of rabbits evaluated.

medium was studied (2). Prostaglandin accumulation by the tissue increased rapidly up to 30 minutes and then remained relatively constant. At this time radioactivity commenced to be found in the external medium, and then increased linearly with time. The translocation of [^{3}H] $PGF_{2\alpha}$ was significantly greater in pseudopregnant than in pregnant animals on day 6, whereas that of [^{3}H] PGE_2 was significantly higher in pregnant than in pseudopregnant animals on day 6.8 (i.e., six days and 20 hours). In pregnant animals, both [^{3}H] $PGF_{2\alpha}$ and [^{3}H] PGE_2 were translocated to the exterior more rapidly at day 6.8 than at days 5 or 6. Translocation of [^{3}H] PGE_2, but not that of [^{3}H] $PGF_{2\alpha}$ was increased by increased intraluminal concentrations of PGE_2. In control experiments the translocation of [^{14}C] sucrose and [^{14}C] urea was studied. Transport of [^{14}C] sucrose was much slower than that of [^{14}C] urea, which was greater than the fastest rates exhibited by the PGs. These experiments demonstrated that PGs could be transported from the uterine lumen across the uterine wall and could thus reach the stroma. Furthermore, the uterus exhibits a certain selectivity in handling these PGs in relation to the stage of pregnancy or pseudopregnancy (i.e., where blastocysts are absent from the uterus but the hormonal conditions are otherwise similar).

In untreated rabbits an increase from 1 to 1.16 ± 0.06 in the capillary permeability ratio of tissue underlying the site of blastocyst implantation is usually seen by day 6.5 of pregnancy (15). To determine how long blastocysts must actually reside in the uterus to induce such an increase, blastocysts recovered from control donor rabbits were transferred to the uteri of day 6 or 6.5 pseudopregnant recipients. The normal time required for a detectable increase was approximately 9 hours (20). A permeability ratio of 1.38 ± 0.14 (n = 6 rabbits with 18 sites) was recorded when day 6.5 blastocysts were transferred to day 6.5 uteri and the permability ratio measured nine hours later. Similar blastocysts left for four or six hours had ratios not different from 1, the control value. Similar blastocysts were recovered from animals that had received indomethacin (10 mg/kg subcutaneously twice daily starting on day 4.5 of pregnancy). These blastocysts were depleted of PGs. Normal day 6.5 blastocysts contained 3.58 ± 1.18 and 20.24 ± 6.02 ng/blastocyst of PGE and PGF, respectively, when measured by radioimmunoassay (RIA). Day 6.5 blastocysts from indomethacin-treated donors contained significantly less PGs, only 0.04 ± 0.01 and 0.02 ± 0.01 ng/blastocyst of PGE and PGF, respectively. For day 7 blastocysts the same trend was seen: normal values were 8.77 ± 2.00 and 23.10 ± 6.57 ng/blastocyst for PGE and PGF, respectively, and after indomethacin treatment only 0.64 ± 0.19 and 0.77 ± 0.24, respectively. Such day 6.5 depleted blastocysts when transferred to uteri of day 6.5 pseudopregnant recipients were unable in nine hours to induce an increase in the capillary permeability ratio over the control value. This result indicated that it is the blastocyst PGs that induce the permeability changes underlying the blastocyst site, because in all other respects the recipients were normal and could produce their own complement of uterine PGs. Indeed, PG levels of the depleted blastocysts were significantly increased after nine to 11 hours in vivo in day 6.5 pseudopregnant animals, although the contents were still lower than normal. If such depleted blastocysts were incubated in vitro in buffer for one hour at 37°C prior to transfer to the untreated recipients, and PG levels measured nine to 11 hours later, then the PG contents were not different from control. The one-hour in vitro incubation presumably removed the residual effect of indomethacin.

Given the propensity of blastocysts to accumulate PGs, day 6.5 PG-depleted blastocysts were incubated in vitro for one hour in buffer containing either 1 μmol/L PGE_2 and 1 μmol/L $PGF_{2\alpha}$ or 1 μmol/L PGE_2 alone. Such blastocysts then containing a normal content of PGE_2 and $PGF_{2\alpha}$ or PGE_2 only *before transfer* were transferred as before to day 6.5 pseudopregnant recipients. Contrary to expectation, the blastocysts containing PGE_2 failed to induce the anticipated increase in capillary permeability in the uteri at the sites underlying the blastocysts in the usual nine to 11 hours. This experimental result did not support the hypothesis that blastocyst PGE_2 released in a controlled fashion was the key mediator of the capillary permeability changes at the implantation site. However, further experiments revealed that indomethacin has very significant effects on the ability of blastocysts to retain PGs and may make them leaky, thus destroying their ability to store PGs. This action could explain the failure of the blastocysts, depleted of PGs by indomethacin, to induce an increase in capillary permeability after PG replenishment. Nonetheless, the fact that indomethacin-treated blastocysts failed to induce a normal increase in uterine vascular permeability strongly suggests that it is indeed blastocyst rather than uterine-derived PGs that are the key factors.

It is been suggested (22) that PGE_2 is the most likely candidate for this role of mediator of increased capillary permeability underlying the implanting blastocyst (13,15), although PGI_2 has also been considered because it is produced in large amounts by uterine homogenates (10) and at the implantation site (26). However, RIA of the various fractions separated by HPLC from extractions of day 6 rabbit blastocysts showed mean $\pm$ SEM contents of 0.24 ± 0.06, 1.05 ± 0.09, and 0.03 ± 0.02 ng/blastocyst (n = six pools of eight to 14 blastocysts) of PGE_2, $PGF_{2\alpha}$, and 6-keto-$PGF_{1\alpha}$ (the major metabolite of PGI_2). The values for PGE_2 and $PGF_{2\alpha}$ correspond well with our previously published values for day 6 rabbit blastocysts (11). Thus, day 6 blastocysts contain eightfold more PGE_2 than 6-keto $PGF_{1\alpha}$ and fourfold more $PGF_{2\alpha}$ than PGE_2. Furthermore, in other unpublished studies we found that [^3H]-Iloprost, a stable analog of PGI_2, was not taken up by blastocysts in the same manner as [^3H] PGE_2 or [^3H] $PGF_{2\alpha}$, reaching an apparent maximal uptake at one hour. $PGF_{2\alpha}$ acts as a vasoconstrictor, unlike PGE_2 and PGI_2, which act as vasodilators and induce blastocyst swelling in vitro.

Conclusions

Consequently, taking all the evidence together, we feel that the most likely explanation for the induction of increased capillary permeability at the implantation site in the rabbit is as follows: Blastocysts as they expand, accumulate increasing quantities of PGs, especially PGE_2 and $PGF_{2\alpha}$. These PGs are stored in an unmetabolized state, and then commencing on day 6, PGE_2 is preferentially released and transported across the luminal epithelium underlying the apposed blastocyst to the stroma. There, perhaps in conjunction with other factors, e.g., platelet-activating factor (PAF) whose concentration in the uterus has been shown to decrease dramatically at the site of blastocyst attachment between days 6 and 7 of pregnancy (Angle, Jones, McManus, Pinckard and Harper, unpublished data), PGE_2 induces increased capillary permeability, which is a required component for normal embryonic development.

References

1. Bito LZ, Davson H, Salvador EV: Inhibition of *in vitro* concentrative prostaglandin accumulation by prostaglandins, prostaglandin analogues and by some inhibitors of organic anion transport. *J Physiol* 1976;256:257–271.
2. Cao Z-d, Jones MA, Harper MJK: Prostaglandin translocation from the lumen of the rabbit uterus in vitro in relation to day of pregnancy or pseudopregnancy. *Biol Reprod* 1984;31:505–519.
3. Castracane VD, Saksena SK, Shaikh AA: Effect of IUDs, prostaglandins and indomethacin on decidual cell reaction in the rat. *J Reprod Fertil* 1974;46:101–104.
4. Chepenik KP, Smith JB: Synthesis of prostaglandins by mouse embryos. *IRCS J Med Sci* 1980;8:783.
5. Davis DL, Pakrasi PL, Dey SK: Prostaglandins in swine blastocysts. *Biol Reprod* 1983;28:1114–1118.
6. Dey SK, Chien SM, Cox CL, et al: Prostaglandin synthesis in the rabbit blastocyst. *Prostaglandins* 1980;19:449–453.
7. Dickmann Z, Spilman CH: Prostaglandins in rabbit blastocysts. *Science* 1975;190: 997–998.
8. Eling T, Warnock R, Dick D, et al: Separation of prostaglandins, thromboxane, hydroxy fatty acids and arachidonic acid by high pressure liquid chromatography. *Prostaglandins Med* 1980;5:345–355.
9. Evans CA, Kennedy TG: The importance of prostaglandin synthesis for the initiation of blastocyst implantation in the hamster. *J Reprod Fertil* 1978;54:255–261.
10. Fenwick L, Jones RL, Naylor B, et al: Production of prostaglandins by the pseudopregnant rat uterus, *in vitro*, and the effect of tamoxifen with the identification of 6-keto-prostaglandin $F_{1\alpha}$ as a major product. *Br J Pharmacol* 1977;59:191–199.
11. Harper MJK, Norris CJ, Rajkumar K: Prostaglandin release by zygotes and endometria of pregnant rabbits. *Biol Reprod* 1983;28:350–362.
12. Hoffman LH: Antifertility effects of indomethacin during early pregnancy in the rabbit. *Biol Reprod* 1978;18:148–153.
13. Hoffman LH, DiPietro DL, McKenna TJ: Effects of indomethacin on uterine capillary permeability and blastocyst development in rabbits. *Prostaglandins* 1978;15:823–828.
14. Holmes PV, Gordashko BJ: Evidence of prostaglandin involvement in blastocyst implantation. *J Embryol Exp Morphol* 1980;55:109–122.
15. Hoos PC, Hoffman LH: Temporal aspects of rabbit uterine vascular and decidual responses to blastocyst stimulation. *Biol Reprod* 1980;23:453–459.

16. Hyland JH, Manns JG, Humphrey WD: Prostaglandin production by ovine embryos and endometrium *in vitro*. *J Reprod Fertil* 1982;65:299–304.
17. Jones MA, Harper MJK: Rabbit blastocysts accumulate [^{3}H]prostaglandins *in vitro*. *Endocrinology* 1984;115:817–823.
18. Jones MA, Cao Z-d, Harper MJK: *In vivo* accumulation of [^{3}H]prostaglandins (PGs) by rabbit blastocysts. *Biol Reprod* [*Suppl*] 1985; Abstract No 204, p 139.
19. Jones MA, Anderson W, Turner TG, et al: Storage *in vivo* of [^{3}H]prostaglandins by rabbit blastocysts. *Endocrinology* 1985;116:993–997.
20. Jones MA, Cao Z-d, Anderson W, et al: Capillary permeability changes in the uteri of recipient rabbits after transfer of blastocysts from indomethacin-treated donors. *J Reprod Fertil* 1986;78:261–273.
21. Kennedy TG: Evidence for a role for prostaglandins in the initiation of blastocyst implantation in the rat. *Biol Reprod* 1977;16:286–291.
22. Kennedy TG: Prostaglandins and increased vascular permeability resulting from the application of an artificial stimulus to the uterus of the rat sensitized for the decidual cell reaction. *Biol Reprod* 1979;20:560–566.
23. Kennedy TG: Prostaglandins and the endometrial vascular permeability changes preceding blastocyst implantation and decidualization. *Prog Reprod Biol* 1980;7:234–243.
24. Kennedy TG, Armstrong DT: The role of prostaglandins in endometrial vascular changes at implantation, in Glasser SR, Bullock DW (eds): *Cellular and Molecular Aspects of Implantation*. New York, Plenum Press, 1981, pp 349–358.
25. Kennedy TG, Barbe GJ, Evans CA: Prostaglandin I_2 and increased endometrial vascular permeability preceding the decidual cell reaction, in Kimball FA (ed): *The Endometrium*. New York, SP Medical and Scientific Books Inc, 1980, pp 331–342.
26. Kennedy TG, Zamecnik J: The concentration of 6-keto-prostaglandin $F_{1\alpha}$ is markedly elevated at the site of blastocyst implantation in the rat. *Prostaglandins* 1978;16:599–605.
27. Lau IF, Saksena SK, Chang MC: Pregnancy blockade by indomethacin, an inhibitor of prostaglandin synthesis: Its reversal by prostaglandins and progesterone in mice. *Prostaglandins* 1973;4:795–803.
28. Lewis GS, Thatcher WW, Bazer FW, et al: Metabolism of arachidonic acid in vitro by bovine blastocysts and endometrium. *Biol Reprod* 1983;27:431–439.
29. Melendez E, Reyes JL: Renal handling of indomethacin and its relationship with the secretory pathways of prostaglandins. *J Pharm Pharmacol* 1982;34:648–652.
30. Niimura S, Ishida K: Histochemical studies of prostaglandin dehydrogenase in hamster eggs. *Jpn J Anim Reprod* 1978;24:16–18.
31. Oettel M, Koch M, Jurishko A, et al: A direct evidence for the involvement of prostaglandin $F_{2\alpha}$ in the first step of estrone-induced blastocyst implantation in the spayed rat. *Steroids* 1979;33:1–8.
32. Racowsky C, Biggers JD: Are blastocyst prostaglandins produced endogenously? *Biol Reprod* 1983;29:379–388.
33. Rajkumar K, Harper MJK: Prostaglandin (PG) production by rabbit blastocysts and endometrium. Society for Gynecologic Investigation, 27th Annual Meeting, 1980, Scientific Abstracts, No 292, p 174.
34. Rankin JC, Ledford BE, Jonsson HT Jr, et al: Prostaglandins, indomethacin and the decidual cell reaction in the mouse. *Biol Reprod* 1979;20:399–404.

35. Saksena SK, Lau IF, Chang MC: Relationship between oestrogen, prostaglandin $F_{2\alpha}$ and histamine in delayed implantation in the mouse. *Acta Endocrinol* 1976; 81:801–807.
36. Samuelsson B, Granstrom E, Gréen K, et al: Prostaglandins. *Ann Rev Biochem* 1975;44:669–695.
37. Sananès N, Baulieu E-E, LaGoascogne C: Prostaglandin(s) as inductive factor of decidualization in the rat uterus. *Mol Cell Endocrinol* 1976;6:153–158.
38. Sharma SC: Temporal changes in PGE, PGF_{α}, oestradiol-17β and progesterone in uterine venous plasma and endometrium of rabbits during early pregnancy, in Crastes de Paulet A, Thaler-Dao H, Dray F (eds): *Les Colloques de l'Inserm, Prostaglandines et Physiologie de la Reproduction*. 1980, Paris, Inserm, pp 243–263.
39. Siegel MI, McConnell RT, Cuatrecasas P: Asprin-like drugs interfere with arachidonate metabolism by inhibition of the 12-hydroperoxy-5,8, 10, 14-eicosatetraenoic acid peroxidase activity of the lipoxygenase pathway. *Proc Natl Acad Sci USA* 1979; 76:3774–3778.
40. Torbert JA: A study of the possible role of prostaglandins in decidualization using a non-surgical method for the instillation of fluids into the rat uterine lumen. *J Reprod Fertil* 1976;47:391–393.
41. Vane JR: Inhibition of prostaglandin synthesis as a mechanism of action of aspirin-like drugs. *Nature*, [*New Biol*] 1971;231:232–235.
42. Wu C, Matthews KP: Indomethacin inhibition of glutathione S-transferases. *Biochem Biophys Res Commun* 1983;112:980–985.

CHAPTER 15

Immunogenetic Control of Pregnancy and Development

Thomas J. Gill III

The reproductive process is an inefficient one, and there is a considerable amount of fetal wastage in a variety of species. This phenomenon had been best studied in the human, rat, and mouse: 48% to 74% of human conceptuses are lost (4); in rats bearing the major histocompatibility complex (MHC)-linked *grc* genes, 20% to 25% of *grc* homozygotes die in the immediate postnatal period (15); and in the CBA × DBA/2 mating combination in mice, approximately 30% of the conceptuses abort spontaneously (1,13). We have tried to organize the various immunological and genetic factors that influence reproduction into an integrated hypothesis that focuses primarily on the effects of MHC and MHC-linked genes (5,6). This hypothesis is outlined in Figure 15.1. The most important factor in the control of normal fetal development is genetic compatibility between the mating partners. Once this critical condition has been met, a variety of immunological factors may modulate the implantation process under certain conditions. Genetic compatibility requires that the mating partners contribute no recessive lethal genes that could act alone or epistatically in the developing fetus. The immunological response in normal pregnancies is not destructive, because it is not directed against the classical class I transplantation antigens but rather against a unique placental class I antigen(s) that has broadly shared (public) specificities only and not haplotype-specific (private) specificities. In the cases where immunological modulation has been shown to affect implantation, the mechanisms are not yet known.

In humans, women who lose their conceptuses without forming antipaternal antibodies may do so because of the primary action of recessive lethal or semi-lethal genes. In women who lose their conceptuses and who make lymphocytotoxic antibodies directed against paternal antigens, the mechanism for repressing the expression of the *classical* class I transplantation antigens on the placental surface may be defective. These transplantation antigens would then elicit a destructive immune response that would lead to spontaneous abortion. There is some evidence from animal studies and from clinical observations that the genes affecting embryological development may also lead to the development of congenital anomalies and to an increased susceptibility to cancer.

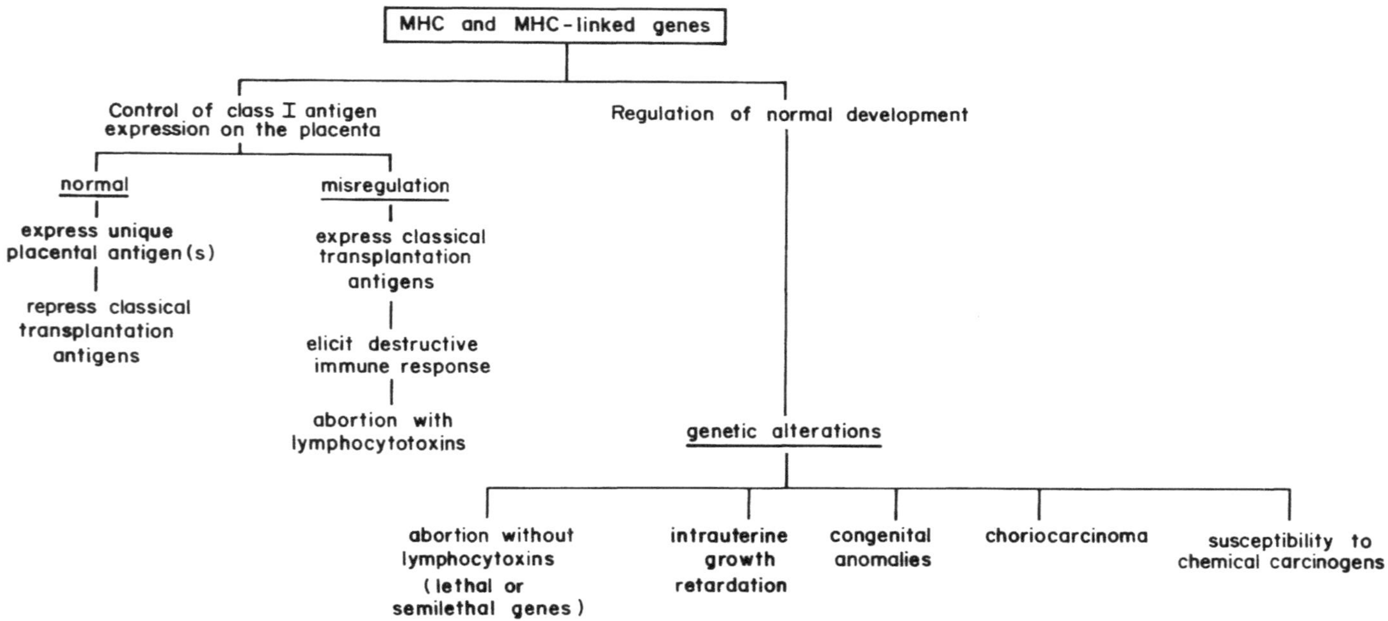

Fig. 15.1. Outline of the hypothesis regarding the role of MHC and MHC-linked genes in reproduction and their possible effects on developmental abnormalities and on susceptibility to cancer.

Studies in Experimental Animals

Studies on the immunogenetics of development have been done in a variety of animals, but the most extensive studies have used either the mouse or the rat. Each species has certain advantages for these types of studies, and the work discussed here focuses primarily on the rat. Among its advantages are a very limited MHC polymorphism; a well-defined MHC; the presence of genes affecting growth and development linked to MHC (the *grc*); the availability of a substantial number of genetically characterized inbred, congenic and recombinant strains; and a size that allows a variety of surgical procedures and that provides large amounts of tissue.

There are a variety of recessive lethal genes in different parts of the genome, for example, the *Tal* and H^{re} genes in the rat. However, the most extensively studied lethal genes are those linked to the MHC in the rat (*grc*) and in the mouse (*t*-haplotypes). Thus, the segment of genome that carries the MHC and the *grc* plays an important immunological and genetic role in affecting development.

The acceptance of the semiallogenic fetus in mammalian reproduction is in apparent contradiction to the laws of transplantation. The mechanism by which this is achieved appears to involve the absence of class II antigens on the placenta and the expression on the placental surface of only those class I antigens that carry broadly shared antigenic determinants. Classical class I transplantation antigens are also present in the basal trophoblast, which is that part of the trophoblast in contact with the maternal tissue (Table 15.1), but they do not appear to be expressed on the placental surface. This lack of expression on the placental surface has been inferred from the observation in rats that only the unique Pa antigen elicits an antibody response (2,3), whereas the classical class I transplantation antigen $RT1.A^a$ does not (11). Studies in the human (12) and in the mouse (23) and preliminary studies in the rat (22) showed that there was a marked decrease in the level of placental mRNA compared with that in lympho-

Table 15.1. Localization of MHC antigens in the placenta.[a]

Species	Antigen	Type of antigenic determinant	Localization in the placenta
Rat	Pa	Public	Basal trophoblast and chorioallantoic membrane
	$RT1.A^a$	Private	Basal trophoblast and chorioallantoic membrane
Mouse	H-2K,D	Public and private	Spongiotrophoblast
Human	HLA-A,B,C	Public	Extravillous trophoblast
	HLA-D,DR	Public and private	Absent from trophoblast
	TLX	Public and private	Syncytial trophoblast and cytotrophoblast

[a] For detailed listing of references, see Gill (4,5), Macpherson et al (16), and Stranick et al (22). Recent work (14) suggests that a whole system of public determinants is present on the placental surface.

Table 15.2. Litter sizes at weaning in matings of *grc* heterozygotes.[a]

Number of offspring of each genotype					
Females			Males		
grc/grc	*grc/+*	+/+	*grc/grc*	*grc/+*	+/+
203	1079	542	229	976	531
237	444	–	233	458	–

[a] Gill et al (9).

cytes or liver. These observations suggest that the production of the classical class I transplantation antigen may be down-regulated in the placenta.

A prime example of fetal loss due to a direct genetic effect is that due to the action of the *grc* genes in the rat (15). This gene complex contains semilethal genes that cause death in approximately 20% to 25% of homozygous offspring in the immediate postnatal period. They do not, however, cause an alteration in the male:female ratio (Table 15.2) nor do they cause segregation distortion (Table 15.3). In the latter case, they differ from the *t*-haplotypes in the mouse, which are characterized by their preferential transmission ratios (10). These genes may act alone or epistatically with other lethal genes (Table 15.4). Epistatic interaction involves genes on different chromosomes that acted in concert; in this case, they cause death of the embryo. This phenomenon in mammals was discovered in the rat (8,21), and there is some evidence that it may also be operative in women having recurrent spontaneous abortions (26). The mechanism by which the *grc* exerts its effect has been studied extensively by biochemical, immunochemical, and morphological approaches. The evidence is consistent with the prime defect's being in membrane biosynthesis. This defect could then lead to abnormal interactions among cells during development and could also make these cells more susceptible to various perturbations, such as those caused by exposure to mutagenic chemicals (5).

Table 15.3. Test for segregation distortion in matings of *grc* heterozygotes.[a]

		Ratio of haplotypes *grc/+*:+/+		
Type of mating[b]	Sex	Expected	Observed	Significance of difference (P value)
grc/+ × *grc/+*	F	2:1	1.99:1	> 0.5[c]
	M	2:1	1.84:1	> 0.1[c]

[a] Gill et al (9).
[b] Data from Table 15.2.
[c] Not significant.

Table 15.4. Lethal epistatic interaction in rats and in humans.[a]

Species:	Rat		Rat		Human	
System						
MHC:	grc		grc		HLA	
nonMHC:	Tal		H^{re}		C3	
	Genotype	Fetal mortality (%)	Genotype	Fetal mortality (%)	Genotype	Antigen sharing in chronic spontaneous abortion (%)
	grc/grc	25	grc/grc	25	HLA/HLA[b]	62
	grc/+	0	grc/+	0	HLA/+[c]	22
	Tal/Tal	100	H^{re}/H^{re}	100	C3/C3	8
	Tal/+	0	$H^{re}/+$	0	C3/+	23
	grc Tal / grc +	70[d]	grc H^{re} / grc +	68[d]	HLA C3 / HLA C3	86[d]
	grc Tal / + +	0	grc H^{re} / + +	0	HLA C3 / HLA +	63

[a] Schaid et al (21); Gill et al (8); Weitkamp and Schacter (26).
[b] Symbol for sharing ≥ 1 HLA antigen.
[c] Symbol for unmatched HLA antigens.
[d] Lethal epistatic interaction.

Studies in Humans

The studies in experimental animals provide the conceptual framework within which the problem of defective development, and possibly of carcinogenesis also, may be examined in humans. These studies form the basis for the hypothesis presented in Figure 15.1. It is critical to combine the results of experimental and clinical studies, because each approach brings a different perspective to the problem and provides data that are complementary. For example, the findings in experimental animals that the placenta expresses only class I antigens and that these antigens carry broadly shared (public) determinants rather than the haplotype-specific (private) ones helped to put into perspective the many studies in humans on the presence and location of HLA antigens in the placenta (Table 15.1).

The unique nature of the class I antigens on the placental surface suggests that one cause of recurrent spontaneous abortion may be the abnormal expression of the classical class I transplantation antigens on the surface of the placenta due to genetic misregulation, as discussed above. Such women are characterized by a high prevalence of cytotoxic antibody against their husbands' lymphocytes (Table 15.5), but they do not have an increased sharing of HLA antigens with their husbands (Table 15.6). By contrast, another population of women has an increased prevalence of HLA sharing with their husbands (Table 15.6), but they do not have cytotoxic antibodies directed against their husbands' MHC antigens (Table 15.5). The latter is consistent with a primarily genetic mechanism that does not

Table 15.5. Immunological differences among normally fertile women, recurrent aborters who do not form antibodies, and recurrent aborters with antipaternal cytotoxic antibodies.[a]

Status	No. of recorded spontaneous abortions	HLA sharing (≥ 2 antigens)		Prevalence of cytotoxic antibodies	
		No./total	Percentage[b]	No./total	Percentage[c]
Normally fertile	0	16/115	13.9	2/12	16.7
Aborters with lymphocytoxins	2-13	6/34	17.6	31/34	91.2
Aborters without antibodies	2-10	57/147	38.8	2/143	1.4

[a] References 17–20,24,25.
[b] 22% in literature (6).
[c] 20% in literature (6).

operate via the immune system: it suggests that such women and their husbands both have MHC-linked lethal or semilethal genes that combine in the fetus to cause its death.

As predicted from animal studies (5), the same genes that influence reproductive loss may also predispose to the development of cancer. Data on HLA sharing in recurrent spontaneous abortion and in choriocarcinoma summarized from the literature and from current work from our laboratory (Table 15.6) provide support for this hypothesis. The data gathered from the literature on HLA sharing between women who developed choriocarinoma and their husbands were obtained prior to the development of precise tissue-typing techniques, and the identification of many of the antigenic specificities at the various loci is quite incomplete. This may be the reason that the aggregate data for HLA sharing in

Table 15.6. Sharing of HLA antigens between husbands and wives in women having recurrent spontaneous abortions and in women developing choriocarcinoma.

Disease	Affected couples			Normally fertile couples		Significance of differences (*P* value)
	Antigen	No./total	Percentage	No./total	Percentage	
Recurrent spontaneous abortion[a]	≥ 2 HLA	125/253	49.4	40/185	21.6	< .001
	HLA-B[b]	156/498	31.3	107/467	22.9	< .001
Choriocarcinoma						
Literature[a]	≥ 2 HLA	27/129	20.9[c]			> .5[d]
	HLA-B	89/303	29.4			< .05
Author's laboratory[e]	≥ 2 HLA	2/3	66.6			< .001
	HLA-B	3/3	100			< .001

[a] Gill (6).
[b] Difference is significant at HLA-A and HLA-DR loci also.
[c] See text.
[d] Not significant.
[e] Gill, Buchsbaum, Rabin, Kristofferson and Klionsky, unpublished data, 1987.

women with choriocarcinoma is not different from that in normal women. Nonetheless, there appears to be an increased sharing at the HLA-B locus in such women, just as there is in recurrent spontaneous abortion. Preliminary data from our laboratory (Table 15.6) suggest that there is a high level of HLA sharing in women who develop choriocarcinoma, but the number of patients examined is, as yet, small. Similar kinds of data suggest that there is an increased presence of congenital anomalies in the offspring of couples who have an increased incidence of HLA sharing (5,6).

Conclusions

A model that summarizes our hypothesis about the mechanisms by which immunological and genetic factors influence reproduction is shown in Figure 15.2. Normally, the expression of the haplotype-specific (private) class I MHC antigens on the placental surface is suppressed. When this control mechanism is defective and these antigens are expressed, the mother mounts a destructive

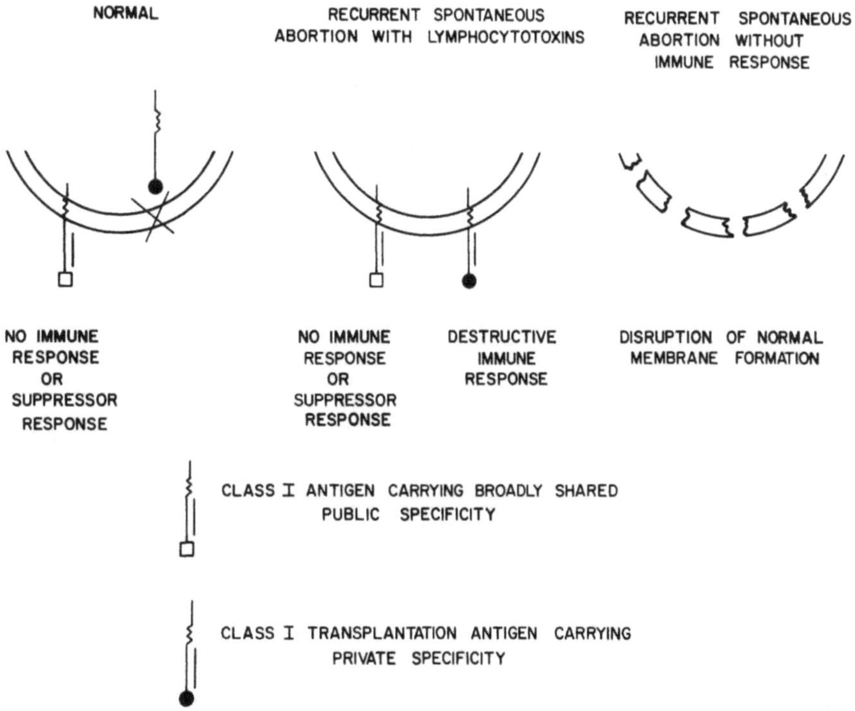

Fig. 15.2. Genetic hypothesis regarding the causes of fetal loss secondary to misregulation of class I MHC antigens on the placental surface or to defective cell membrane formation due to the presence of lethal or semilethal genes.

immune response against her placenta, and this immune response causes fetal loss. When the husband and wife each carry recessive lethal or semilethal genes, the integrity of the cell membrane may be disrupted. This leads to improper cell-cell interactions, and a defective embryo would be formed and aborted. This type of membrane defect in a less severe form could allow the development of a viable fetus that has some developmental abnormalities. This type of defect could also predispose to the development of cancer induced by exposure to chemical carcinogens or to viruses. These agents would perturb cells that already had some structural defects and that were more susceptible to the loss of the control mechanisms regulating growth.

Each component of this hypothesis can be tested explicitly from the clinical to the molecular levels. At best, it may provide a comprehensive scheme of the reproductive process and its aberrations; at worst, it will stimulate clearly focused experiments that will provide valuable information.

Acknowledgment. This work was supported by grants from the National Institutes of Health of the US Public Health Service (HD 08662 and HD 09880) and by grants from the Tim Caracio Memorial Cancer Fund and the Beaver County Cancer Society.

References

1. Clark DA, McDermott MR, Szewczak MR: Impairment of host-versus-graft reaction in pregnant mice. II. Selective suppression of cytotoxic T cell generation correlates with soluble suppressor activity and with successful allogeneic pregnancy. *Cell Immunol* 1980;52:106–118.
2. Ghani AM, Gill TJ III, Kunz HW et al: Elicitation of the maternal antibody response to the fetus by a broadly shared MHC class I antigenic determinant. *Transplantation* 1984;37:187–194.
3. Ghani AM, Kunz HW, Gill TJ III: Pregnancy-induced monoclonal antibody to a unique fetal antigen. *Transplantation* 1984;37:503–506.
4. Gill TJ III: Immunological and genetic factors influencing pregnancy and development. *Am J Reprod Immunol Microbiol* 1986;10:116–120.
5. Gill TJ III: Immunological and genetic factors influencing development and susceptibility to cancer. *Genetica*, 1987, in press.
6. Gill TJ III: Immunological factors influencing pregnancy, in Knobil E, Knobil JD, Neill JD (eds): *Physiology of Reproduction*. New York, Raven Press, 1987, in press.
7. Gill TJ III: Maternal antibody responses and genetic factors influencing the outcome of pregnancy, in Clark D, Croy BA (eds): *Reproductive Immunology 1986*. Amsterdam, Elsevier, 1986, pp 77–84.
8. Gill TJ IV, Gill TJ III, Kunz HW, et al: Genetic and morphometric studies of the heterogeneity in the testicular defect of the H^{re} rat and the interaction between the H^{re} and *grc* genes. *Biol Reprod* 1984;31:595–603.
9. Gill TJ III, Macpherson TA, Ho HN, et al: In Gill TJ III, Wegmann TG (eds): *Immunoregulation and Fetal Survival*. New York, Oxford University Press, 1987, pp 137–155.

10. Gill TJ III, Siew S, Kunz HW: Major histocompatibility complex (MHC)-linked genes affecting development. *J Exp Zool* 1983;228:325–345.
11. Ho HN, Macpherson TA, Kunz HW, et al: The ontogeny of expression of the Pa and A^a antigens on the rat placenta and on fetal tissues by immunohistochemical localization using biotinylated Fab antibody fragments. *Am J Reprod Immunol Microbiol* 1987;13:51–61.
12. Kawata M, Parnes JR, Herzenberg LA: Transcriptional control of HLA-A,B,C antigens in human placental cytotrophoblast isolated using trophoblast- and HLA-specific monoclonal antibodies and the fluoresecence-activated cell sorter. *J Exp Med* 1984;160:633–651.
13. Kiger N, Chaouat G, Kolb J-P, et al: Immunogenetic studies of spontaneous abortion in mice. Preimmunization of females with allogeneic cells. *J Immunol* 1985;134:2966–2970.
14. Konoeda Y, Terasaki PI, Wakisada A, et al: Public determinants of HLA indicated by pregnancy antibodies. *Transplantation* 1986;41:253–259.
15. Kunz HW, Gill TJ III, Dixon BD, et al: The growth and reproduction complex in the rat: genes linked to the major histocompatibility complex which affect development. *J Exp Med* 1980;152:1506–1518.
16. Macpherson TA, Ho HN, Kunz HW, et al: Localization of the Pa antigen on the placenta of the rat. *Transplantation* 1986;41:392–394.
17. McConnachie PR, McIntyre JA: Maternal antipaternal immunity in couples predisposed to repeated pregnancy losses. *Am J Reprod Immunol* 1984;5:145–150.
18. McIntyre JA, Faulk WP: Laboratory and clinical aspects of research in chronic spontaneous abortion. *Diag Immunol* 1985;3:163–170.
19. McIntyre JA, Faulk WP, Nichols-Johnson VR, et al: Immunologic testing and immunotherapy in recurrent spontaneous abortion. *Obstet Gynecol* 1986;67:169–175.
20. McIntyre JA, McConnachie PR, Taylor CG, et al: Clinical, immunologic and genetic definitions of primary and secondary recurrent spontaneous abortions. *Fertil Steril* 1984;42:849–855.
21. Schaid DJ, Kunz HW, Gill TJ III: Genic interaction causing embryonic mortality in the rat: epistasis between the *Tal* and *grc* genes. *Genetics* 1982;100:615–632.
22. Stranick KS, Ho HN, Macpherson TA, et al: Regulation of the maternal-fetal response by trophoblast antigen expression, in Clark D, Croy BA (eds): *Reproductive Immunology 1986*. Amsterdam, Elsevier, 1986, pp 45–52.
23. Tanaka K, Ozato K, Jay G, et al: Control of H-2 antigen and β_2-microglobulin gene expression in mouse trophoblast cell clones. *Proc Natl Acad Sci USA* 1983;80:5597–5601.
24. Taylor C, Faulk WP: Prevention of recurrent abortion with leucocyte transfusions. *Lancet* 1981;i:68–69.
25. Taylor CG, Faulk WP, McIntyre JA: Prevention of recurrent spontaneous abortions by leucocyte transfusions. *J Roy Soc Med* 1985;78:623–627.
26. Weitkamp LR, Schachter BZ: Transferrin and HLA: spontaneous abortion, neural tube defects and natural selection. *N Engl J Med* 1985;313:925–932.

CHAPTER 16

Cell-Mediated Immune Mechanisms in Recurrent Spontaneous Abortion

Joseph A. Hill and Deborah J. Anderson

Approximately 20% of all conceptions end in detectable spontaneous abortion. Recurrent abortion is a common and important clinical problem affecting approximately 1 in 300 pregnancies (14). The risks for recurrent abortion have been calculated by Warburton and Frazier (54) to be 24% after one abortion, 26% after two abortions, and 32% after three consecutive pregnancy losses before 20 weeks gestation. Many instances of recurrent spontaneous abortion have identifiable etiologies such as chromosomal (5), mullerian (38), endocrinological (26,55), and infectious (12,29) abnormalities. However, recurrent abortion was unexplained in as many as 40% of couples studied (52). An immunologic etiology is suspected in women with subclinical autoimmune disease such as systemic lupus erythematosus, who have an increased incidence of otherwise unexplained recurrent abortion (11,13,15,39). The transplacental passage of maternal immunoglobulins that are cytotoxic to trophoblastic or fetal tissue is presumed to be a pathologic mechanism for abortion in such cases (32). Recurrent abortion is also associated with immunity to allogenic (husband's) lymphocytes, sperm, and trophoblast antigens (3,25,33). Our current evidence indicates that the mechanisms of immunologic abortion in women may not only be antibody mediated (humoral response) but could also be cellularly mediated (cellular response) due to cytotoxic effects of lymphokines and monokines released by reproductive antigen (sperm and/or trophoblast) or other foreign (i.e., microbial) antigen-sensitized lymphocytes and macrophages in reproductive tissues.

Our understanding of the interactions between the immune and reproductive systems is rudimentary. Accumulating evidence, however, suggests that the fetal-placental allograft is afforded protection by local immunomodulating factors (8), and that immunologic recurrent abortion may result from an imbalance or breakdown in the mechanisms responsible for immune homeostasis. Immunologic factors mediating abortion may operate at any stage of reproduction since spermatozoa, embryos, trophoblast, and reproductive tract tissues themselves are all capable of stimulating an immune response due to their paternally inherited gene products and tissue-specific differentiation antigens.

Rapid advances in immunoreproduction are occurring due to major advances in the field of immunology. It is now possible to perform in-depth studies of immunoregulatory mechanisms. Numerous biochemical mediators of immunological responses (lymphokines and monokines) have been defined and mass produced in both culture purified and recombinant form for research use. These mediators enable the definition of cellular interactions within the immune system, and between the immune system and reproductive tissues. New monoclonal antibodies have also been produced that identify many of these mediators and leukocyte markers associated with reproductive/immune function. Until recently, immunologic studies have focused on the humoral immune response and the role of antibodies in recurrent abortion. It is now possible using these new immunological reagents to probe the cellular immune mechanisms and immunoregulatory events that also appear to affect reproductive failure both before and after implantation. In this chapter we will present evidence of cellular immune mechanisms that potentially underlie abortion.

Leukocyte Populations Residing in Reproductive Tissues

The human endometrium under hormonal stimulation, cyclically prepares to receive a fertilized ovum in a process known as the menstrual cycle (18). Since the classic histological description of Noyes, Hertig and Rock (40) on cyclic changes in the normal human endometrium, investigators have questioned the nature and activity of lymphoid, myeloid, and other antigen-presenting cells that migrate to the human endometrium and decidua (6,7,8,27,28,46). In the human female, the genital tract is well endowed with macrophages and other immunologically competent cells (46). These cells can process antigens for recognition by the host's immune system. Monoclonal antibodies have been used to identify T-lymphocyte subpopulations in nonpregnant normal and abnormal human endometrial biopsies (56). Both T-helper/inducer and T-suppressor/cytotoxic subpopulations were detected in histologically normal (in phase) biopsies and were present in significantly higher concentrations in proliferative than in secretory phase endometria. In normal biopsies, helper/inducer T cells outnumbered suppressor/cytotoxic T cells throughout the cycle except late in the secretory phase and during menstruation, when suppressor cells predominated. In abnormal (out of phase) biopsies, as revealed by histologic dating, there was a tendency for increased ratios of T-helper cell subpopulations as compared with normal biopsies. In abnormal biopsies from some women with a history of recurrent abortion there were extremely high helper:suppressor cell ratios suggesting that the uterus is indeed an immunologically dynamic tissue and that immunologic mechanisms may be operative in spontaneous abortion. Clark and McDermott (8) have also shown that a granular lymphocyte migrates to the endometrium during the mid to late secretory phase of the menstrual cycle and postulated that this cell type could play an immunoregulatory role in the maintenance of early pregnancy. Lymphoid cells of bone marrow origin are recruited to the uterine decidua at implantation and are shown to possess suppressor activity and are functionally hyporesponsive to paternal antigens (8,28,47). It is also known that human trophoblast (17) and fetal lymphocytes (24) that are in human decidua produce immunosuppressive factors. Abortion in mice has been associated with a lack of decidual suppressor cells (9), giving rise to the concept that such cells locally regulate maternal immune responses directed against fetal tissues and play a critical role in pregnancy maintenance. Other studies have implicated circulating "blocking factors" in the immunological acceptance of gestational tissue and have reported that these factors are absent in women experiencing recurrent abortion (3,45,50). Blocking factors have not been thoroughly characterized, but are presumed to be antibodies. We recently proposed that they could also be lymphokines or monokines that are cellular mediators of immunity (20).

Microbial infections such as chlamydia and mycoplasma may induce an inflammatory infiltration in the reproductive tract (16). It has also been demonstrated in experimental animals (45) and recently in women (43) that semen may be chemotactic, attracting leukocytes to the female reproductive tract shortly after insemination. Leukocytes are also present in human semen (41). A significant percentage of lymphocytes in semen from normal fertile males are of the helper/inducer T-cell phenotype (2). Thus, the genital tracts of both sexes are immunologically dynamic tissues containing a myriad of antigen-presenting and other immune response cells. In response to microbial, seminal, or trophoblast antigens these cells may become activated to either facilitate or hinder reproductive processes.

Regulation of the immune system involves a complex interplay between many diverse cell populations. A central concept is that cytokines (lymphokines and monokines derived from activated lymphocytes and macrophages, respectively) mediate the proliferation and differentiation of immune response cells (48). Many of these factors also mediate cytotoxic effects on a diverse array of nonimmunized cell types including neoplastic and virally infected cells (30,42,51). Macrophages are antigen-presenting cells, and upon activation by foreign antigens such as microbial or reproductive tissues or secretions, produce a monokine, interleukin-1 (Il-1), which plays a fundamental role in initiating the immune response by stimulating the production of the lymphokine interleukin-2 (Il-2) by helper T cells and thus initiating the lymphokine cascade (31) as depicted in Figure 16.1. The spectrum of Il-1 extends beyond T-lymphocyte activation. Il-1 may affect a variety of target cells by increasing their ability to bind and be lysed by natural killer cells (19). Il-2 production by helper T cells promotes T-lymphocyte (particularly cytotoxic T-lymphocyte (CTL) differentiation and proliferation (37). Soluble suppressor factors obtained from the decidua of allopregnant mice inhibit Il-2 responses (10) and may therefore prevent immune activation at the fetal-maternal interface. The T-helper lymphocyte-derived lymphokine, B-cell growth factor (BCGF), can induce B-lymphocyte proliferation and differentiation into antibody secreting plasma cells (35). The interferons (IFN) are another family of cytokines produced during immune activation. Gamma-IFN markedly enhances the expression of class I major histocompatibility (MHC) antigens on a variety of cell types including trophoblast (1); since class I MHC antigens are important intercellular signals in immune responses, and

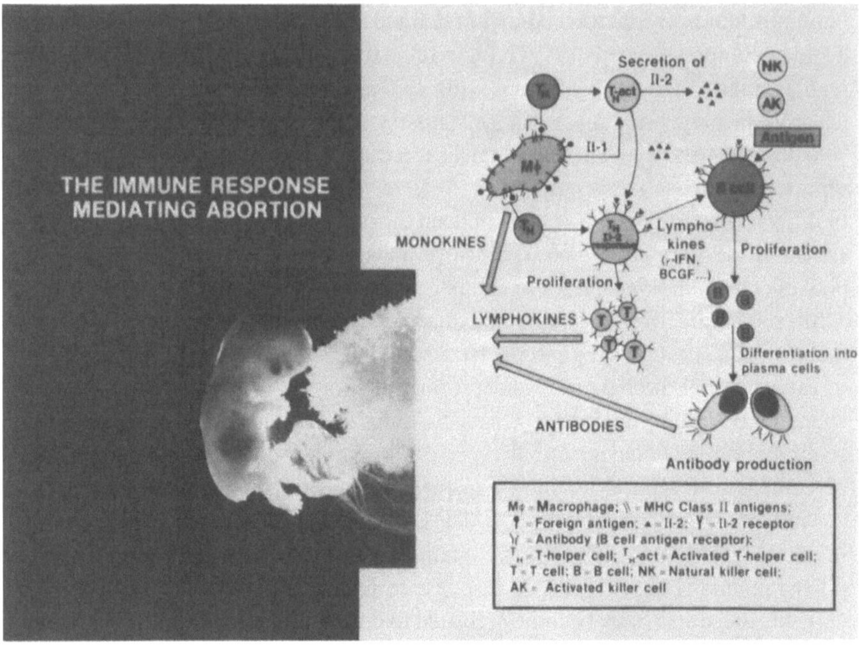

Fig. 16.1. The immune responses mediating abortion.

play a prominent role in recognition of target cells by cytotoxic T cells (49), a dramatic increase in class I MHC expression induced by gamma-IFN could culminate in abortion. Tumor necrosis factor (TNF) is produced by activated macrophages and is cytotoxic to neoplastic cells in vitro and in vivo (34). TNF acts by suppressing transcription and protein biosynthesis (53). Another cytokine, colony stimulating factor (CSF), has been reported to produce cytocidal effects in leukemia cell lines by inducing terminal differentiation (36). Clearly, such biologically active factors could affect reproductive processes.

Effects of Mitogen and Allogenic-Activated Leukocyte Products and Purified Cytokines on In Vitro Reproductive Performance

All of the currently known cytokines were originally characterized from mitogen and allogenic-stimulated peripheral blood mononuclear cells (PBMC). To investigate our hypothesis that activated leukocyte products mediate abortion, we activated PBMC with the mitogens Concanavalin-A (Con-A, a T-cell mitogen) and Pokeweed (PWM, a B- and T-cell mitogen) and in an allogenic (MLC) system. Each person tested served as their own control by using a 24-hour unstimulated leukocyte culture supernatant for control. Activated supernatants and purified cytokines (Table 16.1) were individually added to two in vitro models of

Table 16.1. Summary of well characterized lymphokines

	Principal Producing Cell	Principal Cell Type Affected	Effect	Molecular Weight
Lymphokine				
Interleukin 2 (Il-2)	T lymphocytes	Many cell types, notably T lymphocytes	T-lymphocyte proliferation; cytotoxic T-cell differentiation; lymphokine-activated killer cell (LAK) activation	14,500
Interleukin 3 (Il-3)	Mouse T lymphocytes	T lymphocytes, mast cells	T-lymphocyte proliferation; mast cell proliferation; 20 hydroxysteroid dehydrogenase induction	28,000–41,000
α-Interferon	Virally infected leukocytes	Virus-infected cells, tumors, other lymphocytes, endocrine cells	Cytolysis, hormone-like effects (ACTH, TSH, endorphine)	18,000–20,000
β-Interferon	Virally infected fibroblasts	Diverse cell populations, macrophages	Inhibition of motility and proliferation	23,000
γ-Interferon	Activated T-lymphocytes,	Virally-infected cells, tumors, trophoblast, allografts	Enhanced phagocytosis	20,000–25,000
B-Cell Growth Factor	T lymphocytes	B cells	Induces differentiation and proliferation of B cells	17,000
GM-Colony Stimulating Factor	Leukocytes	Bone marrow, macrophage and granulocyte precursors	Induces differentiation of granulocyte and macrophage lineage cells	17,000
Monokine				
Interleukin 1 (Il-1)	Macrophages (and others)	Lymphocytes, other cell types	Lymphocyte activation, enhanced antibody production, toxic to certain tumor cells, fibrosis, enogenous pyrogen	17,000 (major stable form)
Tumor Necrosis Factor	Macrophage	Tumor cells	Toxic effect on tumor cells	17,000

reproductive performance. Two-cell mouse embryos were used as a preimplantation loss model (21) and human choriocarcinoma cell lines were used as a trophoblast model for postimplantation loss (4). The percentage of mouse blastocyst development was assessed on day 4 and trophoblast proliferation was measured by ^{3}H-thymidine incorporation after three days in culture.

Activated leukocyte supernatants from Con-A-, PWM-, and MLC-stimulated cultures significantly arrested embryo development. Of the purified lymphokines, BCGF and IFN significantly inhibited blastocyst development over a wide concentration range (64.7–0.000647 and 150,000–15 units/ml, respectively). Of the monokines, CSF, had the most pronounced effect on mouse embryo development (25,000–25 units/ml); TNF affected embryos in our culture system only at high concentrations (7.5×10^6 units/ml), and Il-1 β (5,000–0.5 units/ml) exhibited only a slight effect (22).

In the postimplantation abortion model, both Con-A- and MLC-activated-leukocyte supernatants significantly inhibited trophoblast proliferation (4). Over a wide concentration range, IFN (1,500–150,000 units/ml) and TNF (100–10,000 units/ml) significantly inhibited trophoblast proliferation. CSF, Il-1, and Il-2 only affected trophoblast proliferation at high doses with CSF (510 units/ml) inhibiting and Il-1 and Il-2 (500 units/ml) stimulating proliferation. BCGF stimulated proliferation at low concentrations (13–130 units/ml) but inhibited trophoblast proliferation at the highest concentration (650 units/ml). The potential significance of adverse effects seen at high concentrations cannot be excluded as cell-to-cell contact in vivo may assure that very high concentrations of mediators are delivered to target cells.

These data indicate that activated leukocytes produce soluble factors that are toxic to reproductive cells in vitro, and that lymphokines and monokines may mediate abortion.

Effects of Trophoblast-Antigen-Stimulated Leukocyte Supernatants From Women With Recurrent Abortion on In Vitro Reproductive Performance

In further studies to determine whether cell-mediated immune mechanisms may be operative in women with recurrent abortion, trophoblast antigens were used to stimulate PBMC from women with a history of recurrent abortion of unknown etiology, presumed etiology, and fertile controls. Culture supernatants were added to 2-cell mouse embryos (22) and trophoblast cells in culture (23). The percentage of blastocyst development was significantly lower in supernatants of women with unexplained recurrent abortion as compared to the supernatants of women with presumed etiology for their abortion and fertile controls (27% vs. 51% and 63%, respectively). Trophoblast proliferation was also significantly inhibited by supernatants of women with unexplained recurrent abortion. Unstimulated leukocyte supernatants had no effect on trophoblast proliferation. These data suggest that some women with otherwise unexplained recurrent abortion may be stimulated by trophoblast antigens to produce toxic factors that could mediate their abortions.

Conclusions

The acceptance and maintenance of pregnancy is a biological enigma because the conceptus expresses paternal and embryonic antigens that are foreign to its maternal host yet is not normally rejected. It is increasingly apparent that reproductive tissues that are capable of eliciting autoimmune or allograft rejection reactions in their natural hosts, are normally afforded protection by immunomodulating factors produced by the reproductive tissues themselves, and by activities of lymphoid and myeloid cells associated with or activated by reproductive tissues. Although the immunologic interactions are complex, evidence is accumulating that immunologic recurrent abortion can result from cell-mediated immune responses. Leukocytes residing in reproductive tissues may be activated by a variety of antigens. The conceptus may be killed as an innocent bystander to leukocyte activation triggered by infection, or specifically, as a result of cell-mediated immune responses directed against embryonic/trophoblast antigens in women with immunologic reproductive failure.

Acknowledgment. The authors acknowledge the research contributions of Dr. Florina Haimovici, Dr. Ross Berkowitz, and Heidi Faris, and the invaluable assistance of Deirdre Meehan in the preparation of this manuscript.

References

1. Anderson DJ, Berkowitz RS: γ-interferon enhances expression of Class I MHC antigens in the weakly HLA and human choriocarcinoma cell line BeWo, but does not induce MHC expression in the HLA-choriocarcinoma cell line Jar. *J Immunol* 1985;131:540.
2. Anderson DJ, Hill JA: CD4⁺ (T4⁺) lymphocytes in human male reproductive tissues and semen: Implications for the venereal transmission of AIDS. *N Fertil Steril* (in press).
3. Beer AE, Quebbeman JF, Ayers JWI, Haines RF: Major histocompatibility complex antigens maternal and paternal immune responses and chronic habitual abortions in humans. *Am J Obstet Gynecol* 1981;141:987.
4. Berkowitz RS, Hill JA, Kurtz CB, Anderson DJ: Influence of lymphokines and monokines on the growth of JEG3 human gestational choriocarcinoma (CGA) cells in vitro. *Am J Obstet Gynecol* (in press).
5. Boue J, Boue A, Lazor P: Retrospective and prospective epidemiological studies of 1500 karotyped spontaneous abortions. *Teratology* 1975;12:11.
6. Bulmer JN, Johnson PM: Antigen expression by trophoblast populations in human placenta and their possible immunologic relevance. *Placenta* 1985;6:127.
7. Bulmer JN, Sunderland LA: Immunohistological characterization of lymphoid cell populations in the early human placental bed. *Immunol* 1984;52:349.
8. Clark DA, McDermott M: Active suppression of host vs. graft reaction in pregnant mice: III. Developmental kinetics, properties, and mechanisms of induction of suppressor cells during first pregnancy. *J Immunol* 1984;127:1276.
9. Clark DA, Slapys RM, Croy BA, Rossant J: Suppressor cell activity in uterine decidua correlates with success or failure of murine pregnancies. *J Immunol* 1983;131:540.
10. Clark DA, Chaput A, Walker C, Rosenthal KL: Active suppression of host vs. graft reaction in pregnant mice: VI. Soluble suppressor activity obtained from decidua of allopregnant mice blocks the response to Il-2. *J Immunol* 1985;134:1659.

11. Couchock S, Dehoratius RD, Wapner RJ, Jackson LG: Subclinical autoimmune disease and unexplained abortion. *Am J Obstet Gynecol* 1984;150:367.
12. Craea E, Botez D, Ioanid L, Constantinesca S, Georges CU, Braila M, Sirbu P, Nigulescu M, Rocovita V: Genital mycoplasmas and chlamydia in infertility and abortion. *Arch Roum Pathol Exp Microbiol* 1981;40:107.
13. DeWolf F, Carreras LO, Moenman P, Verrylen J, Van Assche A, Renner M: Decidual vasculopathy and extensive placental infarctions in a patient with repeated thromboembolic accidents, recurrent fetal loss, and a lupus anticoagulant, *Am J Obstet Gynecol* 1982;142:829.
14. Edmonds DK, Lindsay KS, Miller JF, Williamson E, Wood PJ: Early embryonic mortality in women. *Fertil Steril* 1982;38:447.
15. Firkin BG, Howard MA, Radford N: Possible relationship between lupus anticoagulant and recurrent abortion in young women. *Lancet* 1980;2:366.
16. Friberg J: Mycoplasmas and ureaplasmas in infertility and abortion. *Fertil Steril* 1980;33:351.
17. Hamaoka R, Matsuzaki N, Itoh K, Tsuji Y, Isumi Y, Fufimara H, Ono S: Human trophoblast and tumor cell-derived immunoregulatory factor. In: *Reproductive Immunology* S. Isojima and W.D. Billington, (eds.) Elsevier Science Publishers, 1983.
18. Healy DL, Hodgen GD: The endocrinology of human endometrium. *Obstet Gynecol Surv* 1983;38:509.
19. Herman J, Dinarello CA, Ken MC, Robson AR: The role of interleukin-1 (Il-1) in tumor NK cell interactions: correction of defective NK cell activity in cancer patients by treating target cells with IL-1. *J Immunol* 1985;135:2882.
20. Hill JA, Anderson DJ: Blood transfusions for recurrent abortion: Is the treatment worse than the disease? *Fertil Steril* 1986;46:152.
21. Hill JA, Haimovici F, Anderson DJ: Products of activated lymphocytes and macrophages inhibit mouse embryo development in vitro. *J Immunol* (in press).
22. Hill JA, Wickersham C, Schiff I, Anderson DJ: Activated lymphocytes produce factors that inhibit mouse embryo development *in vitro*: A possible mechanism for immunologic infertility and recurrent spontaneous abortion, *Fertil Steril* 1986b; 10(suppl):103.
23. Hill JA, Haimovici F, Schiff I, Anderson DJ: Trophoblast antigens induce production of toxic factors by leukocytes from a subgroup of women with unexplained recurrent abortion (RA): Further evidence for a cellular immune mechanism in spontaneous abortion. *Society for Gynecologic Investigation*. (Abstract), Atlanta, GA, 1987.
24. Jacoby DR, Olding LB, Oldstone MBA: Immunologic regulation of fetal-maternal balance. *Adv Immunol* 1984;35:157.
25. Johnson PM, Barnes RMR, Hart CA, Francis WJA: Determinants of immunological responsiveness in recurrent spontaneous abortion. *Transplantation* 1984;38:280.
26. Jones GES, Delfs E: Endocrine patterns in term pregnancies following abortion. *JAMA* 1951;146:1212.
27. Kabawat SE, Mostoufi-Zadeh M, Driscoll SG, Bhan AK: Implantation site in normal pregnancy. A study with monoclonal antibodies. *Am J Pathol* 1985;118:76.
28. Kearns M, Lala PH: Bone marrow origin of decidual cell precursors in the pseudopregnant mouse uterus. *J Exp Med* 1982;155:1537.
29. Kundsin RB, Driscoll SG, Ming PL: Strain of mycoplasma associated with human reproductive failure. *Science* 1967;157:1573.
30. Lachman LB, Dinarello CA, Clansa ND, Fidler IJ: Natural and recombinant human

interleukin 1-B is cytotoxic for human melanoma cell lines. *J Immunol* 1985;135:3962.
31. Larson EC, Iscove NN, Coutinbo A: Two distinct functions are required for induction of T-cell growth. *Nature* 1980;283:664.
32. Lockshin MD, Pruzin ML, Goei S, Qamar T, Magid MS, Jovanovic L, Ference M: Antibody to cardiolipin as a predictor of fetal distress or death in patients with systemic lupus erythematosus. *N Eng J Med* 1985;313:152.
33. Mathur S, Baker ER, Williamson HO, Derrick FC, Teague KJ, Fundenberg HH: Clinical significance of sperm antibodies in infertility. *Fertil Steril* 1981;36:486.
34. Matthews N: Effect on human monocyte killing of tumor cells of antibody raised against an extra cellular monocyte cytotoxin. *Immunol* 1983;48:321.
35. Mehta SR, Conrad D, Sandler R, Morgan J, Montagna R, Maizel AL: Purification of human B cell growth factor. *J Immunol* 1985;135:3298.
36. Metcalf S, Nicola NA: *Intl J Cancer* 1982;30:782.
37. Morgan DA, Rusatti FW, Gallo RC: Selective *in vitro* growth of T lymphocytes from normal human bone marrows. *Science* 1976;193:1007.
38. Musich JR, Behrman ST: Obstetric outcome before and after metroplasty in women with uterine anomalies. *Obstet Gynecol* 1978;52:63.
39. Nilsson IM, Astedt B, Hedner U, Berezin D: Intrauterine death and circulating anticoagulant ("anti thromboplastin"). *Acta Med Scand* 1975;197:153.
40. Noyes RW, Hertig AT, Rock J: Dating the endometrial biopsy. *Fertil Steril* 1950; 1:3.
41. Olsen GP, Shields JW: Seminal lymphocytes, plasma and AIDS. *Nature* 1984; 309:116.
42. Onozaki K, Matsushima K, Kleinerman E, Soito T, Oppenhein JJ: Human interleukin-1 is a cytocidal factor for several tumor cell lines. 1985;135:314.
43. Pandya IS, Cohen J: The leukocytic reaction of the human uterine cervix to spermatozoa. *Fertil Steril* 1985;43:417.
44. Phillips DM, Mahler S: Leukocyte emigration and migration in the vagina following mating in the rabbit. *Anat Res* 1977;189:45.
45. Rocklin RE, Kitzmiller JL, Carpenter CB, Garvoy MR, David JR: Maternal-fetal relation: Absence of an immunologic blocking factor from the serum of women with chronic abortions. *N Eng J Med* 1976;22:1209.
46. Sen DR, Fox H: The lymphoid tissue of the endometrium. *Gynecologia* 1967; 163:371.
47. Slapys RM, Clark D: Active suppression of host vs. graft rejection in pregnant mice. IV. Local suppressor cells in decidua and uterine blood. *J Reprod Immunol* 1983;4:354.
48. Smith KA: Lymphokines regulation of T cell and B cell function, in: *Fundamental Immunology* p 559, W.E. Paul,(ed.) Raven Press, NY 1981.
49. Steeg PS, Moore RN, Johnson HM, Oppenheim JJ: Regulation of murine macrophage Ia antigen expression by a lymphokine with immune interferon activity. *J Exp Med* 1982;156:1780.
50. Stimson WH, Strachon AF, Shepard A: Studies on the maternal immune response to placental antigens: Absence of blocking factors from the blood of abortion-prone women. *Br J Obstet Gynecol* 1979;86:41.
51. Sugarman BR, Aggarual B, Haas P, Figari I, Palladino M, Shepard H: Recombinant tumor necrosis factor: effects on proliferation of normal and transformed cells *in vitro*. *Science* 1985;230:943.

52. Tho PT, Byrd JR, McDonough PG: Etiologies and subsequent reproductive performance of 100 couples with recurrent abortion. *Fertil Steril* 1979;32:389.
53. Torti FM, Dieckmann B, Beutler B, Cerami A, Ringold GM: A macrophage factor inhibits adipocyte gene expression: An *in vitro* model of cachexia. *Science* 1985; 229:867.
54. Warburton F, Frazer FC: Spontaneous abortion risks in man: Data from reproductive histories collected in a medial genetics unit. *Am J Human Genet* 1964;16:1.
55. Winikoff D, Malinek M: The predictive value of thyroid "Test Profile" in habitual abortion. *Br J Obstet Gynecol* 1975;82,760.
56. Xu C, Hill JA, Anderson DJ: Identification of T lymphocyte subpopulations in normal and abnormal human endometrial biopsies. *Society for Gynecologic Investigation* (Abstract), Atlanta, GA, 1987.

Part V
Birth Control Vaccines

CHAPTER 17

Birth Control Vaccines Inducing Antibodies Against Chorionic Gonadotropin

G.P. Talwar and Om Singh

Introduction

The possibility of inducing antibodies against reproductive tract antigens by deliberate immunization was indicated as early as 1899 by the pioneering work of Landsteiner (14) and Metchinikoff (15). These leaders of immunology did obtain antisera toxic to sperm. However, the practical utility of this lead for control of fertility could not materialize owing to several factors. Crude tissue extracts induced antibodies reactive to unwanted tissues with potential toxicity hazards. Not enough was known of the immune processes and the manner in which these could be modulated to advantage. Furthermore, for antigens localized on sperm, it would be necessary to ensure the right type of immune reactions in the genital tract for preventing fertilization; antibodies in systemic circulation alone may not suffice.

Progress in development of immunological approaches to contraception had to await advances in immunology, a better understanding of the reproductive system, and for refined techniques of purification of antigens, peptide synthesis, and gene cloning. At present immunocontraception is an area of active research. A variety of birth control vaccines are under development. Table 17.1 gives the principal candidates. Many of these are discussed in greater detail in other chapters of this volume. It is interesting to note that at times moderate immunization against a hormone leads to promotion of fertility instead of its curtailment. Immunization against androstenedione is observed to increase the frequency of twinning in sheep (13).

Multiple Sites of Intervention

Broadly speaking, experimental immunocontraception approaches are directed at inactivation of either one or more hormones, gametes, or early embryo antigens. The reproductive system is exquisitely sensitive to hormones, and several are involved in a chain-like manner in the process of gametogenesis, in accessory reproductive organ functions, in preparation of the reproductive tract for

Table 17.1. Fertility control vaccines under development.

Category	Target	Remarks
Hormones	Prolactin release inhibitory factor	Recent identification of GnRH associated peptide (GAP) of 56 amino acids; efficacy for fertility control in vivo under study.
	GnRH	Applicable to both sexes; veterinary applications for control of estrus in dogs; medical application in precocious puberty and sex-steroid-hormone dependent cancers in male and female.
	FSH	Being developed for fertility control vaccine for male.
	HCG	Experimental and safety studies completed; three vaccines under phase I clinical trials.
	LH partial and CG	Effective for fertility control in female monkeys.
	Androstenedione	Twinning in lambs.
Gametes	Sperm antigens	Several identified by monoclonal and polyclonal antibodies; few specific to sperm; several immunologically cross-reactive to sperm of other species.
	Zona pellucida	Fertility control demonstrated in rabbits and primates; serious hormonal disturbances; ovarian follicular atresia in rabbits.
Embryo	Stage antigens; trophoblast constituents	Under study
	Riboflavin and thiamine carrier protein	Deprivation of nutrition to embryo.

implantation, and in sustenance of pregnancy. Figure 17.1 is a schematic representation of the mammalian reproductive processes. The gonadotropin-releasing hormone (GnRH), a decapeptide synthesized and secreted by the hypothalamus, stimulates the secretion of follicular-stimulating hormone (FSH) and luteinizing hormone (LH). GnRH, FSH, and LH are made by *both* male and female, and inactivation of the bioactivity of these can influence male as well as female fertility. The gonadotropins act on the male gonads to generate sperms and testosterone; in the female the gonadotropins are involved in follicular development, ovulation, and production of the female sex hormones, estradiol (E_2) and progesterone (P). Sex steroid hormones influence the respective male and female reproductive organs: in the male they are the epididymis, seminiferous tubules, and prostate; in the female they are the uterus, cervix, and vagina. In the first half of the cycle E_2 dominates. The cervical mucus during this phase, and in particular around the day of ovulation is permissive of the ready passage of sperms. In the postovulation phase, dominated by P, the cervical mucosal glycoproteins have altered conformation, hindering the passage of sperms. Progesterone is important for uterine receptivity for implantation of the embryo and for the continuity of pregnancy. Compounds or antibodies interfering in the production or action of this hormone during early pregnancy cause abortion. Data will be presented on vaccines aiming to impede the action(s) of FSH for control of

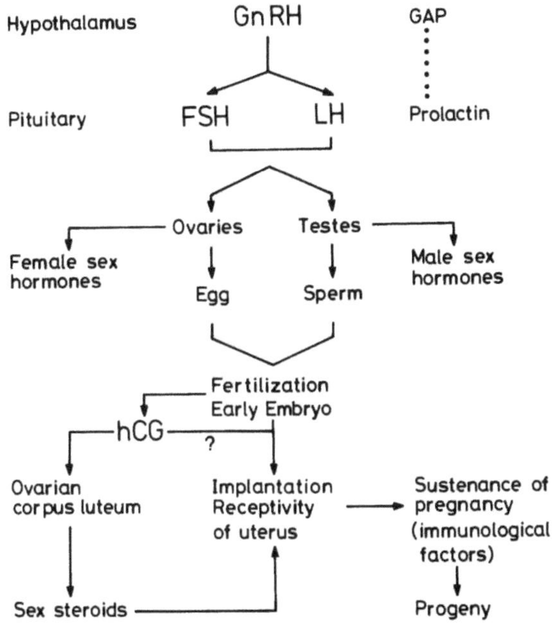

Fig. 17.1. Schematic outline of the reproductive processes and points of possible intervention for control of fertility.

male fertility (Moudgal et al, Chapter 26), on LH for control of female fertility (Thau, Chapter 23) and human chorionic gonadotropin (HCG) (this chapter).

GnRH

It may be appropriate to summarize briefly the current status of immunological approaches against GnRH, as this is not reviewed in other chapters. Immunization against GnRH produced a marked reduction of pituitary LH together with atrophy of gonads in rabbits (1). Fraser (7) immunized female stump tailed macaque monkeys with GnRH linked to bovine serum albumin (BSA). The animals became infertile with diminution of sex steroids. Using the same immunogen, control of fertility was reported in marmosets (12). In studies mentioned above, immunization was carried out with Freund's complete adjuvant (FCA). This adjuvant is not permissible for use in humans and may not even be acceptable for applications in animal husbandry and veterinary medicine. Shastri et al (25) observed that tetanus toxoid (TT) was a better carrier than BSA. Bioeffective immune response could be elicited in monkeys and baboons employing GnRH linked to TT, administered with alum or other permissible adjuvants (34). High antibody titers were induced in these primates by booster injection after 4 to 5 months of primary immunization, and the antibody response was of long duration. E_2 and P levels were significantly low in these animals during the period of high anti-GnRH antibodies. Male rats could also be rendered infertile with this conjugate following a similar immunization schedule; testosterone was significantly reduced in immunized rats (22).

A number of monoclonals have been developed against GnRH. These are useful in better understanding of the epitopes on this molecule, which are both conformational as well as sequential (27,35). A single injection of a small amount of ascites-derived monoclonal antibodies suppresses the progression of estrus in female dogs (35).

Though methods have been developed to produce high antibody titers against GnRH employing acceptable adjuvants, this approach may not be readily acceptable for contraception, as coincident with the blockage of fertility is the reduction of sex steroids. Immunization against GnRH may, however, be useful for control of precocious puberty as well as in patients of sex-hormone-dependent cancers such as carcinoma of the prostate and breast cancer.

Immunization Against Pregnancy Hormone, HCG

HCG is a product of trophoblasts. With the exception of trophoblastic and some other cancers in which oncofetal genes are expressed, the rise in HCG in blood is indicative of pregnancy. It has traditionally served as a reliable index for diagnosis of pregnancy.

The synthesis and secretion of HCG starts fairly early following the fertilization of egg. Human embryo developed from in vitro fertilized eggs in culture secretes HCG (6). Whether HCG has a role in successful implantation of the embryos is not known. Preliminary experiments show that marmoset embryos exposed to anti-βHCG antibodies failed to implant (11).

Following implantation, the levels of HCG rise steeply, roughly doubling everyday during the 5 to 7 weeks of pregnancy. One of the demonstrated roles of HCG is the rescue of corpus luteum and continued secretion of progesterone, which in turn sustains the endometrium and prevents its menstrual shedding. Sustained ovarian steroidogenesis under the influence of HCG is a requisite of pregnancy for about 7 weeks, as ovariectomy during this period results in termination of pregnancy (3). The requirement of HCG for steroidogenesis during this period is also demonstrated by another clinical study. In women undergoing medical termination of pregnancy, administration of HCG but not of estradiol-17β kept up the blood P levels (8).

From the foregoing, it is evident that HCG has a critical role in early pregnancy. The rationale of the vaccine directed against this hormone is to produce antibodies with ability to inactivate HCG bioactivity. As HCG synthesis is established by day 8 following ovulation, annulation of HCG signal to the ovaries would prevent the maintenance of P levels. Hence normal menses will occur even though the cycle is fertile.

Chemical Anatomy of HCG

HCG is a glycoproteinic hormone composed of two dissimilar subunits, α and β, linked to each other noncovalently. The α-subunit is common to three other pituitary hormones, thyroid-stimulating hormone (TSH), LH, and FSH. The β-subunit in each case accords the hormonal specificity. β-Subunit of HCG, a glycoprotein of 23 kDa, is made of 145 amino acids with six carbohydrate residues, four of which are localized on carboxyterminal portion of the subunit. βHCG has large homology of amino acid sequence with β-subunit of HLH, but the former has a unique carboxy terminal portion of 30 amino acids. βHCG has 12 half cystines but no titratable free sulphydryl (SH) group, suggesting the existence of six intrachain disulfide bonds. The disulfide linkages are important for the native conformation of the molecule; reduction and alkylation lead to drastic alterations in the ability of the molecule to recognize receptors on target issues.

Choice of the Antigenic Moiety

Although the native hormone has the best conformation to generate antibodies capable of preventing its binding to the tissue receptors, the entire hormone can not be employed for fear of undesirable cross-reactivity with TSH and FSH, which have the same α-subunit. One has to employ the β-subunit or its subparts for immunization. Both have been used with distinct advantages and disadvantages. The carboxy terminal peptide (CTP) of βHCG generates antibodies devoid of cross-reactivity with HLH. Recent observations of Dr. Noel Rose and co-workers (Chapter 24), however, show that such antibodies do have cross-reaction with striated muscle, somatostatin-secreting cell of the pancreatic islets, and a few cells of anterior pituitary. The cross-reaction is due to CTP and not the carrier, as it is absorbed by the former. Antibodies induced by CTP are probably sequences reading type and it is quite possible that similar sequences exist on other cells demonstrating the reaction. In contrast, the total βHCG generates primarily conformation-reading antibodies. Whether the total lack of reaction with LH is advantageous or otherwise will be evident when efficacy studies are carried out with CTP-based vaccines. Extensive studies of Thau and co-workers (Chapter 23) show the lack of side effects and toxicity of having antibodies reacting with LH. In fact, these confer additional benefit for control of fertility.

Major disadvantages of CTPs are their poor immunogenicity and that the antibodies produced are of low affinity and of relatively poor bioneutralization capacity (2,20). Studies with 31, 35, 45, and 53 CTP of βHCG linked to TT did not give satisfactory results to serve as vaccine (19,21). However, a vaccine based on 37 amino acids CTP linked to diphtheria toxoid (DT) has been developed (29,30), which is administered with muramyl dipeptide derivative, Arlacel A and squalene. Phase I clinical trials have started with this vaccine in one center in Australia on 30 women.

The Use of Carrier

βHCG being a "self" peptide is not expected to be immunogenic in women. To render it immunogenic, it requires modification either by haptenic groups (28) or by linkage to immunogenic carriers (31). We have opted for the latter not merely for its perceived superiority but also for achieving another purpose, namely, concomitant immunoprophylaxis against a health hazard. Tetanus is a major killer in several developing countries. In the first prototype vaccine, βHCG was linked to TT in defined proportions. Improved formulation employs two carriers, TT and cholera toxin chain B (CHB). The rationale of using a second carrier is to provoke responsiveness in hyporesponders to TT.

Counter Gonadotropin Vaccines

βHCG-TT

Conjugate of βHCG and TT (βHCG-TT) adsorbed on alum induced the formation by anti-HCG and anti-TT antibodies in a variety of animal species including the subhuman primates (32). After due toxicology studies (4,9,16,17,23), the vaccine was used for probing phase I clinical trials in women. These were conducted

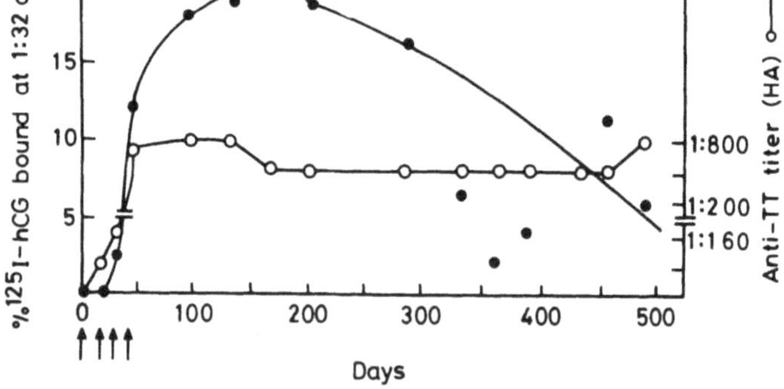

Fig. 17.2. Induction of anti-HCG and anti-TT titers in a human subject (AM) following immunization with βHCG-TT. Four intramuscular injections of the vaccine containing 80 μg βHCG were given at two weeks' interval. The kinetics of the antibody response to HCG and TT are given. ■ Indicates menstrual bleeding. Progesterone was measured in luteal phase. Endometrial biopsy (EB) on days indicated confirmed normal ovulatory cycles. [From Talwar et al (37).]

in six centers located in five countries. The vaccine was given four times at fortnightly interval. Figure 17.2 gives the kinetics of the antibody response in a woman of reproductive age group. Anti-HCG and anti-TT were measurable in circulation. Immunization did not disturb the regularity of the menstrual cycles. Luteal phase P and endometrium biopsies were indicative of normal ovulation. The HCG titers declined after 300 to 500 days, showing the reversibility of the procedure. The antibodies were competent to recognize HCG. Clearance studies were carried out with known amounts of HCG administered in a single or double dose on two consecutive days. Figure 17.3 summarizes the representative results on two subjects. Administration of HCG led to a decline of anti-HCG titers. The antibodies were, however, adequate to completely scavenge the administered amount of HCG. The antigen-antibody complexes were biologically inactive in Leydig cell assay system. The antibody titers returned to near about the same as prevailing before HCG challenge within 25 days, thereby showing the capability of coping up with HCG in two consecutive cycles in case both happen to be fertile. Administration of HCG alone did not result in a booster response. The lack of boosting by HCG alone suggests that pregnancies would by themselves not boost the immune response induced by βHCG-carrier conjugates.

Probing trials were carried out in 63 women. Of them, 61 subjects made anti-HCG antibodies with the vaccine administered at a single dose following a single immunization schedule. In each case the response was reversible. No disturbance was observed in reproductive functions. Women kept ovulating with regularity of menstrual cycles. No other notable side effect was observed on other body functions. The main limitation indicated by these trials was the variability in the antibody titers of women receiving the vaccine. Those with low titers were not protected from pregnancy.

Efficacy of Immunization for Control of Fertility

Immunization with βHCG has been reported to be effective in controlling fertility in baboons (28,33) and marmosets (10). Figure 17.4 is a representative data of a baboon immunized with βHCG-TT. This baboon, otherwise fertile, did not become pregnant during the period of high anti-HCG titers in spite of repeated matings with male of proven fertility. Additional evidence for the ability of antibodies generated by the vaccine βHCG-TT to control fertility is furnished by passive immunization experiments. Figure 17.5 gives data on a baboon with regular ovulatory cycles. She conceived in the second cycle, as shown in the figure. Pregnancy was confirmed by appearance of measurable chorionic gonadotropin in circulation. At this point she was given globulins from a monkey immunized with βHCG-TT. Administration of antibodies resulted in a steep decline of chorionic gonadotropin levels, a fall in P, and heavy bleeding terminating pregnancy. The immediate next cycle, although a bit delayed, was ovulatory, suggesting lack of residual effects on fertility of animals. This was further confirmed by mating her in a subsequent cycle, which resulted in pregnancy. Administration of globulins from a nonimmunized monkey in the same amount

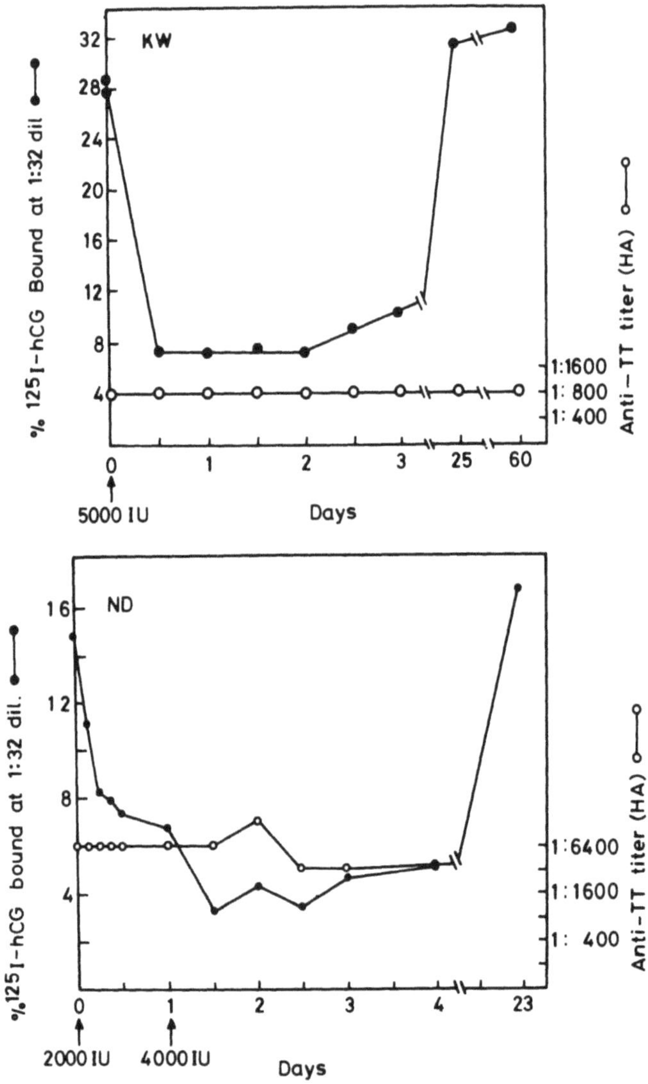

Fig. 17.3. Effect of HCG administration on two human subjects (KW and ND) immunized with βHCG-TT. HCG was given at 5,000 IU or in two doubling doses of 2,000 and 4,000 IU, which resulted in temporary decline of anti-HCG titers but not that of anti-TT. Anti-HCG titers returned to pre-HCG challenge levels on day 23–25. [From Talwar et al (31) and Ramakrishnan et al (18).]

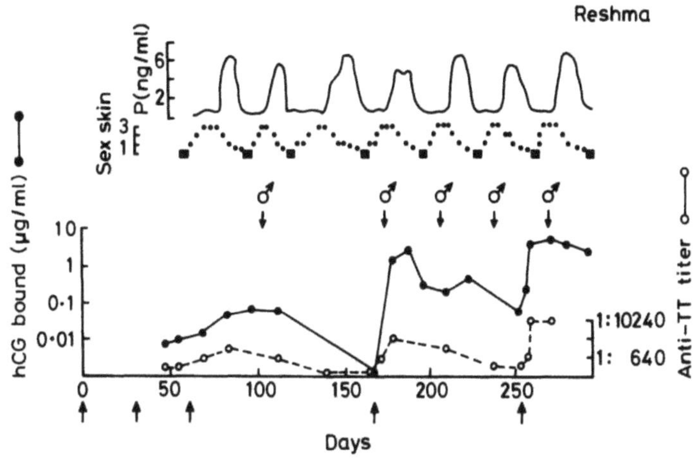

Fig. 17.4. Effect of immunization with βHCG-TT on fertility in a baboon Reshma of proven fertility. Arrows on abscissa indicate immunizations with βHCG-TT. Repeated matings on days indicated did not result in pregnancy. The cycles were ovulatory as evidenced by the P profiles shown in upper part of the figure, which also indicate the changes in sex skin. [From Talwar et al (33).]

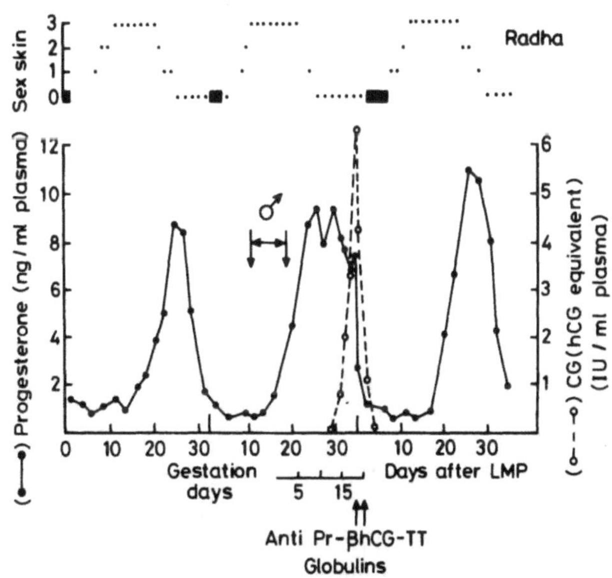

Fig. 17.5. Effect of administration of anti-βHCG-TT globulins to pregnant baboon. Anti-βHCG-TT globulins were administered in two equal doses on the 18th and 19th day of gestation. Pregnancy was confirmed by presence of baboon CG in circulation, which declined with treatment. The preceding cycle and the cycle following treatment are also shown in the figure along with the P levels and sex skin profile. ■ Represents vaginal bleeding. [From Tandon et al (38).]

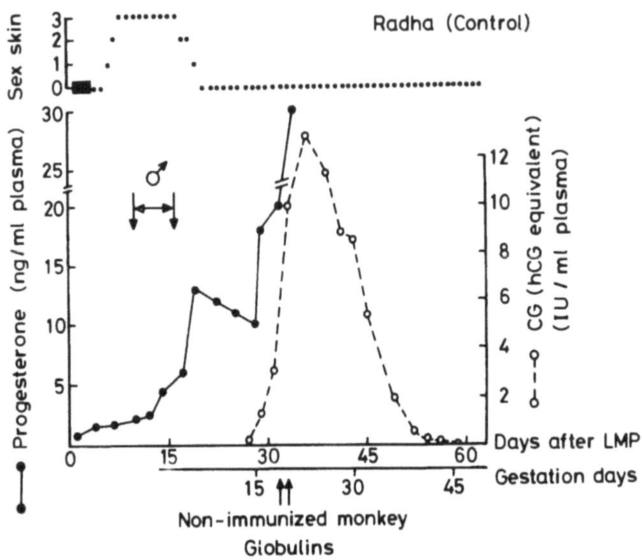

Fig. 17.6. Effect of administration of globulins from non-immunized control monkey serum on pregnancy. Baboon Radha was mated again in a cycle subsequent to those shown in Fig. 17.5 and was given globulins on days 18 and 19 on gestation, which did not abrogate pregnancy. [From Tandon et al (38).]

as in the treatment cycle did not terminate pregnancy, which proceeded to term (Fig. 17.6). The progeny born to such subjects have been observed to be normal in every respect.

Potentiation of Antibody Response

Investigations were carried out on a number of adjuvants to see whether the titers could be enhanced beyond those obtained with alum alone. Detoxified preparation of lipopolysaccharide (LPS) [sodium phthalyl derivative of LPS (SPLPS)] from gram-negative bacteria was found to nearly double the antibody response (26). SPLPS is at least 500 times less pyrogenic than the parent compound. Some derivatives of muramyl dipeptide (MDP) were also found to be of interest. These include the nor-MDP and a derivative of nor-MDP. MDP derivatives, however, require to be given as water-in-oil emulsion whereas SPLPS is usable in aqueous phase. SPLPS has been employed clinically in the United States without apparent side effects (5).

Optimized Conformation Antigen for HCG

HCG is a two subunit protein; the two subunits have a high affinity to bind with each other noncovalently to generate a conformation that is recognized by the target-tissue receptors. The dissociation into the subunits leads to a drastic loss of bioactivity. The sites of association of α-subunit with the β-subunit are conserved and heterospecies α-subunit can bind to βHCG. Thus, α-ovine LH (αoLH) anneals readily to βHCG; the associated molecule has two- to three-fold higher steroidogenic potency than the native HCG composed of homologous α- and β-subunits. The heterologous dimer produces antibodies devoid of cross-reactivity with human FSH and TSH. In the improved formulation of anti-HCG vaccine, the antigen proposed is αoLH annealed to βHCG. This formulation is more immunogenic than βHCG-TT in monkeys (24) and rats (Fig. 17.7).

Mixed Conjugates of βHCG and βoLH

βoLH induces antibodies cross-reactive with HCG and HLH. Immunization of monkeys with βoLH rendered them infertile without any side effects (Chapter 23). The antibodies shortened slightly the luteal phase with diminished P production capacity of corpus luteum. It was observed that a mixture of βoLH and βHCG both linked to TT produced in monkeys higher antibody titers than individual subunit conjugates (36). The spectrum of antibodies generated by the mixture could also be expected to act at a site additional to the abrogation of HCG support to corpus luteum. The anti-oLH antibodies were noted to react with the embryonic antigens at a much earlier stage of development than anti-HCG antibodies (39).

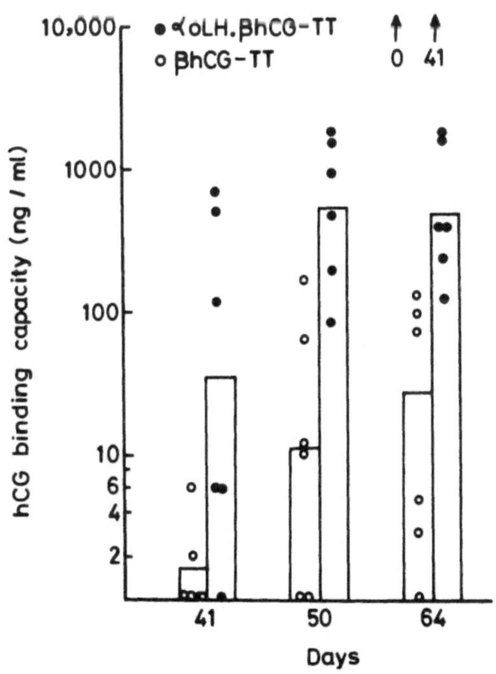

Fig. 17.7. Anti-HCG titers in rats on immunization with αoLH.βHCG-TT(●) or βHCG-TT (○). Two injections were given on days 0 and 41. Bleeds were analyzed on days 41, 50, and 64. Points give the titers in each animal and bars the mean (geometric). [From Talwar et al (37).]

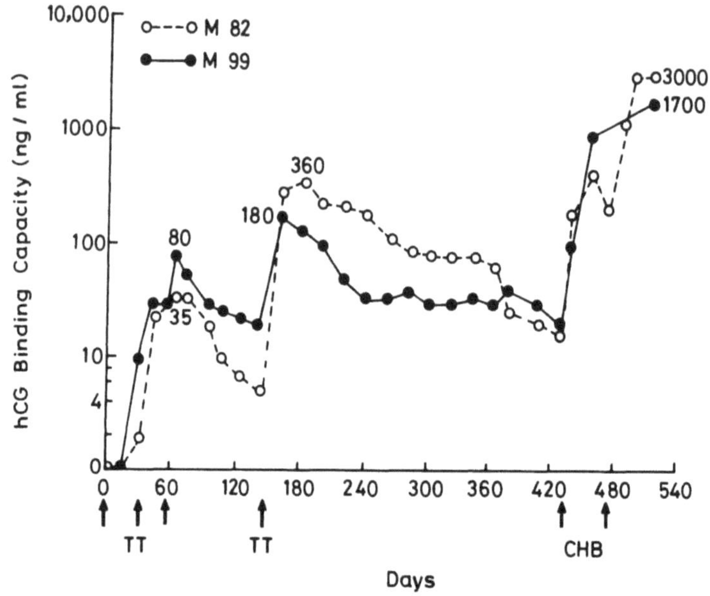

Fig. 17.8. Effect of immunization with gonadotropin conjugated to CHB in two bonnet monkeys producing low anti-HCG titers with TT conjugate. [From Talwar et al (37).]

Diversification of the Carrier

Tetanus toxoid is no doubt a good carrier, enhancing immunogenicity and conferring useful immunoprophylaxis. It was observed, however, that all monkeys did not produce high antibody titers to HCG on immunization with the gonadotropin subunits linked to the carrier TT. Genetic influences on immune response to any given antigen should be expected in an outbred population. This could be a serious limitation to the efficacy of a birth control vaccine where linkage to a carrier is demanded to evoke responsiveness. It was hypothesized that presentation of the antigen to an alternative carrier may evoke better response from those who are hyporesponders to TT. This was indeed found to be the case in a few monkeys investigated in this connection (Fig. 17.8).

New Formulation of Vaccines Under Clinical Trials

Two vaccines have been proposed based on experimental studies. These are 1) αoLH.βHCG-TT/CHB and 2) βoLH-TT/CHB + βHCG-TT/CHB. Toxicology studies have been completed on these formulations in two species of animals, including one subhuman primate. Clearances have been given by the Indian Council of Medical Research and the Drug Controller of India to start Phase I clinical trials on the above mentioned vaccines. The objectives of the trials are to determine the initial safety of immunization, the magnitude and duration of the immune response generated by this immunization procedure. Trials would be

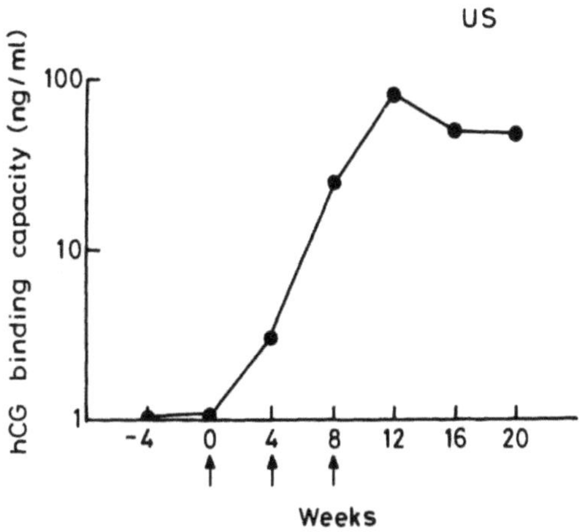

Fig. 17.9. Phase I clinical trials: anti-HCG response in a human subject immunized with a mixture of αoLH.βHCG-TT and αoLH.βHCG-CHB. (αoLH.βHCG-TT/CHB).

conducted in 105 women of reproductive age who have completed their families and have regular ovulatory cycles. Five centers for trials are located at New Delhi (the All India Institute of Medical Sciences and the National Institute of Health and Family Welfare), Bombay (KEM Hospital and BYL Nair Hospital), and Chandigarh (Post Graduate Institute of Medical Education and Research). The protocol consists of three injections of the vaccine given at 4 weeks' interval followed by a booster at 24 weeks. During the first 2 months, basal studies are carried out on hormonal profiles, blood chemistry, and organ functions. After this period immunization starts. By end-September 1986, 76 women were enrolled for the study. Twelve had completed the 3-injection schedule, 21 received two injections, and 18 only one. The vaccine was well tolerated. No side effect of any consequence was reported from any of the Centers. All women immunized had anti-HCG antibodies after the first injection, which were measurable in the blood taken after 4 weeks of immunization (Fig. 17.9).

References

1. Arimura A, Sato H, Kumasaka T, et al: Production of antiserum to LH-releasing hormone (LH-RH) associated with gonadal atrophy in rabbits: Development of radioimmunoassay for LH-RH. *Endocrinology* 1973;93:1092–1103.
2. Chen HC, Matsuura S, Ohashi M: Limitations and problems of hCG specific antisera, in Segal SJ (ed): *Chorionic Gonadotropin*. New York, Plenum Press, 1980, pp 231–251.
3. Csapo AI, Pulkkinen MO, Ruttner B, et al: The significance of the human corpus luteum in pregnancy maintenance 1. Preliminary studies. *Am J Obstet Gynecol* 1972;112:1061–1067.
4. Dubey SK, Sharma NC, Talwar GP: Survivals of animals injected with Pr-β-hCG-TT. *Contraception* 1976;13:195–200.
5. Elin RJ, Wolff SM, McAdam KPWJ, et al: Properties of reference *Escherichia coli* endotoxin and its phthalylated derivative in humans. *J Infect Dis* 1981;144:329–336.

6. Fishel SB, Edwards RG, Evans CJ: Human chorionic gonadotropin secreted by preimplantation embryos cultured *in vitro*. *Science* 1984;223:816–818.
7. Fraser HM: Active immunization of stump tailed macaque monkeys against luteinizing hormone releasing hormone and its effects on menstrual cycles, ovarian steroids and positive feed back. *J Reprod Immunol* 1983;5:173–183.
8. Garner PR, Armstrong DT: The effect of human chorionic gonadotropin and estradiol-17β on the maintenance of human corpus luteum of early pregnancy. *Am J Obstet Gynecol* 1977;128:469–477.
9. Gupta L, Dubey SK, Talwar GP: Investigations on pharmacopeal safety, microbiol sterility and pyrogens of Pr-β-hCG-TT. *Contraception* 1976;13:183–187.
10. Hearn JP: Long term suppression of fertility by immunization with βhCG subunit and its reversibility in female marmoset monkeys, in Talwar GP (ed): *Recent Advances in Reproduction and Regulation of Fertility*. Amsterdam, Elsevier/North Holland, 1979, pp 427–438.
11. Hearn JP: Animal research in understanding human reproduction: An overview, in Soon TE, Rathnam SS, Min LS (eds): *Proceedings of the 12th World Congress on Fertility and Sterility 1986*. Lancaster, Parthenon Press, in press.
12. Hodges JK, Hearn JP: Long term suppression of fertility by immunization against LHRH and its reversibility in female and male marmoset monkeys, in Talwar GP (ed): *Recent Advances in Reproduction and Regulation of Fertility*. Amsterdam, Elsevier/North Holland, 1979, pp 87–96.
13. Hoskinson RM, Scaramuzzi RJ, Campbell BK, et al: Effects of antibodies to steroid hormones on reproductive events of sheep and cattle, in Talwar GP (ed): *Immunological Approaches to Contraception and Promotion of Fertility*. New York, Plenum Press, 1986, pp 351–366.
14. Landsteiner K: Zur kenntnis der specifisch auf blutkorperchen wirkenden sera. *Zentralb Bacteriol* 1899;25:546.
15. Metchinikoff E: Etudes sur la resorption des cellules. *Ann Inst Pasteur (Paris)* 1899;14:577.
16. Nath I, Gupta PD, Bhuyan UN, et al: Autopsy report on rhesus monkeys immunized with Pr-β-hCG-TT vaccine. *Contraception* 1976;13:213–224.
17. Prasad CR, Srimal RC, Dhawan BN: Acute toxicity and pharmacology of beta human chorionic gonadotropin conjugated to tetanus toxoid (Pr-βhCG-TT). *Contraception* 1976;13:189–193.
18. Ramakrishnan S, Dubey SK, Das C, et al: Influence of hCG and tetanus toxoid injections on the antibody titres in a subject immunized with Pr-β-hCG-TT. *Contraception* 1976;13:245–251.
19. Ramakrishnan S, Das C, Dubey SK, et al: Immunogenicity of three C-terminal synthetic peptides of the beta subunit of human chorionic gonadotropin and properties of the antibodies raised against 45-aminoacid C-terminal peptide. *J Reprod Immunol* 1979;1:249–261.
20. Ramakrishnan S, Talwar GP: Immunological studies with β-subunit of human chorionic gonadotropin and its subfragments, in Segal SJ (ed): *Chorionic Gonadotropin*. New York, Plenum Press, 1980, pp 213–230.
21. Sahal D, Ramakrishnan S, Iyer KSN, et al: Immunological characteristics of a carboxy terminal 53 aminoacid peptide of beta subunit of human chorionic gonadotropin. *J Reprod Immunol* 1982;4:145–156.
22. Shaha C, Kaul S, Kini M, et al: Immunization against LHRH without the use of Freund's complete adjuvant, in Talwar GP (ed): *Immunological Approaches to Contraception and Promotion of Fertility*. New York, Plenum Press, 1986, pp 143–150.

23. Sharma NC, Goel BK, Bajaj JS, et al: Metabolic, endocrine and organ functions in monkeys immunized with Pr-β-hCG-TT vaccine. *Contraception* 1976;13:201–211.
24. Sharma NC, Singh O, Gaur A, et al: Improved immunogenic formulation for anti-gonadotropin response, in Talwar GP (ed): *Immunobiological Approaches to Contraception and Promotion of Fertility.* New York, Plenum Press, 1986, pp 37–41.
25. Shastri N, Manhar SK, Talwar GP: Important role of the carrier in the induction of antibody response without Freund's complete adjuvant against a "self" peptide hormone LHRH. *Am J Reprod Immunol* 1981;1:262–265.
26. Singh O, Singh V, Talwar GP: Studies on adjuvants for contraceptive vaccines. *J Reprod Immunol* 1983; [*Suppl*] 105–106.
27. Singh V, Singh O, Kaul S, et al: The dominant role of amide group at C-terminus for recognition by antibody in primates against gonadotropin releasing hormone. *J Steroid Biochem* 1985;23:801–802.
28. Stevens VC: Perspectives of development of a fertility control vaccine from hormonal antigens of the trophoblast, in: *Development of Vaccines for Fertility Regulation.* Copenhagen, Scriptor, 1976, pp 93–110.
29. Stevens VC, Cinader B, Powell JE, et al: Preparation and formulation of a hCG antifertility vaccine: Selection of peptide immunogen. *Am J Reprod Immunol* 1981;6:307–314.
30. Stevens VC, Cinader B, Powell JE, et al: Preparation and formulation of a hCG antifertility vaccine. Selection of adjuvant and vehicle. *Am J Reprod Immunol* 1981;6:315–321.
31. Talwar GP, Sharma NC, Dubey SK, et al: Isoimmunization against human chorionic gonadotropin with conjugates of processed β-subunit of the hormone and tetanus toxoid. *Proc Natl Acad Sci USA* 1976;73:218–222.
32. Talwar GP, Dubey SK, Salahuddin M, et al: Kinetics of antibody response in animals injected with processed beta hCG conjugated to tetanus toxoid (Pr-β-hCG-TT). *Contraception* 1976;13:153–161.
33. Talwar GP, Das C, Tandon A, et al: Immunization against hCG: Efficacy and teratological studies in baboon, in Anand Kumar TC (ed): *Non-human Primate Models for Study of Human Reproduction.* Basel, S Karger, 1980, pp 190–201.
34. Talwar GP, Singh V, Singh O, et al: Pituitary and extra pituitary sites of action of gonadotropin-releasing hormone: Potential uses of active and passive immunization against gonadotropin releasing hormone, in Saxena BB, Catt KJ, Birnbaumer L, Martini L (eds): *Hormone Receptors in Growth and Reproduction.* New York, Raven Press, 1984, pp 351–359.
35. Talwar GP, Gupta SK, Singh V, et al: Bioeffective monoclonal against the decapeptide gonadotropin releasing hormone: Reacting determinant and action on ovulation and estrus suppression. *Proc Natl Acad Sci USA* 1985;82:1228–1231.
36. Talwar GP, Singh O, Singh V, et al: Enhancement of anti-gonadotropin response to beta subunit of ovine luteinizing hormone by carrier conjugation and combination with beta subunit of human chorionic gonadotropin. *Fertil Steril* 1986;46:120–126.
37. Talwar GP, Gaur A, Gupta SK, et al: Immunological control of fertility, in Thompson RA (ed): *Recent Advances in Clinical Immunology* 1986;4:183–200.
38. Tandon A, Das C, Jailkhani BL, et al: Efficacy of antibodies generated by Pr-β-hCG-TT to terminate pregnancy in baboon: Its reversibility and rescue by medroxyprogesterone acetate. *Contraception* 1981;24:83–95.
39. Tandon A, Das C, Jailkhani BL, et al: Effects on pregnancy in mice of passive immunization against ovine LH and human chorionic gonadotropin. *J Reprod Fertil* 1984;70:369–377.

CHAPTER 18

Cloning and Expression of Gonadotropin Hormone Genes

Swatantra K. Jain, William W. Chin, and G.P. Talwar

Introduction

Gonadotropins are a group of interrelated glycoprotein hormones. Two of them, luteinizing hormone (LH) and follicle-stimulating hormone (FSH), are produced by the anterior pituitary; the third, chorionic gonadotropin (CG), is made by placenta. These consist of two dissimilar subunits α and β, associated noncovalently with each other. The α-subunit has 89 to 96 amino acids (depending on the species), and the β-subunit has 112 to 121 amino acids. The α-subunit of these hormones, as well as that of thyroid-stimulating hormone (TSH), is identical within a species. The hormonal specificity is conferred by the β-subunit in each case. The β-subunits of different hormones have variable degree of homology in the sequence of amino acids. Furthermore, the α subunit seems to have distant relationship with the β-subunit. These homologies suggest that the β-subunit of this category of hormones may have evolved from a common β-subunit gene; moreover, both α- and β-subunits may have diverged from another common ancestral gene during the course of evolution (1–3).

The gonadotropins play an important role in the regulation of reproductive functions and of pregnancy. LH and FSH are necessary for the normal sexual development and functions, and CG is necessary for the maintenance of corpus luteum during early stages of pregnancy. It is therefore of interest to understand the regulation of their gene expression. The β-subunit of HCG is a component of the antifertility vaccines (4–6), which are currently in phase I clinical trials. It has been observed that the immunogenicity of βHCG is significantly improved when annealed to ovine α-LH (5). Another vaccine under clinical trial is a mixture of βoLH and βHCG linked to carriers (6). For further experimental and clinical studies, substantial amounts of the hormonal subunits would be needed. By conventional methods, relatively small amounts of hormones are obtained by lengthy processing of pituitaries. By recombinant DNA technology, it is possible to produce substantial amounts of these hormones.

In this chapter we describe the molecular cloning of full-length cDNAs for ovine LH and FSH. These are being subcloned in various vectors to get their expression in mammalian cell lines. A similar approach has been used to get the expression of HCG (7) and bovine gonadotropin (8,9).

mRNA Isolation

Fresh sheep pituitaries were obtained from a local slaughter house. Immediately after the sheep were killed, the pituitaries were excised and quickly frozen in liquid nitrogen. These were stored at −70°C until used. Our experience is that the commercially available pituitaries do not produce good mRNA. The pituitaries were homogenized in four volumes of GTC (4 mol/L guanidium isothiocyanate, 50 mmol/L sodium citrate, pH 7.0, 0.1 mol/L mercaptoethanol, 0.5% sarcosyl) (10,11) using a Polytron homogenizer at medium speed with six strokes of ten seconds each. The homogenate was clarified by centrifugation at 10,000 rpm for 20 minutes. It was further centrifuged at 40,000 rpm through a cushion of 5.7 mol/L CsCl for 18 hours, when the pellet of total RNA was obtained. The pellet was washed, dried, redissolved in sterile water, and precipitated with two volumes of 95% ethanol. It was redissolved in binding buffer (0.4 mol/L NaCl, 0.04 mol/L tris-HCl pH 7.6, and 1 mmol/L EDTA) and passed through a column of oligo-dT-cellulose (type T_2, Collaborative Research, Waltham, MA). The bound poly A^+RNA was eluted with TE (10 mmol/L tris-HCl, pH 7.6, 1 mmol/L EDTA). After two cycles of affinity chromatography, the mRNA was precipitated with two volumes of ethanol. Before use, the mRNA was electrophoresed on agarose gels to check its quality.

cDNA Synthesis

The pituitary mRNA was used as template to synthesize the cDNA by conventional methods (12–15) with certain modifications. The method is described in Figure 18.1. In brief, 15 μg of polyA$^+$mRNA was incubated with 100 units of murine leukemia virus reverse transcriptase (MLV-RTase) and the four dNTPs (deoxy nucleotide triphosphates) in a final volume of 50 μL for 45 minutes at 42°. Ten μCi of ^{32}P labeled deoxycytidine triphosphate (dCTP-[^{32}P]) was included to follow the reaction. After reaction was over, 1 μL of reaction mixture was precipitated with trichloro acetic acid (TCA) and counted to determine the TCA precipitable counts incorporated. Another 1 μL of reaction mixture was electrophoresed on 1% agarose gel to see the size distribution of incorporated radioactivity. It is a common practice to use AMV-RTase (avian myeloblastosis virus reverse transcriptase) for the synthesis of the first strand of cDNA. However, it is well known that reverse transcriptases are associated with DNAse. It has been reported that AMV-RTase is associated with more DNase activity that MLV-RTase (16). The DNA-RNA hybrid was used for a second strand cDNA synthesis by RNase H and DNA polymerase I (120 units), in a final volume of 250 μL. Again, dCTP-[^{32}P] was included to follow the reaction. The incubation was for 60 minutes at 12° followed by 60 minutes at 22°. The use of RNase H and DNA polymerase I is another modification over the conventional method, where Klenow is used for the synthesis of second strand, which later on requires the use of S_1 nuclease to remove the hairpin loop. Many times S_1 nuclease may cause

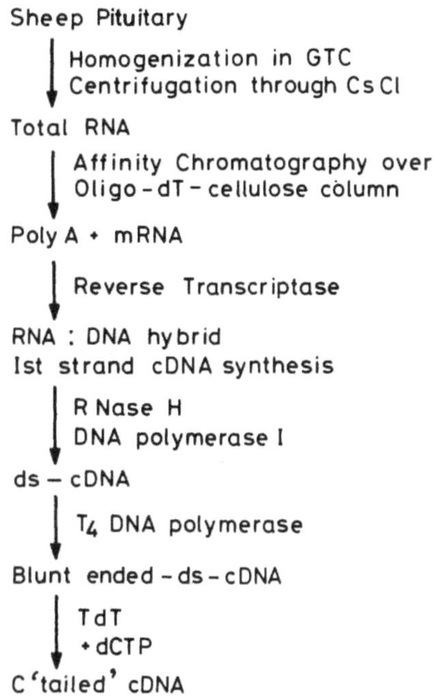

Fig. 18.1. Schematic diagram of double-stranded cDNA synthesis.

nicks and/or degradation of the cDNA. By the modifications discussed, we did not have to use the S_1 nuclease. After the reaction was over, it was heated at 80° for ten minutes and TCA precipitable counts were determined. To look at the quality of cDNA, aliquots were run on nondanaturing agarose gel and on alkaline agarose gel. It showed a distribution between 0.2 kb and 2.0 kb, with 0.5 to 1.0 kb as the predominant size class.

Construction of cDNA Library

The double stranded-cDNA was purified by repeated precipitation with ethanol in presence of 2 mol/L ammonium acetate and treated with T_4DNA polymerase to get the blunt ends. This was then treated with terminal-deoxynucleotidyl transferase (TdT) in presence of dCTP-[^{32}P] to "C tail" the cDNA. Following "C tailing" the cDNA was annealed to PstI digested-"G tailed"-pBR 322 (New England Nuclear, Boston, MA). The vector and the insert were mixed in equimolar ratio (taking average size of cDNA as 500 base pairs, bp) in presence of annealing buffer (0.1 mol/L NaCl, 10 mmol/L tris-HCl pH 7.8, 1 mmol/L EDTA), incubated at 55° for two hours followed by overnight incubation at room temperature (Fig. 18.2).

Escherichia coli MC 1061 cells were used as host. Competent cells were prepared from the exponentially growing culture by $CaCl_2$ treatment (17) and transfected with the annealed mixture as follows. The 0.3 mL cells were incubated

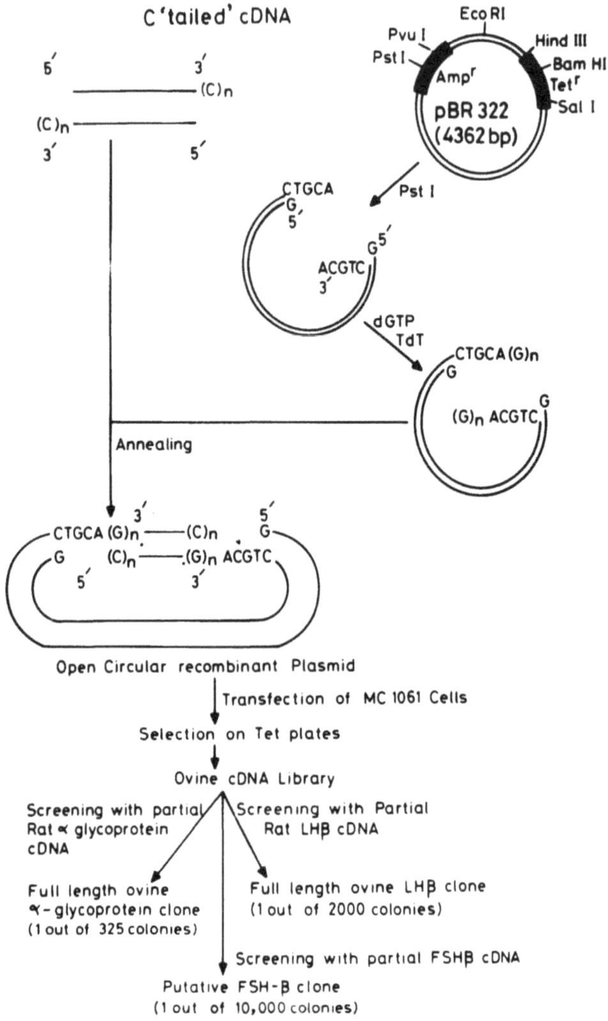

Fig. 18.2. Construction of ovine pituitary cDNA library and its screening.

with annealed mixture for 60 minutes in ice, heat shocked at 42° for two minutes, and mixed with 1.2 mL of L-broth. The mixture was incubated at 37° for 30 minutes and 0.4 mL culture was plated on prewarmed agar plates containing 10 μg/mL tetracycline. The tetracycline-resistant colonies appeared after overnight incubation at 37°. A cDNA library of about 5×10^4 colonies was constructed.

Screening of Library

The library was screened by colony hybridization (18) as follows. The colonies were grown on nitrocellulose filters, and replica filters were prepared by standard procedures. Colonies on the replica filters were lysed by alkali treatment

(15), baked at 80° for two hours, and prehybridized for 24 hours in prehybridization buffer (5X SSC, 1X Denhardt's solution [19], 0.2% sodium dodecyl sulfate [SDS] and 100 μg/mL sonicated salmon sperm DNA [Sigma Chemical Company, St. Louis]). The filters were then hybridized at 65° for overnight with nick-translated (20) probes in the same buffer. Following hybridization, the filters were washed with 5XSSC at 55° and autoradiographed with an intensifying screen (Dupont).

Partial cDNA clones for rat α-glycoprotein (250 bp long, 3′ piece), β-LH (300 bp long, 3′ piece) and bovine FSHβ (250 bp long, middle piece) were used for screening. Although a good homology between rat/bovine probes and respective ovine cDNAs is expected before screening, it was confirmed that they cross-hybridize by the following procedure. A Northern blot of total ovine RNA was hybridized with these probes. In each case the probes selected a single mRNA for α-glycoproteins, LH β and FSH β. The sizes of ovine mRNAs selected by these probes were similar to respective mRNAs for rat gonadotropins. This simple experiment shows that the heterologous probes cross-hybridize with respective sequences of sheep.

The colony hybridization produced several positive clones. The relative abundance of α-glycoprotein clones was very high (about 0.3%–0.4%). LHβ was present in a relatively low abundance of about 0.05%. FHSβ was a minor species with an abundance of only about 0.01%. The colonies from primary screening were rescreened to obtain purified clones.

Northern and Southern Analysis

Northern and Southern analysis were performed by conventional methods. For Northern blot analysis RNA was denatured with 1 mol/L glyoxal and 50% dimethyl sulfoxide (DMSO) and electrophoresed in 1% agarose gel with 10 mmol/L sodium phosphate, pH 7.0 (21). It was transferred to GeneScreen (New England Nuclear, Boston) and baked at 80° for two hours. For Southern blot analysis (22), the DNA was digested with appropriate restriction endonucleases under the conditions recommended by manufacturer and separated on 1% agarose gel in tris-boric acid-EDTA buffer (50 mmol/L Tris, 50 mmol/L boric acid, 1 mmol/L EDTA, pH 8.3). The gel was denatured with alkali, neutralized with Tris/NaCl, blotted to GeneScreen, and baked. The filters were prehybridized and hybridized in the same manner as for the colony hybridization.

Ovine LHβ cDNA

One of the LHβ clones was selected for further analysis. It has an insert band of about 650 bp, which can be excised by digestion with PstI. Southern blot analysis of the gel shows that this band hybridizes with the rat LHβ cDNA probe.

When a Northern blot of total pituitary RNA was hybridized with nick-translated insert, a single mRNA band of about 750 bases was selected. This size

Table 18.1. Properties of ovine LHβ cDNA clone.

1. The abundance in the cDNA library of normal sheep pituitary is about 0.05%. On average, one out of 2,000 clones gives positive hybridization signal.
2. The insert is 650 bp long and can be isolated as a single band by PstI digestion.
3. The clone hybridizes with rat LHβ cDNA probe at a stringency of 5XSSC at 55°.
4. Partial sequencing shows that the clone has full coding region and bears about 80% homology with rat LHβ cDNA.
5. The cross-hybridization studies show that ovine LHβ and α-glycoprotein cDNAs do not cross-hybridize with each other.
6. Northern blot analysis shows that ovine LHβ is coded by a single mRNA of about 700 bases in length.

of mRNA is very similar to the size of rat LHβ mRNA (3). Finally, the identity of clone was confirmed by partial sequencing. For sequencing the PstI and Hae-III, digests of the insert were subcloned in M13 (23) and sequenced by dideoxy chain termination method of Sanger et al (24). The clone bears about 80% sequence homology with rat LHβ cDNA (3) in the region where sequenced. Also it has full coding region. These details are summarized in Table 18.1.

Ovine α-Glycoprotein cDNA

The ovine α-glycoprotein cDNA has two internal PstI sites and gives multiple insert bands upon PstI digestion. However, partial digestion with PstI gives an insert band of 600 bp along with the smaller bands.

The identity of the clone was confirmed by partial sequencing by subcloning in M13 (23) and using Sanger's method (24). In the region where sequenced, it has about 80% homology with rat α-cDNA. The presence of two internal PstI in ovine α-cDNA is in contrast to rat α-cDNA (3), which does not have any internal PstI site. However, the nucleotide sequences bear enough homology to cross-hybridize with each other, at least at low stringency (5X SSC, 55°). Hybridization of a Northern blot with the clone showed that α-glycoproteins are coded by a single species of mRNA. Table 18.2 gives a summary of the characteristics of the α-glycoprotein cDNA clone.

Table 18.2. Properties of ovine α-glycoprotein cDNA.

1. Present in ovine cDNA library at high abundance (0.3%). One out of 325 clones gave positive hybridization signal.
2. The cDNA insert is about 575 bp long. It has two internal PstI sites and digestion with PstI restriction endonuclease gives three insert fragments.
3. As seen by partial sequencing, the clone has full coding region. It has about 80% sequence homology with rat α-chain cDNA and cross-hybridizes with rat cDNA at a stringency of 5XSSC at 55°.
4. The Northern blot analysis shows that a single mRNA of about 800 bases codes for the α-glycoprotein chains.

FSHβ cDNA

The initial screening of the library produced some putative FSHβ clones. The clones have an insert of about 1.4 kb, which is excised as a single band upon PstI digestion. It selects a mRNA of about 1.7 kb in a Northern blot of total pituitary RNA.

Expression of Gonadotropin Hormones

Since these hormones are glycoprotein in nature, posttranslational modifications are needed for synthesizing the bioactive molecules. As bacteria can not glycosylate the proteins, eukaryotic expression vectors are necessary to get the expression. The constructs are being made using various vectors containing metallothionein promotor and poly A region, bovine papilloma virus promotor, along with the intron and the vaccinia virus based vector.

Conclusion

In conclusion, we have isolated the full length cDNA clones for ovine gonadotropins. It is interesting to see that the partial cDNAs from heterologues species could be used for screening the ovine cDNA library to isolate the oLH and oFSH cDNA clones. The striking differences between rat and ovine α-glycoprotein cDNAs, namely, the occurrence of PstI sites, open the issue of the evolution of these genes.

References

1. Stewart M, Stewart F: Constant and variable regions in glycoprotein hormone β subunit sequences: Implications for receptor binding specificity. *J Mol Biol* 1977;116:175–179.
2. Acher R: Molecular evolution of biologically active polypeptides. *Proc R Soc Lond (Biol)* 1980;200:21–43.
3. Chin WW: Organization and expression of glycoprotein hormone genes, in Imura H (ed): *The Pituitary Gland*. New York, Raven Press, 1985, pp 103–125.
4. Talwar GP, Sharma NC, Dubey SK, et al: Isoimmunization against human chorionic gonadotropin with conjugates of processed beta subunit of the hormone and tetanus toxoid. *Proc Natl Acad Sci USA* 1976;73:218–222.
5. Sharma NC, Singh O, Gaur A, et al: Improved immunogenic formulation for antigonadotropin response, in Talwar GP (ed): *Immunological Approaches to Contraception and Promotion of Fertility*. New York, Plenum Press, 1986, pp 37–41.
6. Talwar GP: Immunological approaches to contraception: The need, basic premise and overview, in Talwar GP (ed): *Immunological Approaches to Contraception and Promotion of Fertility*. New York, Plenum Press, 1986, pp 17–24.
7. Khandekar P: Unpublished data.

8. Kaetzel DM, Browne JK, Wondisford F, et al: Expression of biological active bovine luteinizing hormone in Chinese hamster ovarian cells. *Proc Natl Acad Sci USA* 1985;82:7280–7283.
9. Nilson JH, Kaetzel DM: Expression of bovine lutropin genes in Chinese hamster ovary cells: Use of expression vectors to investigate structure requirements for assembly of biologically active LH. *J Reprod Fertil* 1987, in press.
10. Glisin VG, Crkvenjakov R, Byus C: Ribonucleic acid isolated by cesium chloride contrifugation. *Biochemistry* 1974;13:2633–2639.
11. Ulrich A, Shine J, Chirgwind J, et al: Rat insulin genes: construction of plasmids containing the ending sequences. *Science* 1977;196:1313–1319.
12. Efstratiadis A, Kafatos FC, Maxam AM, et al: Enzymatic *in vitro* synthesis of globin genes. *Cell* 1976;7:279–284.
13. Buell GN, Wickens MP, Payvar F, et al: Synthesis of full length cDNAs from four partially purified oviduct mRNAs. *J Biol Chem* 1978;253:2471–2478.
14. Retzel FE, Collet MS, Faras AJ: Enzymatic synthesis of deoxyribonucleic acids by the avian retrovirus reverse trascriptase in vitro. Optimal conditions required for transcription of large ribonucleic acid templetes. *Biochemistry* 1980;19:513–518.
15. Maniatis T, Fritsch EF, Sambrook J: Synthesis and cloning of cDNA, in: *Molecular Cloning, A Laboratory Manual;* Cold Spring Harbor, NY, Cold Spring Harbor Laboratory, 1982, pp 211–246.
16. Gerard G: Cloned moloney murine leukemia virus reverse transcriptese. *Focus* 1985;7:1–3.
17. Mandel M, Higa A: Calcium dependent bacteriophase DNA infection. *J Mol Biol* 1970;53:154–162.
18. Grunstein M, Hogness D: Colony hybridization: A method for the isolation of cloned DNAs that contain a specific gene. *Proc Natl Acad Sci USA* 1975;72:3961–3965.
19. Denhardt DJ: A membrane filter technique for the detection of complimentary DNA. *Biochem Biophys Res Commun* 1966;23:641–646.
20. Rigby PWJ, Dieckmann M, Rhodes C, et al: Labelling deoxy ribonucleic acid to high specific activity *in vitro* by nick translation with DNA polymerase I. *Mol Biol* 1977;113:237–251.
21. McMaster GK, Carmichael GG: Analysis of single and double stranded nucleic acids on polyacrylamide and agarose gels by using glyoxal and acridine orange. *Proc Natl Acad Sci USA* 1977;74:4835–4838.
22. Southern EM: Detection of specific sequences among DNA fragments separated by gel electrophoresis. *J Mol Biol* 1975;98:503–517.
23. Messing J, Crea E, Seaburg PH: A system for shotgun DNA sequencing. *Nucl Acid Res* 1981;9:309–321.
24. Sanger F, Nicklen S, Coulson AR: DNA sequencing with chain terminating inhibitors. *Proc Natl Acad Sci USA* 1977;74:5466–5467.

CHAPTER 19

Hormonal Regulation of Pituitary Gonadotropin Gene Expression

William W. Chin, Soheyla D. Gharib, and Margaret E. Wierman

Introduction

It is well known that the endocrine regulation of the reproductive system is complex, involving multiple hypothalamic, pituitary, and gonadal factors in the coordination of gonadal function. The pituitary gonadotropins, luteinizing hormone (LH) and follicle-stimulating hormone (FSH), play pivotal roles in this regulatory network. They act directly on gonadal tissue to stimulate the production and release of gonadal sex-steroid hormones and to regulate gametogenesis. Hence, appropriate levels of LH and FSH determine oocyte maturation and ovulation in female animals and spermatocyte development in male animals. As a result, the regulation of LH and FSH synthesis and secretion must be a critical element involved in normal reproductive function (Fig. 19.1). Much has been learned over the last few decades concerning the factors that modulate LH and FSH production and secretion. It has been long known that the gonadal sex-steroid hormones, including estrogen, testosterone, and progesterone, regulate the pituitary gonadotropins either directly at the pituitary or indirectly at the hypothalamic levels. In addition, a hypothalamic factor, gonadotropin-releasing hormone (GnRH), has been shown to be the major positive controller of LH and FSH release. Recent data indicate that GnRH is released by neurons in the median eminence of the hypothalamus in a pulsatile fashion that varies in amplitude and frequency, both of which influence the amount of LH and FSH produced and secreted (16). Although these two regulators of gonadotrope function are well established, they are by no means the exclusive ones. In particular, recent studies indicate that there are several gonadal peptides that regulate FSH synthesis and release. They include inhibin, which is present in at least two forms, α and β (9,20,21) and inhibin-like FSH-releasing peptide (FRP) (19,25). As their names suggest, inhibins decrease the release of FSH, and FRP increases the release of FSH from gonadotrope. Finally, recent work by Adelman et al (1) on the characterization of the GnRH gene indicates that the carboxy-terminal portion of the GnRH precursor encodes a protein known as GnRH-associated-protein (GAP), which may also act to stimulate gonadotrope function. Hence, it is increasingly evident that there are several hypothalamic and gonadal major

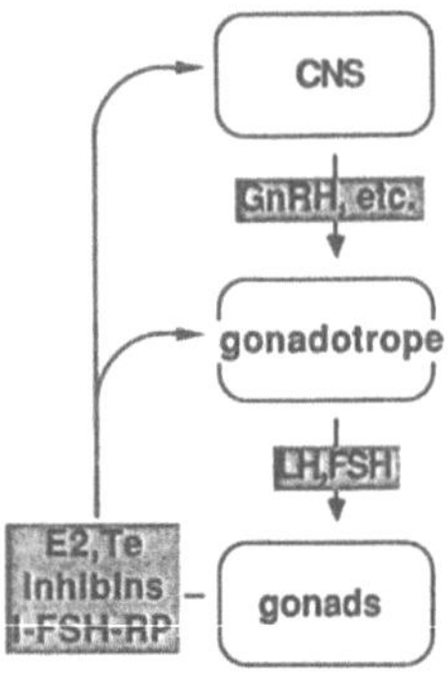

Fig. 19.1. Regulation of pituitary gonadotropin synthesis and release. The gonadotrope in the anterior pituitary gland synthesizes and secretes lutropin (LH) and follitropin (FSH), which then stimulate the gonads to produce sex-steroid hormones [estrogen (E2) and testosterone (Te)] and gametes. The sex-steroids and recently characterized gonadal peptides such as inhibin and inhibin-like FSH-releasing peptide (I-FSH-RP), in turn, modulate the activity of the gonadotrope with differential effects on LH and FSH synthesis and release. The gonadal factors probably also have major effects on the central nervous system to regulate further hypothalamic effectors of gonadotrope function.

factors involved in the modulation of gonadotrope activity, some of which are relatively unexplored.

Pituitary Gonadotropins: LH and FSH

LH and FSH are members of a group of hormones termed the glycoprotein hormone family. They also include pituitary thyrotropin (thyroid-stimulating hormone, TSH) and placental chorionic gonadotropin (CG). Each glycoprotein hormone is made up of two different subunits, α and β, which are associated in a noncovalent fashion (Fig. 19.2). There is a single α-subunit gene in most mammals, encoding a common α-subunit that is shared among all four glycoprotein hormones. In marked contrast, the β-subunits (LHβ, FSHβ, TSHβ, and CGβ) are encoded by separate genes. Hence, the β-subunits impart to glycoprotein dimers specific biologic activity. The α-subunit is approximately 18 to 22 Kd and the β subunits, 17 to 25 Da in size. Each subunit contains N-linked and O-linked

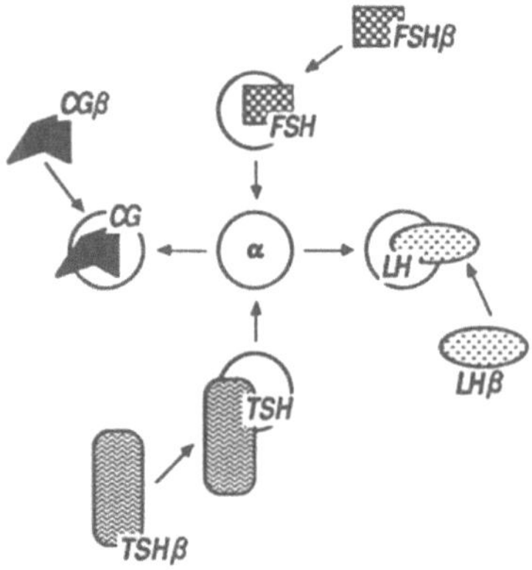

Fig. 19.2. Structure of the pituitary gonadotropins. LH and FSH share structural similarities with two other hormones, thyrotropin (TSH) and chorionic gonadotropin (CG). Each hormone in this family contains two different subunits that are noncovalently associated. The common α-subunit is nearly identical among hormones with this family. In contrast, the LHβ- and FSHβ-subunits are unique and confer specific biological activity upon the dimer. Each subunit is approximately 18 to 25 kDa in size and is encoded by a separate gene located on separate chromosomes.

intramolecular disulfide linkages. These bonds are necessary for each subunit to achieve the appropriate conformation to allow for monomer-monomer interaction to form the biologically active dimer. In most cells that produce the glycoprotein hormones, it has been observed that the α-subunit is often synthesized in molar excess of the dimer and free β-subunits. However, it is yet unclear what the physiologic significance of the excess α-subunit might be. Interestingly, recent work has revealed that the free α-subunit possesses a distinct biochemical structure from its native or combined α-counterpart. This variation is a result of an additional sugar group added to the α-subunit protein backbone (4).

LH and FSH are produced by a group of cells in the anterior pituitary gland which are known as gonadotropes. Recent data indicate that the gonadotrope population is heterogeneous. Specifically, immunologic staining studies have shown that as many as 70% of normal rat pituitary gonadotropes produce both LH and FSH whereas 15% produce only LH and the remainder produce only FSH (3). However, it is also clear that the pituitary gonadotropes are not produced elsewhere in the body. In particular, LH and FSH are not produced in the pituitary thyrotrope, which synthesizes only TSH. Thus, the expression of the β-subunits is apparently tissue-specific, with the genes that encode these β-subunits being expressed only in certain cells. In contrast, it appears that the α-subunit gene is expressed in both gonadotrope and thyrotrope populations within the anterior pituitary gland.

Biosynthesis of Glycoprotein Hormone Subunits

Before we discuss the studies concerning the effects of sex-steroid hormones and GnRH on gonadotropin subunit mRNA levels in the pituitary gland, we would like to review briefly the general flow of information within a cell with regard to gene expression and ultimate formation of a mature secretion peptide such as a glycoprotein hormone subunit (4,8) (Fig. 19.3). Every peptide is encoded by a gene. The gene is transcribed to yield a precursor mRNA that needs to be processed in order to yield functional mRNA. These events include 5′-capping, poly-A addition at the 3′-end, and splicing of intron sequences. The mature mRNA, having been processed rapidly in the nucleus, is then transported to the cytoplasm where it engages the protein synthetic machinery consisting of ribosomes and ribosomal proteins. There, the mRNA is translated into the protein precursor. However, even during the translational process, this polypeptide precursor is altered, including the removal of N-terminal peptide sequences and the addition of N-linked sugar moieties. This partially processed polypeptide is transported through the Golgi-endoplasmic reticulum complex where it undergoes further maturation, including polypeptide folding, formation of disulfide linkages, dimerization, more extensive glycosylation, etc to yield the near mature-polypeptide, which then may be packaged into secretory vesicles and granules. These near-mature proteins then remain in a storage form until the appropriate secretion signals are applied to the cell. Hormonal regulation of

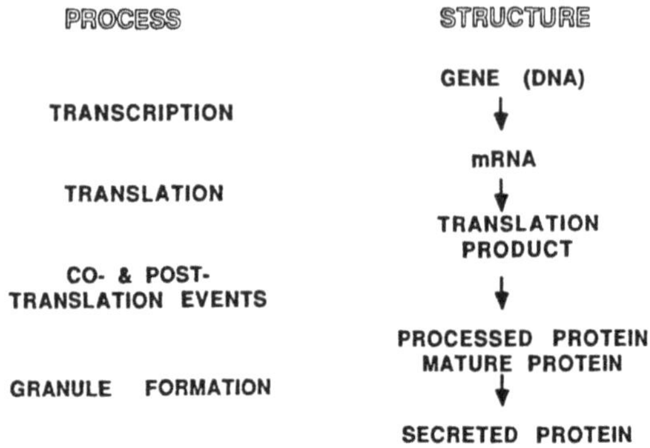

Fig. 19.3. Informational flow within a cell. Each protein produced by a cell is encoded by a single gene. This gene is transcribed in the nucleus to yield a messenger RNA that enters the cytoplasm to encounter the ribosomal complex where protein synthesis occurs. The initial protein product is a precursor protein referred to as the translation product. This immature protein is processed extensively as described in the text via a series of co- and posttranslational steps that lead eventually to the mature product. In the case of the gonadotropins, the near mature subunits combine soon after translation and undergo complex glycosylation steps in the posttranslational modification process.

synthesis of the subunits of glycoprotein hormones may, of course, occur at any of these levels. However, it is clear, as is the case for most proteins, that the major regulation probably occurs at the transcriptional and posttranscriptional levels. Hence, this establishes the rationale for our studies to evaluate the effects of gonadal sex-steroid hormones and GnRH on pituitary subunit mRNA levels.

Molecular Cloning of Gonadotropin Subunit cDNAs and Genes

In order to determine the effects of these various hormones on the subunit genes and their transcripts it has been necessary to obtain molecular clones of mRNAs and genes encoding α-, LHβ-, and FSHβ-subunits. These nucleic acids may then be used to detect subunit-specific mRNAs in whole anterior pituitary glands using blot hybridization analyses. Over the last five years, our laboratory has been successful in obtaining complementary DNAs (cDNA) and genes that encode α-, LHβ-, and FSHβ-subunits in the rat. Using such labeled DNAs, we have shown that the mRNAs for each subunit appear to be singular and that the sizes of the α-, LHβ-, and FSHβ-mRNAs are 800, 700, and 1,700 nucleotides, respectively (5,10,13) (Fig. 19.4). In addition, these studies have allowed us to determine that each rat glycoprotein hormone subunit gene is encoded by a single gene in mammalian genomic DNA, which, in turn, is apparently located on different chromosomes (14,22,26). Hence, the expression of α- and β-subunit genes must be regulated via common mediation factors that then impinge on these genes, which are apparently separated by great distances within the genome.

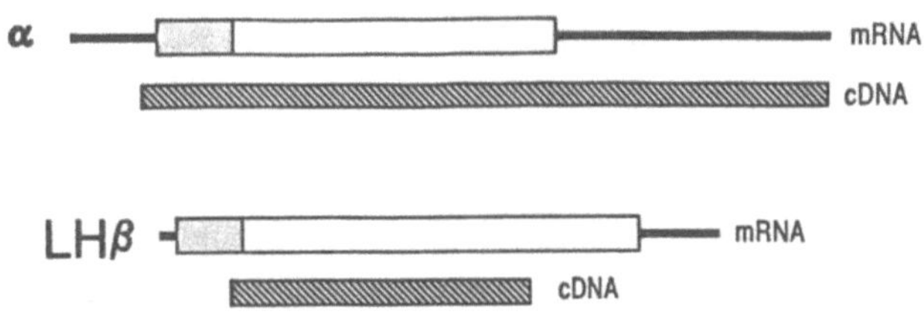

Fig. 19.4. Structure of the mRNAs and cDNAs encoding the α- and β-subunits of LH and FSH. The top figure in each section represents, in schematic form, the mRNAs. The box indicates the coding region for the precursor protein, with the leader peptide depicted by a shaded area. The solid lines to the left and right of the coding region indicate the 5′- and 3′-untranslated regions, respectively. The hatched bar beneath each mRNA indicates a representative cloned complementary DNA or cDNA encoding a portion of the respective RNA. The α- and LHβ-mRNAs are 800 and 700 nucleotides in size, respectively.

Hormone Regulation of LH and FSH Subunit Genes: Sex-Steroid Hormones

It is well known that gonadal sex-steroid hormones negatively regulate the synthesis and secretion of LH and FSH. However, until several years ago, there was little information about such regulatory events at the pretranslational level. Early work using cell-free translational systems suggested that gonadal sex-steroid hormones regulate LHβ and FSHβ mRNAs in a negative fashion. Godine et al (12) examined the pituitary mRNA from normal and castrate rats in the wheat-germ cell-free translation system and demonstrated approximately tenfold elevations of the subunit mRNA levels after castration. Studies with similar results in other species were reported by Landefeld et al (17,18). Counis et al (7) and Corbani et al (6) studied the effect of estradiol and progesterone administration on α- and LHβ-mRNA levels in the pituitary glands of ovariectomized rats. Using cell-free translation and immunoprecipitation techniques, as described previously, they showed that estrogen but not progesterone rapidly decreases the levels of α- and LHβ-mRNAs. Nilson et al (23) studied the effect of estrogen on the α- and LHβ-mRNAs in the anterior pituitary gland of castrate sheep. These studies were the first to use cloned α- and LHβ-subunit cDNAs and blot hybridization techniques to evaluate the changes in the subunit mRNAs. After three weeks of treatment with large doses of estrogen, they showed that the α- and LHβ-mRNAs decreased markedly from control castrate pituitary levels. In addition, Alexander and Miller (2) have known that estrogen decreases FSHβ-mRNA in the pituitary of the castrate ram.

We have studied the effect of gonadal sex-steroid hormones on the biosynthesis of LH and FSH at the pretranslational level in the rat. We have employed two experimental protocols: (a) castration and (b) hormone replacement of castrate animals, in both male and female rats. In the first experimental protocol, groups of sexually mature male and female Sprague-Dawley rats (150–200 g) were gonadectomized at time 0 and killed at intervals of time from 0 to 28 days after surgical castration. The pituitary glands were isolated and plasma LH and FSH levels were determined by RIA. Total RNA was isolated from single pituitary glands using a microtechnique. The typical yield of total RNA from each gland was 50 to 75 μg of RNA. Total RNA samples (5–15 μg) from single pituitary glands were assayed for α-, LHβ-, and FSHβ-mRNAs using blot hybridization techniques. Values were expressed in arbitrary densitometric units relative to β-actin mRNA, which was shown to be invariant with hormonal manipulation. In the female rats, α-, LHβ-, and FSHβ-mRNAs increased gradually within the first week after castration in male and female animals. The subunit mRNA levels peaked at 21 to 28 days with changes in α-, LHβ-, and FSHβ-mRNAs that are 4-fold, 8 to 10-fold, and 4-fold, respectively, at peak levels. In male rats, similar increases in α- and LHβ-mRNAs were observed. One major difference, however, was found in the FSHβ-mRNA. In contrast to the fourfold changes seen after castration in female animals, the pituitary gland of the males showed only a 1.5 to 2-fold increase in FSHβ-mRNA after 28 days. Thus, the levels of LH and FSH subunit mRNAs increase with removal of gonadal sex-steroid hormones with an effect that is greatest for LHβ-mRNA, with effects that are smaller for α-, and FSHβ-mRNA, and with effects that are most modest for FSHβ-mRNA in the male.

In the second experimental protocol, male and female rats that had first been castrated three to four weeks previously were treated with estrogen for female and testosterone for male animals. The dose for estrogen was 10 μg/100 g body weight/day and for testosterone it was 500 μg/100 g body weight/day administered subcutaneously in sesame oil. Animals were killed at 0, 1, and 7 days after replacement with gonadal sex-steroid hormone. The anterior pituitary gland was isolated and plasma LH and FSH determined as described above. The plasma LH and FSH levels declined rapidly, as described previously. The pituitary mRNA levels for α, LHβ, and FSHβ declined rapidly so that by seven days there was possibly 10% to 25% of the original level present. Again the effects of the estrogen

SUBUNIT mRNA	FEMALE	MALE
α	+++	+++
LHβ	++++	++++
FSHβ	++	+/-

Fig. 19.5. Regulation of LH and FSH subunit mRNAs by sex-steroid hormones. This chart summarizes the effects of estrogen and testosterone on LH and FSH subunit mRNAs in the pituitary glands of female and male sexually mature Sprague-Dawley rats. Note that, in general, the sex-steroids decrease the levels of the mRNAs, respectively. The effect is strongest for LHβ-mRNA and weakest for FSHβ-mRNA. Strikingly, the regulation of FSHβ-mRNA in the male rat is minimal in contrast to severalfold changes seen in this mRNA in the female animal.

and testosterone replacement were greatest for the LHβ-subunit and less for α- and FSHβ-mRNAs. In marked contrast, FSHβ-mRNA in the male rat was much more modest. Thus, in general, the decline in the subunit mRNA paralleled the decline in plasma LH and FSH values.

In summary, these data suggest that the gonadal sex-steroid hormones regulate the gonadotropin subunit mRNAs in a negative fashion, with an effect that is greatest for LHβ- than for α- and FSHβ-subunit mRNAs (Fig. 19.5). There exists the possibility that there is a difference between male and female rats in the regulation of the FSHβ-subunit mRNA. Whether this is due to sexual dimorphic changes, differences in the applied sex-steroid hormone, or influence of variable gonadal inhibin and FRP levels is still unclear. Whether these effects occur at the gene transcription or mRNA stability level is also unknown at this time.

Regulation of LH Subunit Genes: GnRH

Finally, preliminary experiments have been performed in collaboration with Dr. Marshall's group at the University of Michigan School of Medicine. These studies involve the effect of GnRH on the levels of α- and LHβ-mRNAs in male rat pituitary glands (24). The experimental paradigm is the partial replacement of acutely castrate male rats with physiologic levels of testosterone. The general approach is to suppress the endogenous GnRH pulse generator so that exogenously administered GnRH will determine the amount of LH and FSH synthesized and released. In these studies GnRH was given at varying doses from 10 to 250 ng per dose but administered every 30 minutes for a 48-hour period immediately after castration and administration of low levels of testosterone. The results can be summarized as follows: both α- and LHβ-mRNAs were shown to increase 1.5 to 2-fold from basal levels. These increases were dose-dependent and were different for α- and LHβ-mRNAs. The peaks of LHβ- and α-mRNAs were achieved with GnRH doses of 25 and 75 ng per dose, respectively. The 25-ng dose also produced peak levels of LH release with a given GnRH dose and of pituitary GnRH receptors. These results interestingly show that the GnRH pulse pattern can indeed result in increases in α- and LHβ-mRNAs, which may be dose-dependent. In addition, there is a differential effect on α- and LHβ-mRNAs by various GnRH doses. This, in combination with recent preliminary data from Jonassen and Roberts (15), which indicate that GnRH may rapidly stimulate the transcription of LHβ gene, suggests that GnRH may regulate the synthesis of LH at the pretranslational level.

Summary

Recent work using the techniques of molecular biology and of recombinant DNA technology has allowed careful analysis of the effects of gonadal sex-steroid hormones and GnRH on the synthesis of LH and FSH at the pretranslational level. Clearly gonadal sex-steroid hormones regulate LH and FSH biosynthesis in a negative fashion at the mRNA level and perhaps transcriptional level. Recent data concerning inhibins and FRP as well as GAP give us reason to believe that there may be other regulatory factors involved in the synthesis of FSH and LH. Studies in these important areas are underway.

Acknowledgments. We thank Drs. Thomas M. Badger, Edward Bernstine, and John C. Marshall for their collaborative efforts. We also thank Aparna Roy for her technical assistance and Nancy Patterson for typing this manuscript. This work is supported in part by National Institutes of Health grant HD 19938 to W.W.C.

References

1. Adelman JP, Mason AJ, Hayflick JS, et al: Isolation of the gene and hypothalamic cDNA for the common precursor of gonadotropin-releasing hormone and prolactin release-inhibiting factor in human and rat. *Proc Natl Acad Sci USA* 1986; 83:179–183.
2. Alexander DC, Miller WL: Regulation of ovine follicle-stimulating hormone β chain mRNA by 17β-estradiol *in vivo* and *in vitro*. *J Biol Chem* 1982;257:2282–2286.
3. Childs GV, Ellison DG, Lorenzen JR, et al: Immunocytochemical studies of gonadotropin storage in developing castration cells. *Endocrinology* 1982;111:1318–1328.
4. Chin WW: Organization and expression of glycoprotein hormones genes, in Imura H (ed): *The Pituitary Gland*. New York, Raven Press, 1985, pp 103–125.
5. Chin WW, Godine JE, Klein DR, et al: Nucleotide sequence of the cDNA encoding the precursor of the β subunit of rat lutropin. *Proc Natl Acad Sci USA* 1983;80: 4649–4653.
6. Corbani M, Counis R, Starzec A, et al: Effect of gonadectomy on pituitary levels of mRNA encoding gonadotropin subunits and secretion of luteinizing hormone. *Mol Cell Endocrinol* 1984;35:83–87.
7. Counis R, Corbani M, Jutisz M: Estradiol regulates mRNAs encoding precursors to rat lutropin (LH) and follitropin (FSH) subunits. *Biochem Biophys Res Commun* 1983;114:65–72.
8. Darnell J: Variety in the level of gene control in eucaryotic cells. *Nature* 1982;297:365–371.
9. Forage RG, Ring JM, Brown RW, et al: Cloning and sequence analysis of cDNA species coding for the two subunits of inhibin from bovine follicular fluid. *Proc Natl Acad Sci USA* 1986;83:3091–3095.
10. Gharib SD, Wierman ME, Badger TM, et al: Regulation of mRNAs encoding the subunits of rat follicle-stimulating hormone (FSH) by gonadal steroids. The Endocrine Society 68th Annual Meeting, 1986, Abstract 903.
11. Gharib SD, Bowers SM, Need LR, et al: Regulation or rat luteinizing hormone (LH) subunit mRNAs by gonadal steroid hormones. *J Clin Invest* 1986;77:582–589.
12. Godine JR, Chin WW, Habener JF: Luteinizing and follicle-stimulating hormones. Cell-free translations of messenger RNAs coding for subunit precursors. *J Biol Chem* 1980;255:8780–8783.

13. Godine JE, Chin WW, Habener JF: α subunit of rat pituitary glycoprotein hormones. Primary structure of the precursor determined from the nucleotide sequence of cloned DNAs. *J Biol Chem* 1982;257:8386–8371.
14. Jameson JL, Chin WW, Hollenberg AN, et al: The gene encoding the β-subunit of rat luteinizing hormone: Analysis of gene structure and evolution of nucleotide sequence. *J Biol Chem* 1984;259:15474–15480.
15. Jonassen JA, Roberts JL: Transcription of the rat beta LH gene is stimulated by GnRH. 14th Annual Meeting of the Society for Neuroscience, 1984, Abstract #308.4.
16. Kalra PS, Kalra SP: Control of gonadotropin secretion, in Imura H (ed): *The Pituitary Gland*. New York, Raven Press, 1985, pp 189–220.
17. Landefeld TD, Kepa J, Karsch FJ: Regulation of α subunit synthesis of gonadal steroid feedback in the sheep anterior pituitary. *J Biol Chem* 1983;258:2390–2393.
18. Landefeld TD, Kepa J, Karsch FJ: Estradiol feedback effects on the α-subunit mRNA in the sheep pituitary gland: Correlation with serum and pituitary luteinizing hormone concentrations. *Proc Natl Acad Sci USA* 1984;81:1322–1326.
19. Ling N, Ying S-Y, Ueno N, et al: Pituitary FSH is released by a heterodimer of the β-subunits from the two forms of inhibin. *Nature* 1986;321:779–782.
20. Mason AJ, Hayflick JS, Ling N, et al: Complementary DNA sequences of ovarian follicular fluid inhibin show precursor structure and homology with transforming growth factor-β. *Nature* 1985;318:659–663.
21. Mayo KE, Cerelli GM, Spiess J, et al: Inhibin A-subunit cDNAs from porcine ovary and human placenta. *Proc Natl Acad Sci USA* 1986;83:5849–5843.
22. Naylor SL, Chin WW, Goodman HM, et al: Chromosome assignment of genes encoding the α and β subunits of glycoprotein hormone in man and mouse. *Somatic Cell Gen* 1983;9:757–770.
23. Nilson JH, Nejedlik MT, Virgin JB, et al: Expression of α subunit and luteinizing hormone β genes in the ovine anterior pituitary. Estradiol suppresses accumulation of mRNAs for both α subunit and luteinizing hormone β. *J Biol Chem* 1983;258:12087–12090.
24. Papavasiliou SS, Zmeili S, Khoury S, et al: Gonadotropin-releasing hormone differentially regulates expression of the genes for luteinizing hormone α and β subunits in male rats. *Proc Natl Acad Sci USA* 1986;83:4026–4029.
25. Vale W, Rivier J, Vaughan J, et al: Purification and characterization of a FSH releasing protein from porcine ovarian follicular fluid. *Nature* 1986;321:776–779.
26. Watkins P, Eddy R, Beck A, et al: Assignment of the human gene for the β subunit of follicle-stimulating hormone (FSHβ) to chromosome 11. *Cytogenet Cell Genet* 1985; 40:773.

CHAPTER 20

Active Immunization of Rhesus Monkeys Against the β-Subunit of Ovine Luteinizing Hormone: Evaluation of Safety and Effectiveness of an Antigonadotropin Vaccine

Rosemarie B. Thau

Introduction

The capacity to interfere with reproductive processes by immunizing against hormones has long been recognized (19). However, more recent studies have aimed at developing a vaccine that would prevent pregnancy while affecting other reproductive functions (such as menstrual cycles and ovulation) as little as possible. Therefore, great efforts have been made to raise antibodies against human chorionic gonadotropin (HCG) and to avoid cross-reactions with the pituitary gonadotropin luteinizing hormone (LH). HCG, which is produced by the implanting embryo, induces the corpus luteum to continue production of the progesterone required for the maintenance of pregnancy. LH regulates (in conjunction with follicle-stimulating hormone, FSH) normal follicular development and induction of ovulation. The α-subunit of HCG is identical to that of LH, FSH, and thyrotropin, while the β-subunit is different for each gonadotropin. The C-terminal peptide (CTP) of HCGβ (being unique to HCGβ) avoids even existing homologies among the β-subunits. The possibility of active immunization with the β-subunit or fragments of HCG as a method of fertility regulation in females has been investigated for almost two decades. Recently, several anti-HCG vaccines have reached the stage of phase I clinical trials (4,9,10,20). Studies in primate models were carried out by actively immunizing baboons, marmosets, and rhesus monkeys against HCGβ and baboons against the CTP of HCGβ (7,1,5). Marmoset chorionic gonadotropin (CG) is highly cross-reactive with HCG (12), whereas baboon CG cross-reacts less than 2% (11) and rhesus (Rh) CG has limited cross-reaction with HCG and no cross-reaction with HCGβ (2). Therefore, anti-HCGβ antibodies cannot be used to neutralize the biological activity of RhCG.

The observation of Hodgen et al (3) that antisera to the β-subunit of ovine LH (oLHβ) cross-react strongly with RhCG but minimally with RhLH suggested that immunizing rhesus monkeys against oLHβ would result in a good animal model for studying many aspects of antigonadotropin vaccines. Choosing an appropriate animal model is a difficult task and, as happens so frequently in research, our results were unexpected. We hoped that after active immunization against oLHβ, the circulating anti-oLHβ antibodies would neutralize the biological activity of RhCG without affecting RhLH function. Surprisingly, the cross-reaction of the oLHβ antibodies with RhLH was significant. In spite of this fact, the females continued to ovulate and to have regular menstrual cycles (14,15). Their fertility was drastically reduced. Thus, the rhesus monkeys immunized against oLHβ provided an opportunity to evaluate the effects of a gonadotropin vaccine that had significant cross-reaction with a pituitary hormone. We proceeded to characterize the anti-oLHβ antisera and to study their endocrine action and long-term effects on safety in animals immunized for more than six years.

In this chapter I will review our findings in 63 rhesus monkeys that were immunized against oLHβ. This constitutes the most extensively studied primate model for pregnancy vaccines based on immunization against gonadotropins. Following initial investigations on effectiveness, antibody characterization, mechanism(s) of action, and reversibility, a comprehensive evaluation of long-term effects on the arterial system and tissue damage due to possible immune complex deposits was carried out. Our observations indicate that a certain degree of cross-reaction with LH not only does not cause any significant tissue damage, but may be a contributing factor to the antifertility properties of the antibodies.

Rhesus Monkey Model

Sixty-three rhesus monkeys were actively immunized in 1975 and 1976 using 50 μg of oLHβ in Freund's complete adjuvant. Booster injections were given in Freund's incomplete adjuvant at approximately six- to 12-month intervals (8,14,15). The investigations on some of the animals were carried through 1984, and three female offspring from immunized mothers are still under observation.

Antibody Response and Characteristics

All immunized monkeys developed measurable antibody titers; however, antibody response varied greatly among animals. The time course of antibody titers for ten monkeys that were studied for more than seven years is shown in Figure 20.1. For routine antibody assessment, a "rapid titer assay" was used in which the percentage of [^{125}I]-HCG bound by 2 μL of serum was determined when approximately 0.2 ng of this tracer was used (23). Six to 12 months after the primary immunization, approximately 75% of all immunized monkeys had developed antibody titers that were effective in preventing pregnancy (Fig. 20.2). Only one of 53 monkeys became pregnant when antibody titers were more than 30% or had a binding capacity higher than 0.2×10^{-7} mol/L (as measured by HCG-displacement analysis) (23). The neutralizing potency of the antisera for HCG and RhCG correlated well with binding capacity (Fig. 20.3) but not with binding affinity (22).

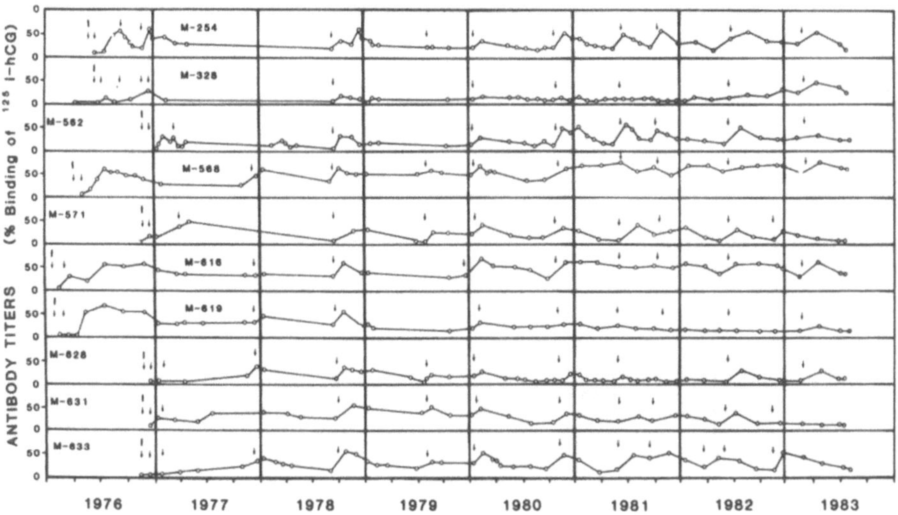

Fig. 20.1. Antibody titers (percent [^{125}I]-HCG bound) in rhesus monkeys immunized for seven years. Arrows indicate initial immunization (I) and booster injections. [From Thau et al (18).]

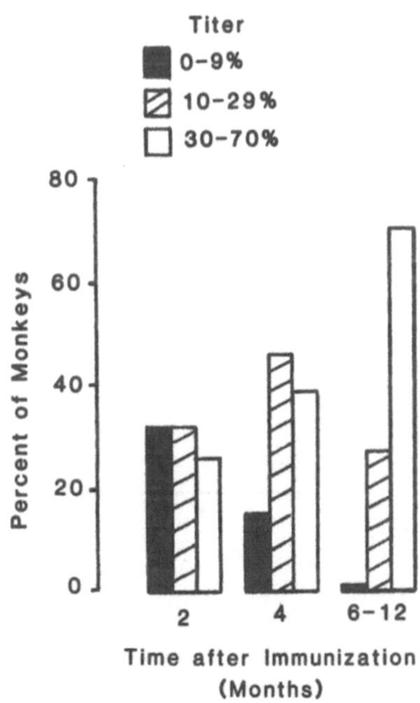

Fig. 20.2. Percentage of monkeys (N=63) reaching effective antibody titers (30%–70%) following primary immunization and booster injections. [From Yamamoto et al (21).]

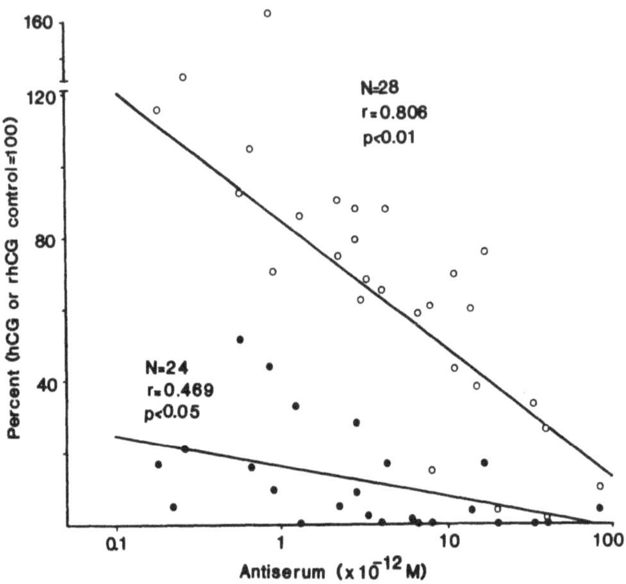

Fig. 20.3. Correlation between neutralizing activity of anti-oLHβ antiserum for HCG (o) and RhCG (●) in the in vivo mouse bioassay and the HCG binding capacity as calculated from in vitro displacement analysis. (From Yamamoto and Thau (22).]

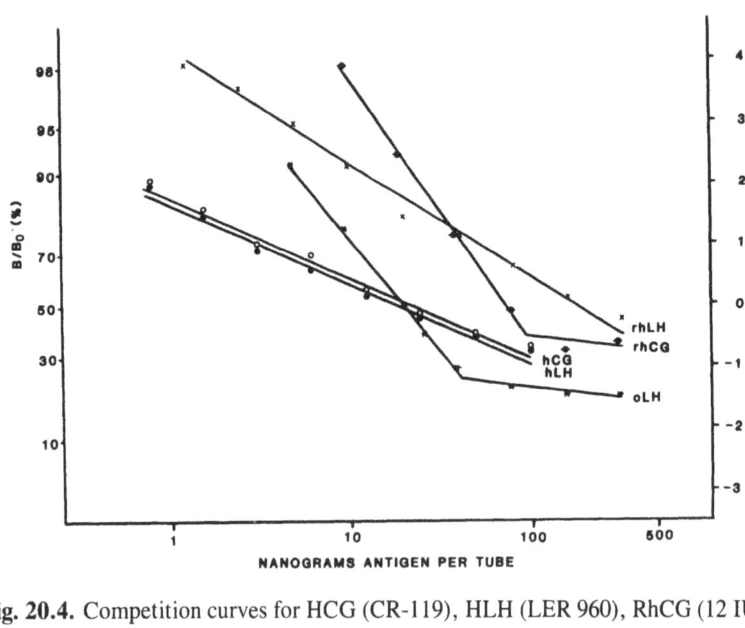

Fig. 20.4. Competition curves for HCG (CR-119), HLH (LER 960), RhCG (12 IU/mg), RhLH (LER 1909-2), and oLH (NIH S18) in an assay system using oLHβ-antiserum from a typical immunized high-titer monkey (M580). B_0, counts bound with labeled ligand alone; B, counts bound in the presence of labeled and unlabeled ligand. Dose-response curves for HCG and HLH were constructed in a homologous radioimmunoassay; iodinated HCG was used for RhCG, RhLH, and oLH. [From Yamamoto et al (21).]

Binding capacity and affinity of the antisera to HCG and HLH were calculated from homologous competition curves. Both binding capacities (per unit of each hormone) and binding affinities were usually higher for HCG than those for HLH (21). Displacement analysis (using [^{125}I]-HCG) showed that oLHβ antibodies cross-reacted extensively with RhLH (Fig. 20.4).

Effects on Fertility

Following an initial study in which 13 animals were immunized and shown to be infertile (8) as long as antibody titers remained high, 50 more rhesus monkeys were immunized and their fertility tested over a four-year period. Among nonimmunized control monkeys the pregnancy rate per mating was approximately 50%. Among immunized animals with titers above 30% binding of [^{125}I]-HCG (rapid titer assay), the pregnancy rate never exceeded 4% and was 2% overall (Table 20.1).

Normal fertility could be restored by either allowing antibody titers to decline (8) or by the administration of a progestational agent between day 4 and 40 of gestation (15). Pregnancies in low-titer animals did not end in abortions (see below).

Endocrinological Effects and Mechanism(s) of Action

Immunization did not affect regular menstrual cycles (Table 20.2), but the length of the cycle was shortened by an average of two days (Fig. 20.5). This decrease was due to shortened luteal phases. In addition, the mean plasma progesterone levels were significantly lower during the luteal phase (Fig. 20.5, cycles I and II), whereas plasma estradiol and 17α-hydroxyprogesterone concentrations were not affected (16). Following mating, there was no evidence of "corpus luteum rescue" (CG-induced prolongation of the corpus luteum) beyond that of the infertile cycle

Table 20.1. Pregnancy rates of control and immunized monkeys over a four-year period.[a]

	No. of matings	No. of pregnancies	Pregnancy rate (%)
Controls	142	72	50
oLHβ			
Titers >30%	53	1	2
Titers <30%	50	12	24
Total	103[b]	13	13

[a] Monkeys were mated at midcycle for 72 hours.

[b] The total number of immunized monkeys was 63. Some monkeys were mated when their titers were low and again several months later when their titers were high, so that the total number of matings exceeds 63.

Table 20.2. Effects of immunization on regularity of the menstrual cycle.

	No. of monkeys	No. of monkeys with regular cycles	No. of monkeys with irregular cycles	% of monkeys with regular cycles
Controls	49	21	28	43
oLHβ				
Titers >30%	27	14	11	52
Titers <30%	11	4	7	36
Total	36	18	18	50

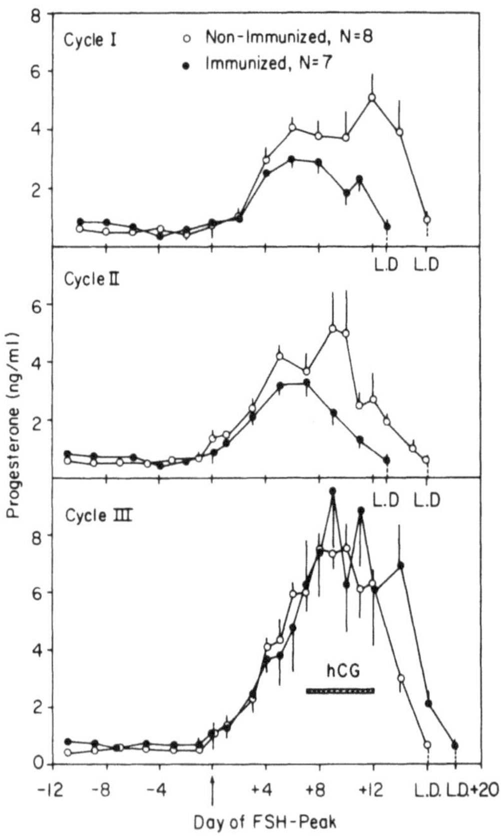

Fig. 20.5. Plasma progesterone concentrations throughout three consecutive menstrual cycles (mean ± SE; L.D., last day of the cycle). The data are synchronized to the day of the FSH peak and to the mean lengths of follicular and luteal phases. HCG was administered during cycle III. Analysis of variance showed significant differences between immunized and control monkeys for progesterone levels during the luteal phase of cycles I and II [$F = 14.6\ (1,12)$; $P < .01$] but not for cycle III. N is number of monkeys; ////, HCG treatment. [From Thau et al (16).]

even in monkeys whose pregnancy was maintained by the administration of medroxyprogesterone acetate (15,13,6).

To determine whether the failure of the corpus luteum to produce normal amounts of progesterone was due to the neutralization of RhCG by circulating oLHβ antibodies or to reduced sensitivity of the corpus luteum to gonadotropin stimulation, HCG was administered to immunized and nonimmunized animals for five days, beginning on day 17 of the cycle. Treatment with HCG increased progesterone levels equally in controls and immunized monkeys and resulted in cycles of similar length (Fig. 20.5, cycle III), demonstrating that the antifertility effect of the immunization was not due to reduced ability of the corpus luteum to respond to gonadotropin.

Follicular steroidogenesis was not impaired in immunized monkeys, as evidenced by normal serum levels of estradiol and 17α-hydroxyprogesterone (16). Serum FSH concentrations during the follicular phase were significantly elevated, which may have caused a change in the serum FSH to LH ratios and subsequently impaired folliculogenesis, leading to the shortened luteal phases. The short luteal phases are likely to be a contributing factor to the antifertility action of this vaccine, since the corpus luteum could have regressed before endogenous CG became available to rescue the corpus luteum and maintain pregnancy.

Long-Term Effects

Contraceptive methods based on long-term immunization must be assessed for their potential to induce adverse side effects, including immunopathological complications. Because of the significant cross-reaction of the anti-oLHβ antibodies with endogenous RhLH, our monkey model provided the opportunity to examine these and other aspects of vaccine safety.

It must be emphasized at this point that all monkeys had been immunized using Freund's complete adjuvant and had undergone repetitive booster injections with Freund's incomplete adjuvant. Thus, these animals had experienced a particular vigorous immunologic challenge and constitute a "worst example" for any possible adverse side effects.

Evidence on Safety From Necropsy Findings

Originally, the major purpose of our studies was the investigation of the effectiveness and mechanism(s) of action of immunization against gonadotropic hormones. Thus, while routine histopathological examinations were carried out on 48 of the 63 immunized monkeys, systematic toxicology evaluations in comparison with age-matched controls were carried out in ten of these animals. The latter had been immunized for seven years (Fig. 20.1) and were fed a hypercholesteremic diet during the last year.

Neither the gross pathology nor the organ weights revealed evidence of adverse effects from the immunization. There was a higher evidence of tuberculosis-like lesions in immunized as compared with control animals. This may have been caused by the fact that the use of Freund's complete adjuvant in the initial vaccination rendered the monkeys tuberculin-positive, so that it was impossible to identify early exposure to tuberculosis.

Histopathological examination showed that follicular cysts were present in four of the ten immunized monkeys on the special atherosclerosis diet, but not in any of the control animals. This finding may be related to the immunization and will require special attention in future studies.

Circulating Immune Complexes and Arterial Changes

A major safety issue in the development of contraceptive methods based on long-term immunization is the possible effect of circulating immune complexes (CIC) on the arterial wall. Because of the significant cross-reaction of the oLHβ-antibodies with serum LH, the serum levels of CICs were examined by Raji and Clq-assays in 24 immunized monkeys. Eleven of these were found to have CIC concentrations that were more than two standard deviations above the mean for controls in at least one of the assays used (Fig. 20.6) (17). There was also an indication that immunization against oLHβ was associated with a moderate thickening of the intima in the arteries in some of the animals (17). This prompted us to carry out a more defined study in long-term immunized monkeys and to manipulate the diet to accentuate arterial damage. A diet containing 0.33 mg/kcal cholesterol was fed for a one-year period to ten nonimmunized controls and ten

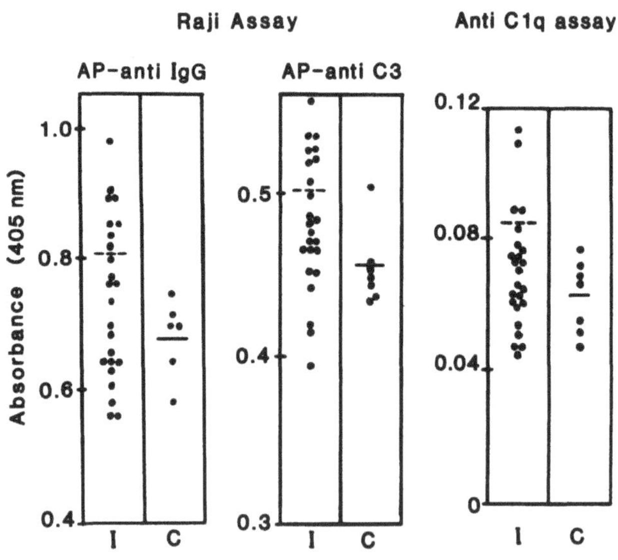

Fig. 20.6. Plasma concentrations of circulating immune complexes (absorbance at 405 nm) in 24 immunized (I) and seven control (C) monkeys. The solid line shows the control mean. The dashed line indicates control mean plus two standard deviations. Alkaline phosphatase (AP) conjugated antisera were used in the Raji assay. [From Thau et al (17).]

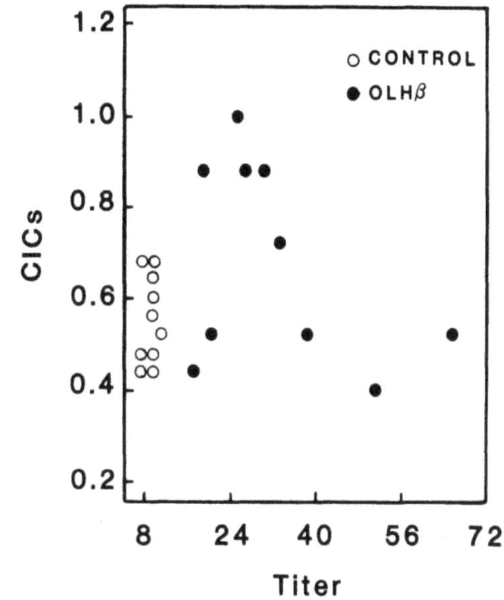

Fig. 20.7. Scatter plot for mean CIC concentrations (absorbance at 405 nm) versus mean antibody titers (percent [^{125}I]-HCG bound) during the one-year elevated cholesterol diet study for ten controls and ten immunized monkeys. [From Thau et al (18).]

monkeys that had been immunized for six years. Serum cholesterol levels increased from a mean of 161 to 394 mg/dL. CIC levels were measured monthly and varied substantially during the year the cholesterol diet was given. However, they were not significantly nor consistently higher in immunized animals as compared with controls. Mean values for CICs and antibody titers for each monkey were used for correlation analysis. No significant correlation between the two variables was observed (Fig. 20.7). Arterial characteristics such as plaque area, maximal plaque thickness, percent lesions, and percent stenosis (which can be used as indicators of the extent and/or severity of atherosclerosis), were measured in the abdominal and thoracic aortas and the coronary and carotid arteries. Statistical analysis showed no significant differences between immunized and nonimmunized groups for any of the parameters examined, indicating that long-term immunization against pregnancy had no adverse effects on the extent and/or severity of atherosclerosis in hypercholesteremic monkeys (18).

Effects on the Pituitary

LH and FSH are produced and secreted by the pituitary. The interaction of the antibodies produced by oLHβ vaccination with endogenous LH raised the question of possible adverse effects on the pituitary. We therefore investigated effects of long-term immunization on pituitary function by comparing responsiveness to the gonadotropin-releasing hormone agonist, buserelin (GnRH-A), in immunized and nonimmunized monkeys and by examining the morphology of pituitaries from both groups. Changes in serum FSH (rather than LH) levels in response to challenge with GnRH-A were used as a measure of pituitary responsiveness, since the circulating anti-oLHβ antibodies interfered with the LH radioimmunoassay in immunized monkeys. The magnitude and pattern of response were similar in immunized and control animals (Fig. 20.8), although the immunized monkeys had higher initial levels of FSH.

Histopathological studies that were carried out following seven years of immunization showed that there was no difference between the pituitaries of immunized and nonimmunized animals. The extent and frequency of immune complex deposits were also not significantly different between the two groups (unpublished observations).

Effects on Progeny

Some of the immunized monkeys became pregnant after antibody titers declined; some pregnancies occurred when titers were below 30%. One monkey had a high titer at the time of conception, and several monkeys were pregnant at the time of the primary immunization. In addition, several pregnancies were maintained in high-titer animals by injections of medroxyprogesterone acetate. Unlike the marmoset monkeys, which were shown to experience repeated abortions occurring progressively later in pregnancy as anti-HCG antibody titers declined (1), the rhesus monkeys in our study did not show any indications of abortions associated with low concentrations of circulating anti-oLHβ antibodies.

A total of 27 pregnancies occurred in immunized rhesus monkeys. Anti-oLHβ-titers at the time of mating ranged from under 10% to over 60%. Twenty-five pregnancies were carried to term. None of the infants had obvious defects at birth, although there was one stillbirth, and a set of twins died after premature delivery. The stillbirth occurred in an animal that had had a stillbirth prior to immunization. During the same time period, four stillbirths and two abortions were observed among 34 pregnant controls.

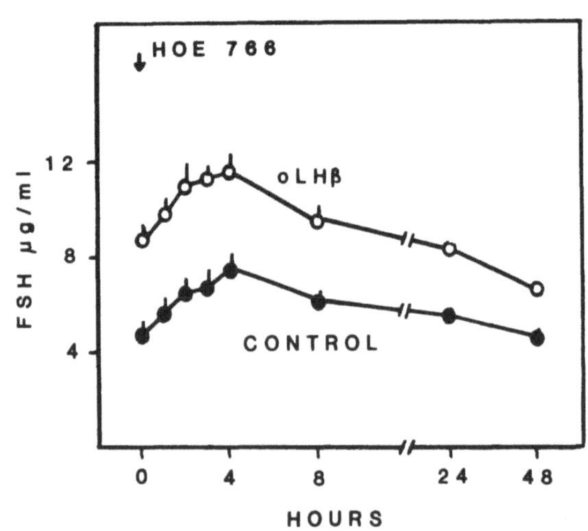

Fig. 20.8. Mean serum FSH concentration (mean ± SE) of six immunized and five nonimmunized monkeys before and following a challenge with buserelin (Hoe 766, GnRH-A).

Table 20.3. Luteal phase progesterone peaks in offspring of control and immunized monkeys.

	Progesterone in serum (ng/mL)					
	Offspring of controls			Offspring of immunized monkeys		
Monkey number	755	756	757	1638	3571	4484
January-May 1984[a]	<2.0	3.1	5.5	3.0	3.3	5.0
	<2.0	3.7	4.7	<2.0	4.0	3.0
	<2.0	5.0	4.2	<2.0	3.2	5.7
	<2.0	3.4	3.3		2.7	7.6
November 1984-January 1985[b]	13.0	8.0	9.0	3.0	7.0	9.0
	12.0	7.0	13.0	<2.0	10.0	7.0
	7.0	7.0	14.0	<2.0	9.0	11.0
September-December 1985[b]	6.1	<2.0	<2.0	<2.0	7.2	<2.0
	8.7	2.3	3.2	<2.0	6.0	<2.0
	5.4	<2.0	4.6	<2.0	5.4	4.3
	9.8	3.8	7.4	<2.0	7.0	3.3

[a] Samples taken for assay two times per week.
[b] Samples taken for assay three times per week.

Three of the offspring (all female) from immunized mothers were observed throughout puberty and into early adulthood. They started ovulatory menstrual cycles at the same time as three age-matched control animals. Two of the three offspring from immunized mothers have cycled at least as regularly as the three control animals. The third animal has shown more irregularity than the other five, but was certainly not outside the range frequently seen in nonimmunized monkeys in our colony. This animal has also shown fewer ovulatory cycles than the other five (as measured by serum progesterone levels; Table 20.3). However, it showed a positive response to a GnRH-A challenge test, indicating that its pituitary is capable to normal production and release of LH, Overall, the above observations lend reassurance that exposure of female monkeys to anti-oLHβ antibodies during life in utero does not drastically affect their endocrine development. The small number of progeny examined precludes more definitive conclusions.

Summary and Conclusions

The animal model described above represents a particularly vigorous immunologic challenge since the antigen was initially administered with Freund's complete adjuvant and, in booster injections, with Freund's incomplete adjuvant.

Since the antibodies elicited interacted strongly with RhLH, the experiment provided evidence on the safety of inducing antibodies against pituitary gonadotropins.

Three-quarters of the 63 monkeys immunized developed titers that effectively prevented pregnancy. The antifertility effect could be reversed by either allowing antibody titers to decline (which usually occurred with six to 12 months) or by providing progestational support during the first few weeks after fertilization. Ovulatory menstrual cycles continued, but luteal phases were shortened and serum progesterone levels were reduced.

The prolongation of progesterone production due to endogenous CG ("corpus luteum rescue") normally seen in pregnant monkeys was not seen in immunized monkeys that became pregnant following treatment with medroxy-progesterone acetate. These results indicate that vaccination with oLHβ impaired luteal function during the second half of the luteal phase and at the time of corpus luteum rescue. The impairment of luteal function and the shortened life span of the corpus luteum may have been caused by cross-reaction of the anti-oLHβ antibodies with RhLH. Either the impaired luteal function secondary to shortened luteal phases, or the lack of corpus luteum stimulation with RhCG, or a combination of both factors are possible mechanisms of antifertility action of the anti-oLHβ antibodies. Successful application of oLHβ as an antigen for contraception in humans may depend on relative cross-reactions of HLH and HCG with anti-oLHβ antibodies. Although findings in the monkey model are promising, only experiments in the human can positively answer the question whether oLHβ will be an effective contraceptive vaccine in the human.

The extensive toxicology studies carried out in oLHβ-immunized monkeys showed that even though several of these animals had slightly elevated concentrations of circulating immune complexes, there was no correlation between antibody titers and CICs. Investigations on the effects of immunization on atherosclerotic lesions in the coronary and carotid arteries and in the aorta showed no differences between immunized and control monkeys with respect to the arterial characteristics that were used as indicators of the extent and/or severity of atherosclerosis.

Pituitary function as measured by GnRH challenge tests was normal. Histopathological studies showed neither a significant effect of oLHβ-immunization on the frequency of immune complex deposits in the pituitary nor a significant difference in the immunocytochemical staining of the gonadotrophs between the two groups. The sexual development of three female progeny from immunized mothers was similar to that of age-matched controls.

We conclude from our findings in oLHβ-vaccinated rhesus monkeys that over the time period studied, the immunization did not cause significant adverse effects on ovarian and pituitary functions. The extent and severity of atherosclerotic lesions were not increased. Tissue damage due to immune complex deposits was not observed.

References

1. Hearn JP: *The Immunobiology of Chorionic Gonadotropin*, Hearn JP (ed): in *Immunological Aspects of Reproduction and Fertility Control*. Baltimore, University Park Press, 1980, pp 229–244.
2. Hodgen GD, Nixon WE, Vaitukaitis JL, et al: Neutralization of primate chorionic gonadotropin in radioimmunoassay and bioassay. *Endocrinology* 1973;92:705–709.
3. Hodgen GD, Tullner WW, Vaitukaitis JL, et al: Specific radioimmunoassay of chorionic gonadotropin during implantation in rhesus monkeys. *J Clin Endocrinol Metab* 1974;39:457–464.
4. Nash H, Talwar GP, Segal S, et al: Observations on the antigenicity and clinical effects of a candidate antipregnancy vaccine: β-subunit of human chorionic gonadotropin linked to tetanus toxoid. *Fertil Steril* 1980;34:328–335.
5. Sharma NC, Goel BK, Bajaj JS, et al: Metabolic, endocrine and organic functions in monkeys immunized with Pr-β-hCG-TT. *Contraception* 1976;13:201–211.
6. Spinola PG, Seidman LS, Sundaram K, et al: Impaired steroidogenesis in the luteal phase of the reproductive cycle and during pregnancy in rhesus monkeys immunized with the β-subunit of ovine luteinizing hormone. *J Steroid Biochem* 1982;16:151–156.
7. Stevens VC: Female contraception by immunization with hCG-prospects and status, in Nieschlag E (ed): *Immunization with Hormones in Reproduction Research*. Amsterdam, The Netherlands, North-Holland Publishing Company, 1975, pp 217–231.
8. Sundaram K, Chang CC, Laurence KA, et al: The effectiveness in rhesus monkeys of an antifertility vaccine based on neutralization of chorionic gonadotropin. *Contraception* 1976;14:639–653.
9. Talwar GP, Sharma NC, Dubey SK, et al: Isoimmunization against human chorionic gonadotropin with conjugates of processed β-subunit of the hormone and tetanus toxoid. *Proc Natl Acad Sci USA* 1976;73:218–222.
10. Talwar GP, Dubey SK, Salahuddin M, et al: Antibody responses to Pr-β-hCG-TT vaccine in human subjects. *Contraception* 1976;13:237–243.
11. Thanavala YM, Hay FC, Stevens VC: Immunological control of fertility: measurement of affinity of antibodies to human chorionic gonadotropin. *Clin Exp Immunol* 1978;33:403–409.
12. Thanavala YM, Hearn JP, Hay FC, et al: Characterization of the immunological response in marmoset monkeys immunized against hCGβ-subunit and its relationship with their subsequent fertility. *J Reprod Immunol* 1979;1:263–273.
13. Thau RB: *In vivo* determination of progesterone production rates, in Alexander NJ (ed): *Animal Models for Research on Contraception and Fertility*. Hagerstown, Harper & Row, 1978, pp 142–150.
14. Thau R, Sundaram K, Thornton YS, et al: Effects of immunization with the β-subunit of ovine luteinizing hormone on corpus luteum function in the rhesus monkey. *Fertil Steril* 1979;31:200–204.
15. Thau RB, Sundaram K: The mechanism of action of an antifertility vaccine in the Rhesus monkey: Reversal of the effects of antisera to the β-subunit of ovine luteinizing hormone by medroxyprogesterone acetate. *Fertil Steril* 1980;33:317–320.
16. Thau RB, Yamamoto Y, Sundaram R, et al: Human chorionic gonadotropin stimulates luteal function in rhesus monkeys immunized against the β-subunit of ovine luteinizing hormone. *Endocrinology* 1983;112:277–283.

17. Thau RB, Witkin SS, Bond MG, et al: Effects of long-term immunization against the beta-subunit of ovine luteinizing hormone on circulating immune complex formation and on arterial changes in rhesus monkeys. *Am J Reprod Immunol* 1983;3:83–88.
18. Thau RB, Bond MG, Witkin SS, et al: Lack of toxicological effects following seven years of active immunization of rhesus monkeys with the β-subunit of ovine luteinizing hormone, in Talwer P (ed): *Immunological Approaches to Contraception and Promotion of Fertility.* New York, Plenum Press, 1986, pp 25–33.
19. Thompson KW: Antihormones. *Physiol Rev* 1941;21:588–631.
20. World Health Organization: Special Programme of Research, Development and Research Training in Human Reproduction. 14th Annual Report, 1985, pp 53–61.
21. Yamamoto Y, Gunsalus GL, Sundaram K, et al: Characterization of anti oLHβ-antibodies acting as contraceptives in rhesus monkeys. I. *In vitro* binding properties. *J Reprod Immunol* 1982;4:295–311.
22. Yamamoto Y, Thau RB: Characteristics of anti oLHβ-antibodies acting as contraceptives in rhesus monkeys. II. In vivo neutralizing ability of gonadotropic hormones. *J Reprod Immunol* 1983;5:195–202.
23. Yamamoto Y, Gunsalus GL, Sundaram K, et al: Antibodies against the β-subunit of ovine luteinizing hormone can decrease the clearance of human chorionic gonadotropin in rhesus monkeys. *Am J Reprod Immunol* 1984;5:164–170.

CHAPTER 21

Safety Evaluation of HCG Vaccine in Primates: Autoantibody Production

Noel R. Rose, C. Lynne Burek, and Julian P. Smith

Introduction

The immediate objective of this investigation was to evaluate the possible production of autoantibodies after immunization of baboons with a vaccine prepared from the C-terminal peptide of the β- chain of human chorionic gonadotropin (HCG). In a broader sense, however, the project has served as a paradigm of the immunological problems that arise following immunization with small-chain synthetic peptides. To produce a protein antigenic stimulus, small peptides must be conjugated to a protein carrier and incorporated into an adjuvant. Each component of this complex vaccine mixture has the potential for inducing autoantibody production or of enhancing the level of preexisting natural autoantibodies.

Materials and Methods

Immunization

A 36 amino acid C-terminal peptide of the β- chain of HCG was used as the antigen in order to reduce possible cross-reactions with other peptide hormones that share sequences with HCG. The peptide was conjugated to diphtheria toxoid and incorporated into an oil vehicle, squalene:arlacel, and mixed with muramyl dipeptide as adjuvant.

Two sets of animals were studied. The first, a short-term experiment, consisted of four groups of eight baboons, each receiving, respectively, saline; vaccine conjugate, oil, and no adjuvant; vaccine conjugate, oil, and adjuvant; and vaccine conjugate, oil, and ten times concentrated adjuvant. Serum samples were collected prior to immunization and at ten weeks and 20 weeks after the start of immunization. The second experiment comprised four groups of baboons, with six animals receiving saline; six animals, oil alone; six animals, diphtheria toxoid, oil, and adjuvant; and 15 animals, vaccine conjugate, oil, and adjuvant. The animals in this study received a total of six injections at two-week intervals. Serum samples were collected before the first injection and at biweekly intervals until termination of the experiment.

Serological Studies

The main test procedure for demonstrating autoantibodies was indirect immunofluorescence using frozen sections of rat or baboon tissue, and antimonkey immunoglobulin conjugated to fluorescein isothiocyanate. All tests were performed in duplicate, and the reactions were evaluated "blindly" on an arbitrary scale from 0 to 4. In order to identify the specific cells with which autoantibodies reacted, a double-labeling technique was used in some experiments. Finally, absorption experiments were carried out by addition of soluble antigen to diluted antiserum or by adding antigen coupled to Sepharose 4B beads.

Results

Nonorgan-Specific Antibodies

Preexisting Antibodies

A major finding in this investigation was the existence of naturally occurring antibodies in baboon sera, especially antibodies to smooth muscle and striated muscle. The frequency of preexisting antibodies was evaluated using both rat and baboon tissue substrates, and results are presented in Tables 21.1 and 21.2. Using the rat tissue substrate, each of the four groups of animals had a small number of suspicious, weak, or even moderate reactions to smooth muscle prior to immunization. Fewer animals showed initial reactivity to striated muscle. The reactivity of preimmunization samples to baboon smooth- and striated-muscle tissue substrate was more pronounced than with the rat substrate. There are at least two possible explanations for this observation: 1) The autoantibodies are somewhat species restricted so that baboon tissue is a more suitable indicator or 2) for technical reasons, negative reactions are harder to distinguish using baboon tissue substrate. Both of these explanations appear to be true in part.

Table 21.1. Frequency of smooth-muscle and striated-muscle antibody in preimmunization baboon sera.

		Rat tissue substrate							
		Smooth-muscle antibody				Striated-muscle antibody			
Group	(N)	Neg	Susp	Weak	Mod	Neg	Susp	Weak	Mod
Saline	(6)	4	1	1		5	1		
Arlacel	(6)	5	1			6			
Diphtheria toxoid	(6)	3	2	1		5			1
HCG peptide	(15)	12	2		1	12	3		

Neg, negative; Susp, suspicious; Weak, weak; Mod, moderate.

Table 21.2. Frequency of smooth-muscle and striated-muscle antibody in preimmunization baboon sera.

		Baboon tissue substrate							
		Smooth-muscle antibody				Striated-muscle antibody			
Group	(N)	Neg	Susp	Weak	Mod	Neg	Susp	Weak	Mod
Saline	(6)	2	4			1	4	1	
Arlacel	(6)	3	3			2	3	1	
Diphtheria toxoid	(6)	5		1		3	2	1	
HCG peptide	(15)	7	5	3		5	8	2	

Neg, negative; Susp, suspicious; Weak, weak; Mod, moderate.

Autoantibodies Produced Following Immunization With HCG Subunit Vaccine

The data reported here were obtained from rat substrate. Experiments were duplicated on baboon substrate, and the results were comparable.

Smooth-Muscle Antibodies

None of the animals injected with saline alone showed a significant rise in antibody during the observation period, although there were occasional random changes in antibody levels as judged by immunofluorescence. Of six animals injected with arlacel and followed over a long period of time, two showed no response whatever (Fig. 21.1A). Two other animals showed slight increases in smooth-muscle antibody after the first or second injection. These increases persisted for the remainder of the study. Two other animals demonstrated a very gradual increase in the antibody level until the fourth or fifth injection, at which time there was substantial rise in antibody activity (Fig. 21.1A).

Of the animals that received diphtheria toxoid as well as squalene:arlacel and muramyl dipeptide adjuvant, only one animal showed no response (Fig. 21.1B); another developed very slight but persistent reactivity to smooth muscle. Four baboons developed increased levels of smooth-muscle reactivity soon after the start of injections. One of them responded weakly, but antibody level soon declined to preinjection status. Three of the four developed moderate levels of smooth-muscle antibodies. Overall, the response of this group was one of quick rise in smooth-muscle antibody during the injections. The antibodies reached a plateau or even declined after cessation of the injections. There are, inter alia, marked individual variations. A sampling of the baboons immunized with complete vaccine comprising arlacel, diphtheria toxoid, muramyl dipeptide, and HCG subunit is illustrated in Figure 21.1C. All the animals exhibited an increase in smooth-muscle reactivity to varying degrees during the course of the experiment. In most cases the antibody tended to decline after the cessation of injections.

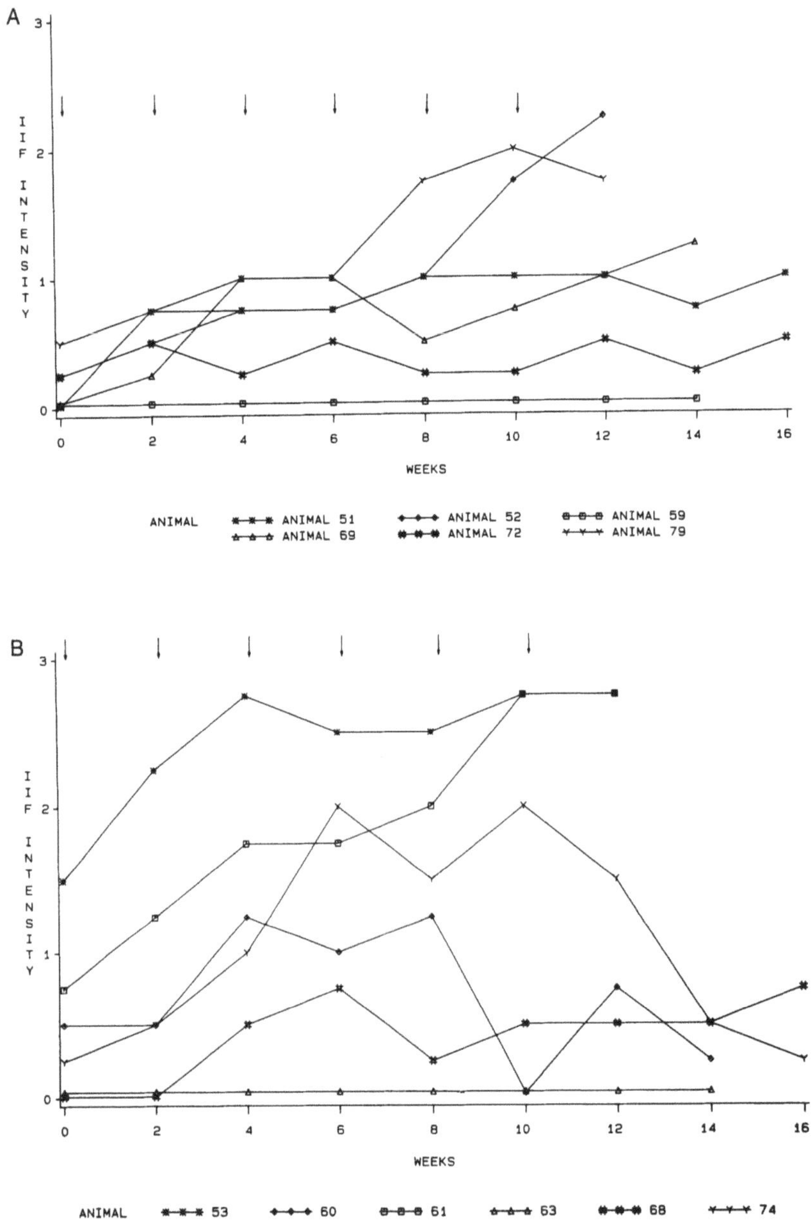

Fig. 21.1. Indirect immunofluorescence test for smooth-muscle antibody on rat tissue substrate. Fluorescence intensity was evaluated for sera at 1:10 dilution. Samples were taken prior to immunization (0 weeks) and then every two weeks until the termination of the experiment. The arrows indicate the time of immunizations. Animal immunizations–**A**: squalene:arlacel; **B**: squalene:arlacel, diphtheria toxoid, and muramyl dipeptide adjuvant; **C**: complete vaccine, consisting of squalene:arlacel, diphtheria toxoid, muramyl dipeptide adjuvant, and HCG peptide.

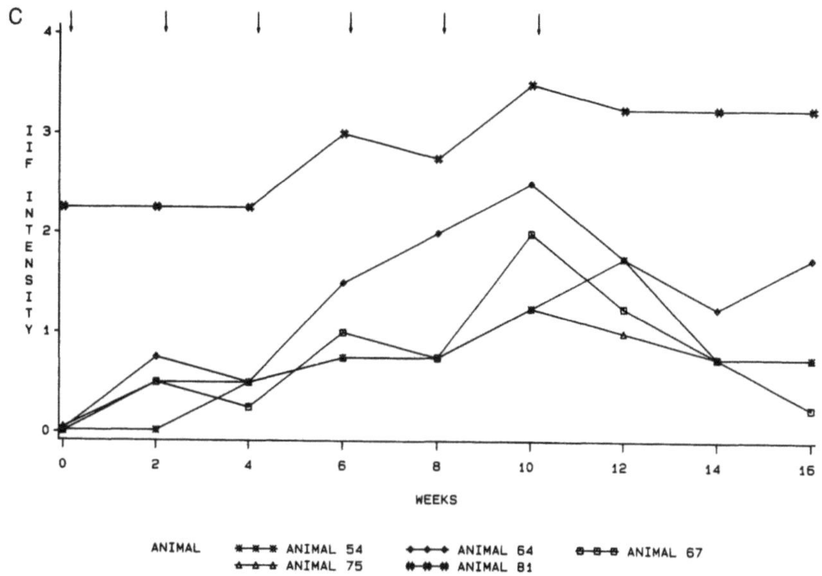

Fig. 21.1. *(Continued)*

Striated-Muscle Antibodies

Cardiac tissue was used as substrate for striated-tissue antibodies. None of the group injected with saline showed any significant rise in reaction with striated muscle. Several of the animals receiving arlacel developed weak reactions, but antibody was not detected until the animals had been given at least three injections (Fig. 21.2A). Among the six animals given diphtheria toxoid with arlacel and adjuvant, two produced virtually no striated-muscle antibody, three exhibited small but definite increases in striated-muscle reactivity, and one animal produced a high level of antibody. Significantly, that animal started with a high initial level (Fig. 21.2B). A sample of five animals receiving the complete vaccine is illustrated in Figure 21.2C. The pattern seems to be a general rise in antibody level for ten or 12 weeks followed by a decline after immunizations were stopped.

Absorption Experiments

Whether carried out with free antigen or antigen coupled to beads, β-HCG peptide reproducibly absorbed striated-muscle antibody from all serum samples. In contrast, smooth-muscle antibody was mainly absorbed by addition of diphtheria toxoid, while the reaction to striated muscle was unaffected. These findings suggest that striated-muscle antibodies are attributable to the HCG subunit component of the vaccine, whereas smooth-muscle antibodies are related to the diphtheria toxoid.

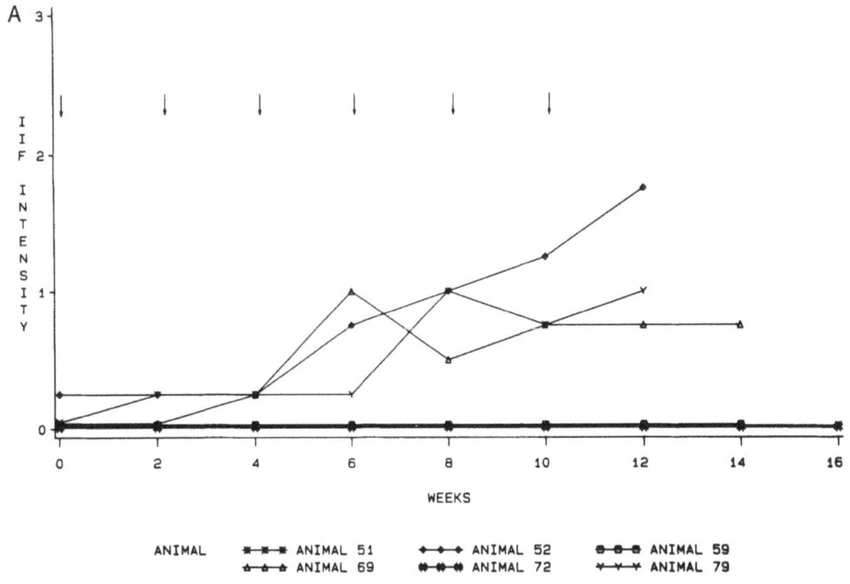

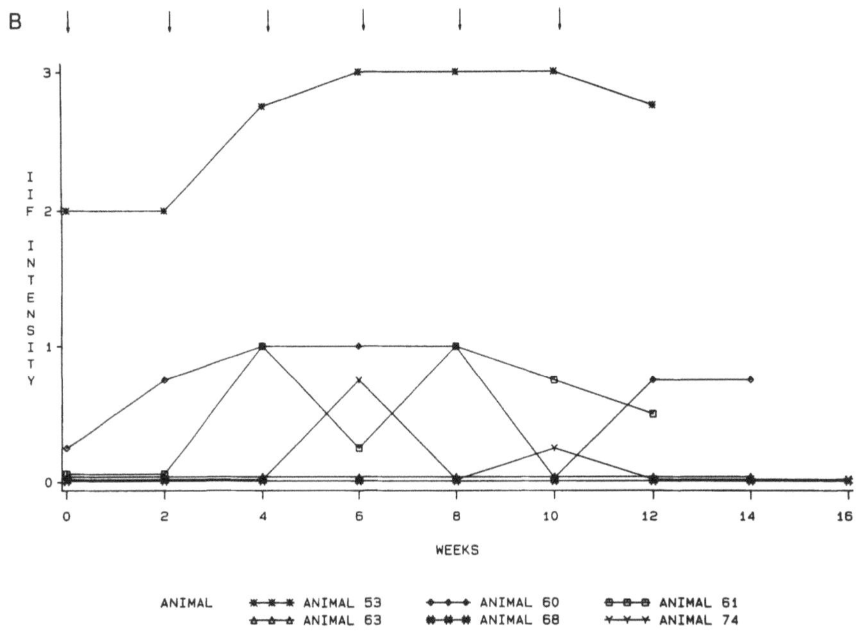

Fig. 21.2. Indirect immunofluorescence test for striated-muscle antibody on rat tissue substrate. Fluorescence intensity was evaluated for sera at 1:10 dilution. Samples were taken prior to immunization (0 weeks) and then every two weeks until the termination of the experiment. The arrows indicate the time of immunizations. Animal immunizations – **A**: squalene:arlacel; **B**: squalene:arlacel, diphtheria toxoid, and muramyl dipeptide adjuvant; **C**: complete vaccine, consisting of squalene:arlacel, diphtheria toxoid, muramyl dipeptide adjuvant, and HCG peptide.

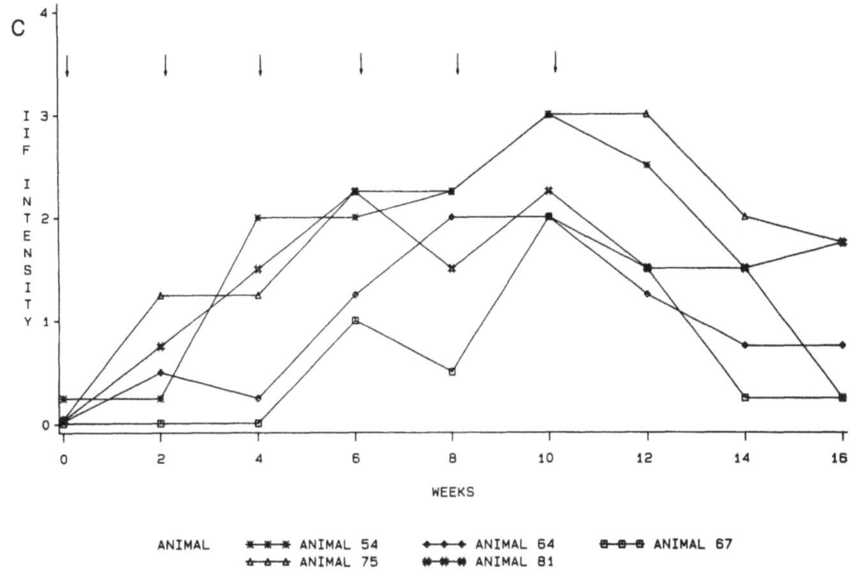

Fig. 21.2. *(Continued)*

The Antinuclear and Mitochondria Antibodies

None of the baboon sera showed any remarkable reaction with nuclear antigens or mitochondria.

Organ-Specific Antibodies

Antibodies reactive to parietal cells of the rat stomach were found in the sera of a few of the animals receiving complete HCG vaccine. However, one animal in the saline group also showed parietal cell antibodies. Therefore, we regard these results as inconclusive. No antibodies to adrenal or ovary were found in any of the animals. A few sera showed reactions with thyroid microsomes using baboon thyroid as substrate but were negative when repeated on alternative baboon substrate. These reactions were weak and did not show an association with any particular component of the vaccine. No antibodies were found to typical salivary gland or epidermal antigens. A radioimmunoassay performed by Dr. Helene Rauch at Wayne State University failed to detect any antibodies to acetyl choline receptor in the sera.

Antibodies to pancreatic islet cells were found in sera from three of the baboons immunized with the complete vaccine. Double-labeling experiments showed that the reactive cells were the somatostatin-producing cells of the pancreatic islets. However, absorption with somatostatin did not reduce the reaction. No reactions were found with the insulin-secreting or glucagon-secreting cells of the pancreas. The HCG peptide removed antibody activity to the islet cells. We conclude, therefore, that the baboon sera contained antibodies to antigen of somatostatin-secreting cells of the pancreas islet but not to the secreted hormone. This antibody was predominantly IgG and could bind complement.

A number of sera exhibited reactivity to baboon pituitary tissue. However, the pituitary cells reacting with the baboon sera have not yet been identified. The reactivity does not appear to be related to prolactin or to growth-hormone-secreting cells.

Discussion

Naturally occurring autoantibodies to many tissues are present in sera of normal adult baboons. Whether this results from genetic predisposition to autoimmune responses or to a response to external cross-reacting stimuli is unknown. However, the presence of such antibodies must be taken into account in interpreting the results of an investigation such as the one described herein.

There is considerable variability among animals in the development of autoantibodies following immunization. The injection of saline by itself apparently induces a weak antibody response in a few animals. It appears that arlacel enhances this response, suggesting that tissue damage produced by the injection itself might trigger a transient autoimmune response in genetically predisposed animals. These autoantibodies may be a mechanism for natural elimination of damaged tissue products.

Absorption studies have shown definitely that the smooth-muscle antibody is distinct from striated-muscle antibody. However, there is a general correlation between smooth- and striated-muscle antibody in individual animals, both with respect to their presence and their relative titers. This finding suggests that certain baboons are more apt to produce autoantibodies of the nonorgan-specific variety. The smooth-muscle antibody develops primarily in response to diphtheria toxoid and can largely be absorbed by diphtheria toxoid. The striated-muscle antibody results primarily from the HCG in the vaccine mixture and is largely absorbed by HCG peptide. Both of these antibodies are effectively absorbed by the complete vaccine conjugate. This observation implies that there are unique sites on the conjugate to which the animals develop antibodies. Presumably the sites represent conformational determinants produced during conjugation of the HCG to the carrier protein.

Antibody levels in most animals decrease once the injections have ceased. This finding indicates that the responses are related to active immunization. In a few animals, responses remain elevated. These animals may be genetically predisposed to autoimmune responses. Our study of baboons immunized in past experiments showed that antibodies to smooth and striated muscle can persist as long as four years.

One of the most important issues in the development of HCG vaccine is the possible appearance of organ-specific antibodies, particularly to hormone or hormone-producing cells. Antibody responses to several tissues were seen in animals receiving the complete vaccine. We found that absorption with the HCG subunit removes virtually all of this antibody activity. The most prominent organ-specific antibody is directed to the somatostatin-secreting cells of the pancreatic islet. This antibody does not react with soluble somatostatin itself. Since there seems to be no sequential homology between the HCG subunit and somatostatin, we assume that this cross-reaction is due to a shared conformational determinant between these two molecules. Finally, we have detected reactions with a few cells of the anterior pituitary. However, we have not been able to identify the cells responsible for this reaction.

This study has taught us many lessons about the possible autoimmune sequelae of immunization with subunit vaccines. We have found that the use of adjuvants by themselves often increases the levels of naturally occurring autoantibodies, particularly those to smooth and striated muscle. The protein carrier, diphtheria toxoid in this case, may also share determinants with autologous components and give rise to more autoantibodies. Finally, the primary antigen, the HCG subunit, induces antibodies to a number of antigens, including both organ-specific and nonorgan-specific antigens. The fact that one cannot predict the target of these reactions is worrisome.

On the other hand we have obtained no evidence that the autoimmune responses described in this chapter are indicators of a disease process. The immunized baboons have been examined thoroughly by both histological and endocrinological methods, with no evidence of induced abnormality found. We do not feel, therefore, that production of autoantibodies by itself is a contraindication to careful testing of an HCG subunit vaccine. It is important, nonetheless, that the presence of autoantibodies be documented in human as well as in animal experiments. The possibility that prolonged immunization in genetically predisposed individuals might lead to immunopathic complications must remain a consideration.

Acknowledgments. The HCG vaccine used in this study was prepared under the direction of Dr. Vernon Stevens of Ohio State University, who also carried out the immunization in baboons. He provided us with periodic serum samples from the immunized animals as well as baboon tissue for immunological study.

This project was supported by a grant from the Special Programme of Research Development and Research Training in Human Reproduction, World Health Organization, Geneva.

CHAPTER 22

Potential of LH Receptor as an Immunocontraceptive Vaccine: Studies in the Baboon (Papio)

Brij B. Saxena and Mukul Singh

Introduction

Among the new approaches, the immunological regulation of fertility has recently received considerable attention (7,14,15). Immunization against antigens present in the ovum, sperm, and placenta (1), especially the β-subunit of the chorionic gonadotropin, has demonstrated a potential of an immunocontraceptive vaccine (15). We have examined the concept of active immunization against lutenizing hormone (LH) receptor (12) on the ovarian function of female baboons (Papio), as shown in Figure 22.1. Our observations have indicated that LH receptor has a potential for the development of a contraceptive vaccine (13).

Materials and Methods

LH Receptor

The LH receptor was isolated from bovine corpora lutea as described in Figure 22.2. The LH receptor is a glycolipid-protein that is soluble in detergent (4,5,10). The LH-receptor polymer consists of repeating units of disulfide-linked oligomers. The molecular assembly of the receptor consists of a dimer of approximate molecular weight ratio (M_r) 280K, which is composed of two identical monomers of M_r 120 to 140K as determined by gel-permeation chromatography and sodium dodecyl sulfate (SDS)-gel-electrophoresis. Each monomer of the LH receptor is dissociated into two nonidentical, disulfide-linked subunits of M_r 85K and 38K (3,11). The LH receptor has a binding capacity range of 2000 to 3000 pm/mg and a dissociation constant (K_d) of approximately 1.0 × 10 liter M^{-1}. The hormone receptor complex is capable of stimulating adenylate cyclase via G-protein to convert adenosine triphosphate (ATP^{32}) to $cAMP^{32}$ in vitro (9,11). The LH receptor is species nonspecific. However, the production of antibodies against bovine LH receptor in other species, such as the rabbit, mouse, and baboon circumvent the barrier of immunological tolerance inherent in the use of isoantigens in the development of a contraceptive vaccine. The antibodies

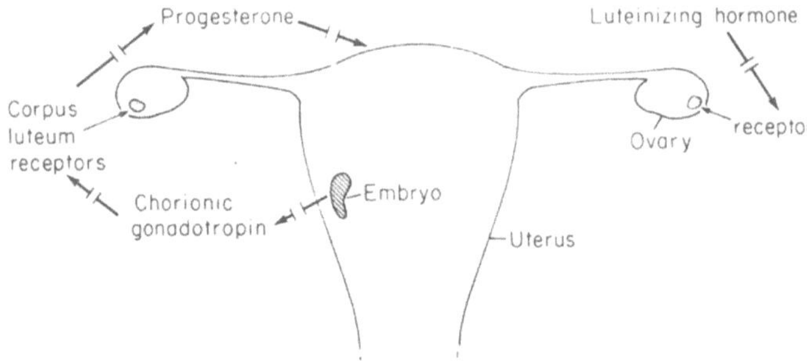

Fig. 22.1. The folliculogenesis and corpus luteal functions in the primate are dependent upon luteotropic stimuli from the pituitary luteinizing hormone (LH) and from chorionic gonadotropin (CG) from the blastocyst, respectively. Hence, an immunological block against LH receptors, which are common with CG sites, can retard follicular growth and *prevent* pregnancy.

against the LH receptor bind to membrane receptor as shown by immunofluorescence studies (8) and thus the antibodies inhibit the binding of ^{125}I-HCG to the gonadal membrane receptor, inhibit the production of testosterone production by HCG-stimulated rat Leydig cells (6), as well as inhibit the production of progesterone by granulosa cells in vitro (2–5,9–11).

LH-Receptor Implants

The implants containing LH receptor were prepared as follows. Silastic tubing was cut into 3-mm × 10-mm pieces. An electric needle heated at 90°C was used to pierce ten holes in the tubing to allow the material to pass through the silastic tube. The inner volume of the silastic tube was calculated to be approximately 250 μL . The receptor solution was concentrated and sterilized by ultrafiltration through an 0.45 filter. The LH receptor was suspended at a concentration to contain 2.5 mg of receptor in 250 μL of 10 mmol/L Tris-HCl buffer of pH 7.5 containing 0.5% Triton X-100. A sufficient number of silastic tubes were subjected to steam sterilization and were then placed in a vial in an upright position tightly against each other (Figure 22.3). The sterile receptor suspension was poured into the vial to cover the silastic tubes. Air bubbles trapped in the tubing were removed by suction with the aid of a syringe to allow the silastic tubes to be completely filled with the receptor solution. The receptor-filled silastic tubes were lypholized under sterile conditions. Each silastic tube thus contained 2.5 mg of the LH receptor. The implants were stored individually in sterile vials at 4°C in a desiccator under sterile conditions.

Implantation of LH-Receptor Implants

Three adult, normally cycling baboons of ages 5 to 8 years and body weight of 10 to 20 kg were immunized against LH receptor. Disposable preloaded Teflon

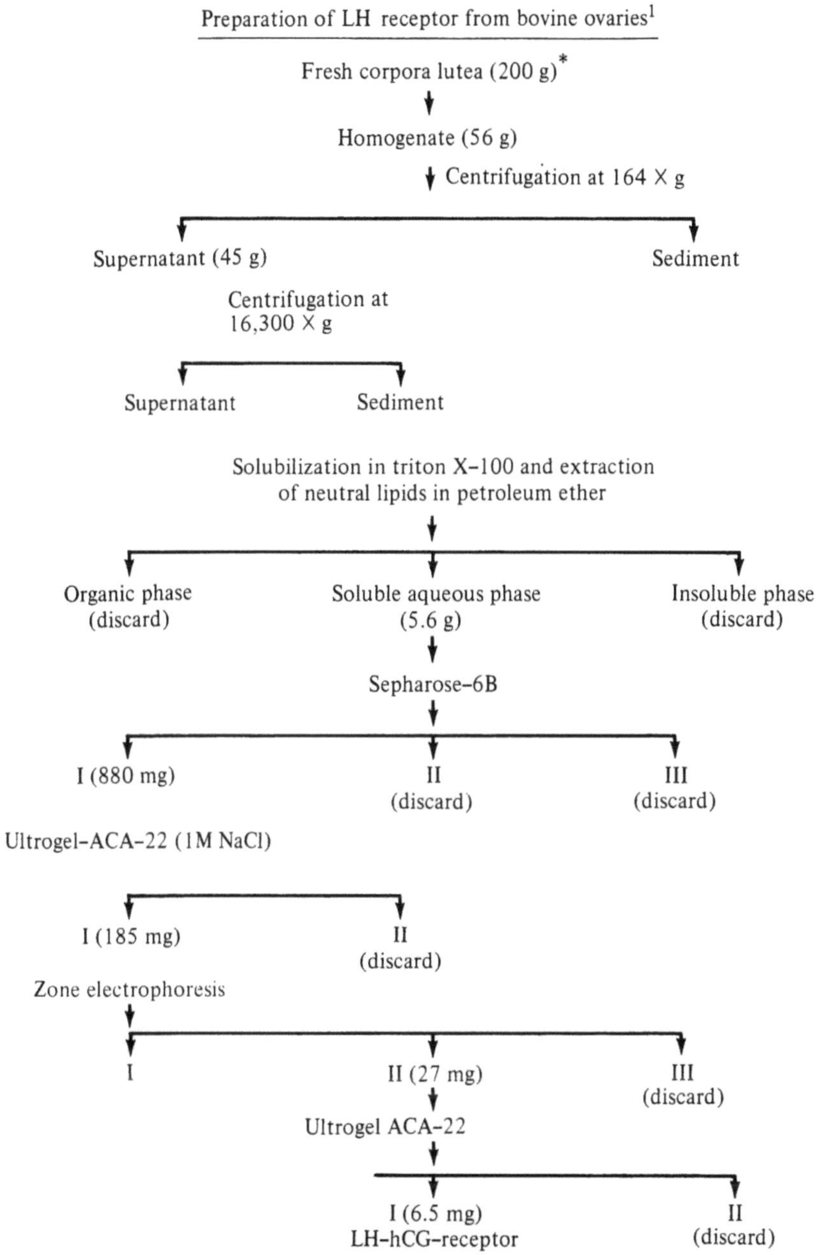

[1] Prepared every two weeks
*Protein

Fig. 22.2. Purification of LH receptor from bovine corpora lutea.

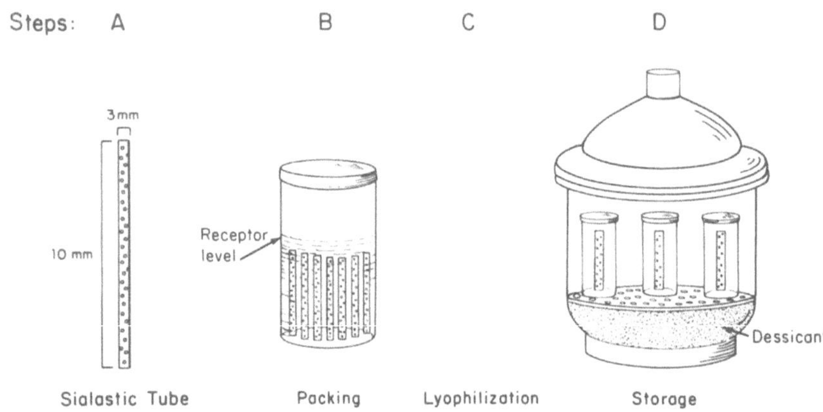

Fig. 22.3. Preparation of LH-receptor implant.

Harman trocars were used to place the implantation of LH receptor under the skin (Figure 22.4). One ml of ketamine was injected to anesthetize the baboons. A portion of the skin on the back of the baboon was shaved and cleaned with poridone-iodine and alcohol. A 3 mm incision was made in the skin. The stylet of the trocar was introduced through the incision, approximately 3 cm under the skin. The silastic implant preloaded in the stylet was inserted by the plunger, and the trocar was withdrawn. The incision was closed by one superficial stitch with 5-0 black braided silk. The implant was removed under general anesthesia. The site of the implant was shaved and prepared as described above. A 3-mm incision was made. The implant was freed of the surrounding tissue and was carefully removed with the aid of forceps. The incision was closed by one stitch as described above. The tissue around the implant was sent for histopathological evaluation at Research Animal Resources Center, Cornell University Medical Center. Aliquots of 10 mL blood were collected by venipuncture biweekly. The antibody titers were determined by the inhibition of testosterone production by

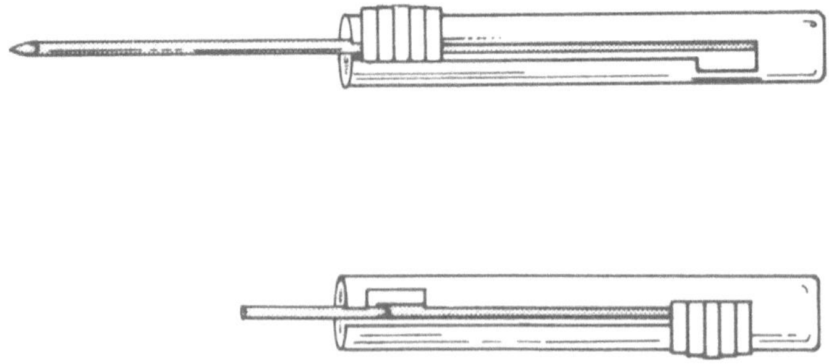

Fig. 22.4. Disposable preloaded Harman injector. (Reproduced with permission of Dr. S. Mitchell Harman, Gerontology Research Center, Baltimore, MD.)

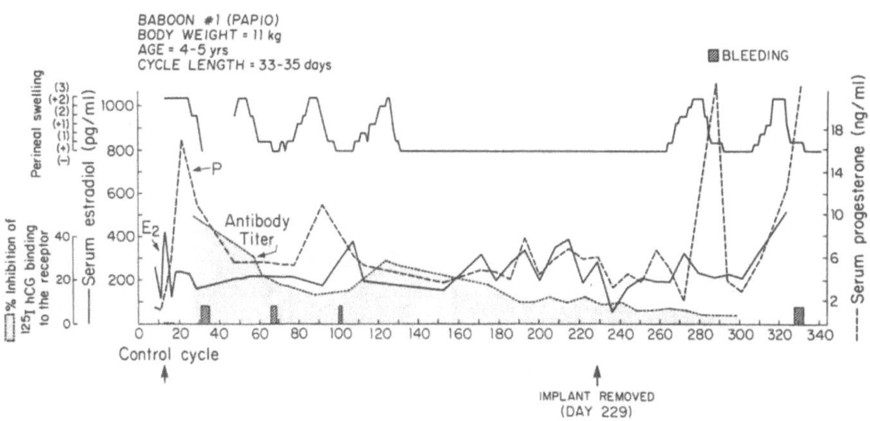

Fig. 22.5. Cyclic changes in the perineal swelling, serum estradiol (E_2), progesterone (P) levels, menstrual bleeding patterns, and antibody titers in baboon #1.

rat Leydig cells stimulated by HCG. Serum levels of estradiol (E_2), and progesterone (P) were determined by radioimmunoassay. Most of the follicle-stimulating hormone (FSH) and LH levels in the serum were determined by radioimmunoassay in the laboratory of Dr. Vernon Stevens at The Ohio State University Hospital (Columbus, Ohio). Serum samples were also analyzed for metabolites before and after immunization. The changes in perineal turgescence, menstrual bleeding patterns, and cycle lengths were recorded daily during the control cycle, after implantation, and after the removal of the LH-receptor implant. The baboons were regularly observed for any adverse effects throughout the study.

Results

Baboon #1, of 4 to 5 years of age with a body weight of 11 kg and cycle length of 33 to 35 days, showed typical midcycle perineal swelling of 3+ consistent with midcycle ovulatory surge of E_2 levels of 420 pg/ml (Figure 22.5). The luteal phase serum P levels of 17.5 ng/mL followed by menstrual bleeding on day 33 to 34 were indicative of a fertile cycle. An LH-receptor implant was placed under the skin on day 13 of the control cycle. The antibody against the receptor was detected on day 25 of the cycle. In the subsequent cycle, the midcycle E_2 surge and luteal phase P levels were significantly lower accompanied by a decrease in the duration of midcycle perineal turgescence. However, baboon #1 had normal menstrual bleeding. There was a consistent decrease in the duration of perineal swelling; E_2 level was at 300 pg/mL and progesterone level at 10 ng/mL serum during the next cycle, which was significantly lower than that in the control cycle. The effects of persistent LH-receptor antibody became more dramatic in subsequent

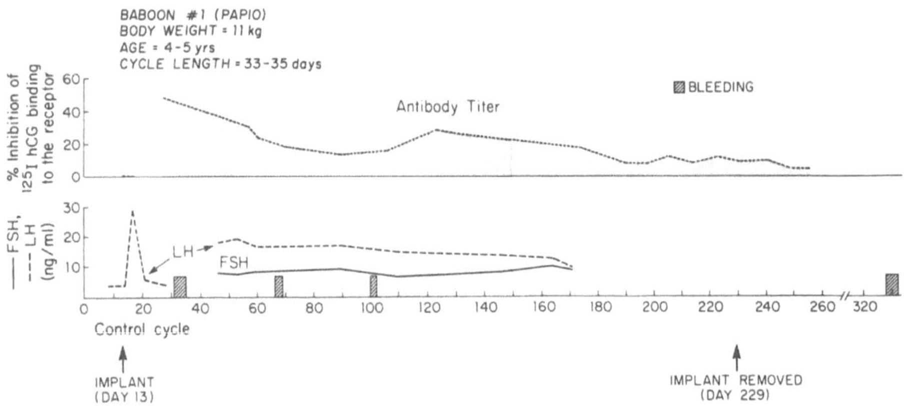

Fig. 22.6. Top: Receptor antibody titers. Bottom: Serum levels of follicle-stimulating hormone (FSH), luteinizing hormone (LH), and menstrual bleeding pattern in baboon #1.

cycles. For example, the perineal turgescence disappeared while the serum E_2 and P levels remained at lower levels and there was no menstrual bleeding during the entire period of implant in situ. These observations were suggestive of a state of infertility. As shown in Figure 22.6, there was a normal midcycle serum LH surge during the control cycle. However, in the presence of LH-receptor antibody, the midcycle LH and FSH surge disappeared and the levels of serum LH and FSH remained slightly elevated and sustained. The presence of steroids suggested follicular growth, however, at a lower rate. It was noteworthy that there was no total

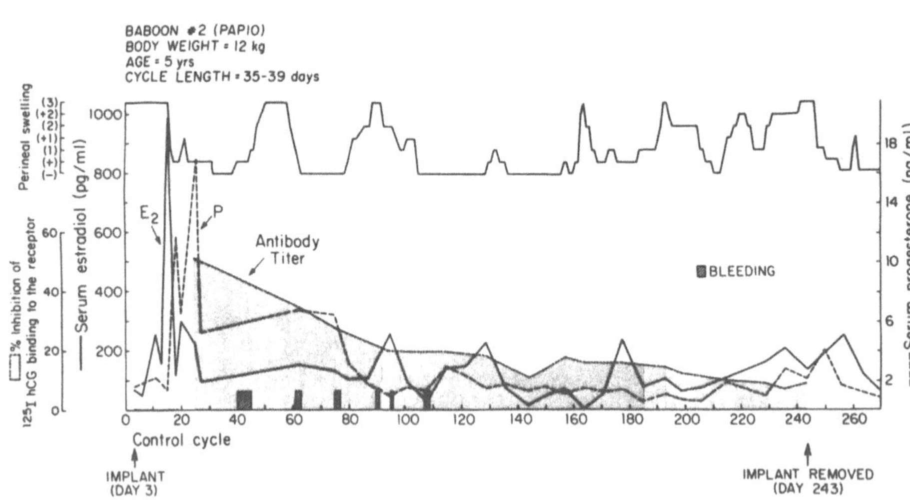

Fig. 22.7. Cyclic changes in the perineal swelling, serum estradiol (E_2) and progesterone (P) levels, menstrual bleeding pattern, and antibody titers in baboon #2.

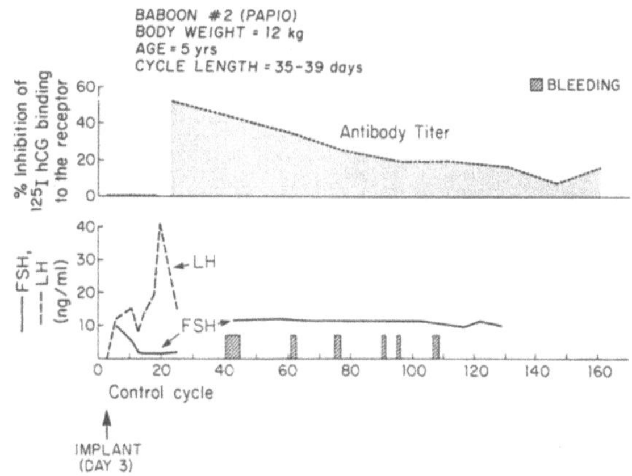

Fig. 22.8. Top: Receptor antibody titer. Bottom: Serum levels of follicle-stimulating hormone (FSH), luteinizing hormone (LH), and menstrual bleeding pattern in baboon #2.

deprivation of ovarian steroids. The absence of menstrual bleeding was indicative of inadequate ovulation due to poor folliculogenesis or anovulation which consequently resulted in the formation of inadequate corpus luteum. The LH-receptor implant was removed on day 229 while the antibody titers were declining. As shown in Figure 22.5, 30 to 35 days after the removal of the implant, the perineal swelling coincident with normal cyclical levels of serum estradiol and progesterone levels began to appear again. The baboon, after 230 days of ammenorrhea, returned to normal menstrual bleeding, suggesting the reversibility of fertility status.

Baboon #2, with a body weight of 12 kg, received the LH-receptor implant in the follicular phase on day 3 of the control cycle with normal 3+ perineal turgescence as well as midcycle preovulatory surge of E_2 and luteal phase elevation of P levels in the serum (Figure 22.7). The baboon had normal menstrual bleeding on day 40. The antibody titers were detected in the serum sample collected on day 25 of the cycle, i.e., 22 days after the LH-receptor implant was placed subdermally. In the subsequent cycle there was a consistent decreased in both the intensity and the duration of perineal turgescence that was also accompanied by lower levels of serum E_2 and P, shorter cycle lengths, and irregular menstrual bleeding until day 60 postimplant. Subsequently the baboon remained in complete amenorrhea with consistently lower levels of serum E_2 and P, as well as impaired perineal turgescence. The implant was removed on day 244 while the antibody titers were in the declining phase. Subsequently the perineal turgescence began to resume the cyclic pattern together with cyclic elevation in the serum E_2 and P levels. The baboon eventually returned to normal menstrual and endocrine pattern typical of a fertile cycle. As shown in Figure 22.8, baboon #2 also showed normal follicular elevation in serum FSH and midcycle LH surge. Subsequent to LH-receptor implant the FSH levels remained elevated until day 130.

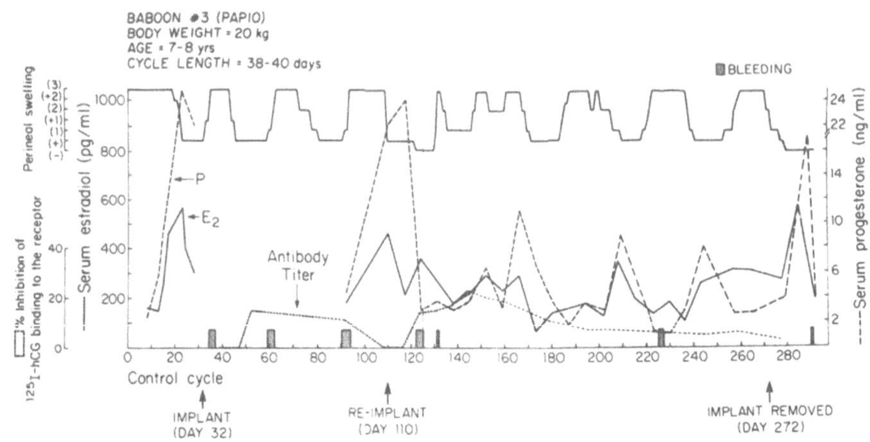

Fig. 22.9. Cyclic changes in the perineal swelling, serum estradiol (E_2) and progesterone (P) levels, menstrual bleeding patterns, and antibody titers in baboon #3.

Baboon #3, with a body weight of 20 kg, received the subdermal LH-receptor implant during the luteal phase on day 32 of the control cycle with normal patterns of perineal swelling, midcycle E_2 surge, and luteal phase elevation of P in the serum (Figure 22.9). The baboon had normal menstrual bleeding on day 38. The antibody was detected in the serum sample collected on day 48. The baboon again had menstrual bleeding on day 60, with slight impairment in the perineal swelling in each cycle. The antibody titers also declined. The third cycle postimplant showed normal perineal swelling and serum levels of E_2 and P. It was discovered that the implant had been pulled out by the baboon; therefore, antibody titers declined and the baboon retained a profile of a fertile cycle. The LH-receptor implant was replaced on day 110. The antibody was again

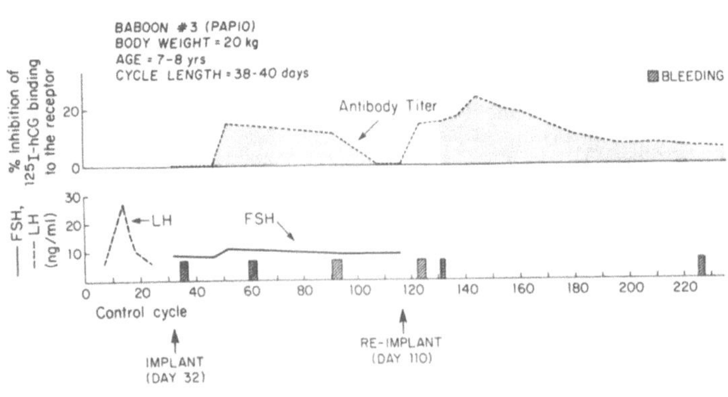

Fig. 22.10. Top: Receptor antibody titer. Bottom: Serum levels of follicle-stimulating hormone (FSH), luteinizing hormone (LH), and menstrual bleeding pattern in baboon #3.

detected in the serum collected on day 122 which was accompanied by a reduction in the intensity and duration of perineal turgescence, the occurrence of intermenstrual bleeding, and a decrease in serum E_2 and P levels. During the subsequent period, baboon #3 showed increase in cycle length with slight impairment in perineal turgescence and slightly lower levels of serum E_2 and P. The implant was removed on day 272 which in the subsequent cycle showed almost normal levels of serum E_2 and P. The baboon had normal menstrual bleeding on day 290 and subsequently returned to a normal pattern of fertile cycle. As shown in Figure 22.10, baboon #3 showed a normal midcycle LH surge in the serum. Levels of serum FSH during the postimplant period were elevated until day 118. Serum metabolites during pre- and postimplant periods did not change significantly (Table 22.1).

Discussion

The effects of immunization against LH receptor via implant on the three baboons were similar with minor individual variations. For example, after placement of the LH-receptor implant during the midcycle in baboon #1, during the follicular phase in baboon #2, and during the luteal phase in baboon #3, the antibody against the receptor was detected in all three baboons between 2½ and 3 weeks, indicating relatively rapid response. There was an initial increase in the production of antibody. For example, antisera of baboons #1 and #2 showed up to 50% of inhibition in the binding of ^{125}I-HCG to the rat Leydig cell receptor in the first serum sample obtained 2½ weeks after placement of the LH-receptor implant. The initial antibody titer in baboon #3 showed only 15% inhibition;

Table 22.1. Serum metabolites (SMAC) in baboon #30025 before and after LH-receptor implant.

Test	Preimplant	Postimplant	Normal range[a]
Phosphate (mg/dl)	5.1	6.2	2.2- 4.6
Protein (gm/dl)	7.2	6.9	6.4- 8.1
Alb:Glob (ratio)	1.15	1.55	1.1- 2.3
Calcium (mg/dl)	8.8	8.0	8.8- 10.5
Glucose (mg/dl)	74.0	97.0	65.0-130.0
Bun (mg/dl)	16.0	21.0	6.0- 23.0
Bilirubin (mg/dl)	0.1	0.06	0.2- 1.6
SGPT (IU/L)	51.0	56.0	5.0- 55.0
SGOT (IU/L)	42.0	50.0	10.0- 50.0
Sodium (mmol/L)	147.0	145.0	134.0-143.0
Potassium (mmol/L)	4.1	4.1	3.5- 5.3
Chloride (mmol/L)	100.0	104.0	96.0-107.0
Creatinine (mg/dl)	1.0	1.3	0.6- 1.3
Cholesterol (mg/dl)	125.0	117.0	120.0-240.0

[a] In humans.

baboon #3 was, however, almost double the weight of baboons #1 and #2. The antibody titers then slowly declined in a steady manner for up to 300 days. It remains to be established whether a relationship between titer and body weight in relation to a constant dose of antigen exists. It is also noteworthy that the effect of antibody on the receptor serum levels of E_2, P, FSH, and LH, and on the cycle length and menstrual bleeding becomes quite visible during the next cycle after the placement of the LH-receptor implant in all three baboons. The early onset of response may be directly related to the initial rise in the production of antibody. There is apparent individual variability in the quantitative response during the immunized period in the intensity and duration of perineal turgescence, in serum levels of E_2, P, FSH, and LH, and in cycle length, and menstrual bleeding. It is not known as to what degree the individual variability is related to body weight, age, or other metabolic factors. An optimum dose remains to be established to overcome individual variation and to attain zero fertility rate without adverse side effects. It is evident that all the baboons exhibited ovarian dysfunction and presumably infertility as a result of the production of antibody against LH receptor in the ovary. The evaluation of contraceptive efficacy of the antibody against the receptor awaits mating of immunized baboons with males of known fertility. Antigen dose titration versus receptor antibody titers that will create zero fertility rate remains to be established. The merits of antibody against LH receptor are:

1. The LH receptor from bovine corpora lutea acts as a heteroantigen
2. The use of adjuvants is not required
3. The antigen is provided as a one-time administration in the form of implant
4. There is sufficient availability of the LH receptor from bovine corpora lutea
5. There is quick immune response to the antigen
6. The antibodies against the LH receptor are effective at low titers
7. There is quick biological response to the antibodies
8. There is convenience of dose-regulation by the replenishment or removal of the antigen from the implant
9. There is neither total steroid deprivation, or unopposed estrogen production, nor an elevation of pituitary gonadotropins indicative of a pathophysiological concern
10. All endocrine, metabolic, and reproductive functions are completely and rapidly reversible to normal
11. There are no observed adverse effects either systematically or at the site of implant.

The possible mechanism of action of LH-receptor antibodies may be a reversible suppression of folliculogenesis, ovulation, and corpora lutea function owing to partial blockade of LH receptor in the ovary during the entire menstrual cycle. These observations in baboons indicate that LH receptor has a potential in the development of a contraceptive vaccine for use in human as well as in animal breeding.

Acknowledgments. This work was supported by grants from the National Cancer Institute of the National Institutes of Health–CA-13908, Andrew W. Mellon Foundation, New York, and The Robert Wood Johnson Charitable Foundation, New York. Part of these studies were presented at The Society for the Advancement of Contraception International Medical and Scientific Conference, Chicago, IL, September, 1986 and The American Fertility Society/Canadian Fertility and Andrology Society International Congress Meeting, Toronto, Canada, September, 1986.

Mukul Singh is a Lalor Foundation Fellow.

References

1. Bahl OP, Muralidhar K: Current status of antifertility vaccines. Dhindsa D, Schumacher E (eds): in *Immunological Aspects of Infertility and Fertility Regulation*. New York, New York, Elsevier North Holland Inc., 1980, pp. 225–257.
2. Channing CP: The use of granulosa cell cultures and short-term incubations in the assay for gonadotropins. Moudgal NR (ed): in *Gonadotropins and Gonadal Function*. New York, Academic Press, 1974, pp. 185–188.
3. Dattatreyamurty B, Rathnam P, Saxena BB: Isolation of the luteinizing hormone-chorionic gonadotropin receptor in high yield from bovine corpora lutea: molecular assembly and oligomeric nature. *J Biol Chem* 1982;258(5):3140–3158.
4. Haour F, Saxena BB: Characterization and solubilization of gonadotropin receptor of bovine corpus luteum. *J Biol Chem* 1974;249:2195.
5. Khan FS, Rathnam P, Saxena BB: Purification and properties of human chorionic gonadotropin-lutropin receptor from plasma-membrane and soluble fractions. *Biochem J* 1981;197:7–22.
6. Metsikko K, Rajaniemi H: Antibodies to purified luteinizing hormone receptor localized the receptor at the luteal surface. *Endocrinology* 1981;109:1399–1403.
7. Moudgal NR, Sairam MR, Mahoney J: On the immunogenicity of the β-subunit of bovine luteinizing hormone (oLH) and equine chorionic gonadotropin (eCG) in the chimpanzee (Pan troglodytes): Effect of antiserum on monkey cycle and early pregnancy. *Am J Reprod Immunol* 1985;8:120–124.
8. Podesta JE, Sanchez ML, Solano AR: Dynamic studies of free and occupied LH-receptor with a monoclonal anti-LH-receptor antibody. Les colloques de l'Inserm, Developements recents de l'endocrinologre du testicule. Recent progress in cellular endocrinology of the testes. *INSERM* 1984; pp. 277–286.
9. Rojas FJ, Asch RH: Properties of gonadotropin-stimulable adenylyl cyclase of the human corpus luetum: Regulation of hormonal responsiveness by guanyl nucleotide and magnesium ion. *J Clin Endocrinol Metab* 1984;59(2):219–227.
10. Saxena BB: Gonadotropin Receptors. Melvin Blecher (ed): in *Methods in Receptor Research*. New York, Marcel Dekker, Inc., 1976, pp. 251–299.
11. Saxena BB, Dattatreyamurty B, Ota H, et al: Subunits of luteinizing hormone human chorionic gonadotropin receptor from bovine corpora lutea. *Biochemistry* 1986;25:7943–7950.
12. Saxena BB, Landesman R, Gupta GN, et al: New approaches in fertility regulation. *J Obstet Gynecol* 1984; Vol. 4 (Suppl. 1):6–22.

13. Saxena BB, Singh M: Effect of active immunization with LH-receptor on the reproductive functions of female rabbits and baboons, abstract, ed, 1986. The American Fertility Society/Canadian Fertility and Andrology Society. International Congress meeting, Toronto, Canada.
14. Stevens VC: Pre-clinical safety studies on a hCG vaccine, in Gleicher N (ed): *Reproductive Endocrinology.* New York, Allan R Liss, 1981, p. 451.
15. Talwar GP, Singh O, Singh V, et al: Enhancement of antigonadotropin response to the β-subunit of ovine luteinizing hormone by carrier conjugation and combination with the β-subunit of human chorionic gonadotropin. *Fertil Steril* 1986;46:120–126.

CHAPTER 23

Development of a Contraceptive Vaccine for Use by the Human Male: Results of a Feasibility Study Carried Out in Adult Male Bonnet Monkeys (*Macaca radiata*)

N.R. Moudgal, G.S. Murthy, N. Ravindranath, A.J. Rao, and M.R.N. Prasad

Introduction

Among the various methods being evaluated for contraceptive use in the human male, the development of an effective contraceptive vaccine appears attractive. Currently three possible immunogens are considered: two are polypeptide hormones [lutenizing hormone-releasing hormone (LHRH) and follicle-stimulating hormone (FSH)] and the third is a sperm-specific protein [lactic dehydrogenase χ (LDH-X)]. The merits/demerits of using these as a vaccine are provided in Table 23.1. The use of LHRH as a vaccine is in keeping with the clinical observation that both gonadotropins (2), FSH and LH, are required for maintenance of spermatogenesis (2). However, since blockade of LH secretion leads to cessation of testosterone production and consequently loss of libido, immunization with LHRH would require continuous supplementation with exogenous testosterone. The long-term effects of continuous testosterone supplementation remain to be investigated. Although sperm-specific antigens have been shown to be effective immunogens (3), their contraceptive efficacy has been demonstrated by imunizing only the females of the species, and, as such, this does not qualify as a vaccine for the male.

The suggestion that FSH could make an effective contraceptive vaccine in man was made by us a couple of years ago based on conclusive evidence we had earlier provided that chronic neutralization of FSH by passive immunization with a well-characterized FSH antiserum leads to testicular dysfunction, as evidenced by drastic reduction in the quantity and quality of ejaculated sperms voided as well as setting in of infertility (4,5). This is best illustrated by the results presented in Figure 23.1 and Table 23.2.

Our earlier observation on the immunogenicity of ovine FSH (oFSH) and the pronounced cross-reactivity its antibody exhibits toward FSH of other species, including the human (9), was of relevance here as it obviated the need to use the homologous FSH as the vaccine as well as the attended problems it would pose such as development of procedures to couple it to carrier proteins.

Table 23.1. Possible contraceptive vaccines for the male.

Number	Immunogen	Method of presentation	Merits/demerits
1	LHRH	Coupled to a carrier protein; adjuvant essential.	Effective as it blocks both LH and FSH. Libido inhibited unless supplemented with testosterone.
2	Sperm-specific protein (including LDH-X)	No carrier protein needed if heterologous protein used; adjuvant essential.	Effective only if the female of the species is immunized.
3	Ovine FSH	No carrier protein required; use of Alhydrogel as adjuvant adequate.	Effective in inducing infertility; libido and testosterone levels unaffected; process is reversible.

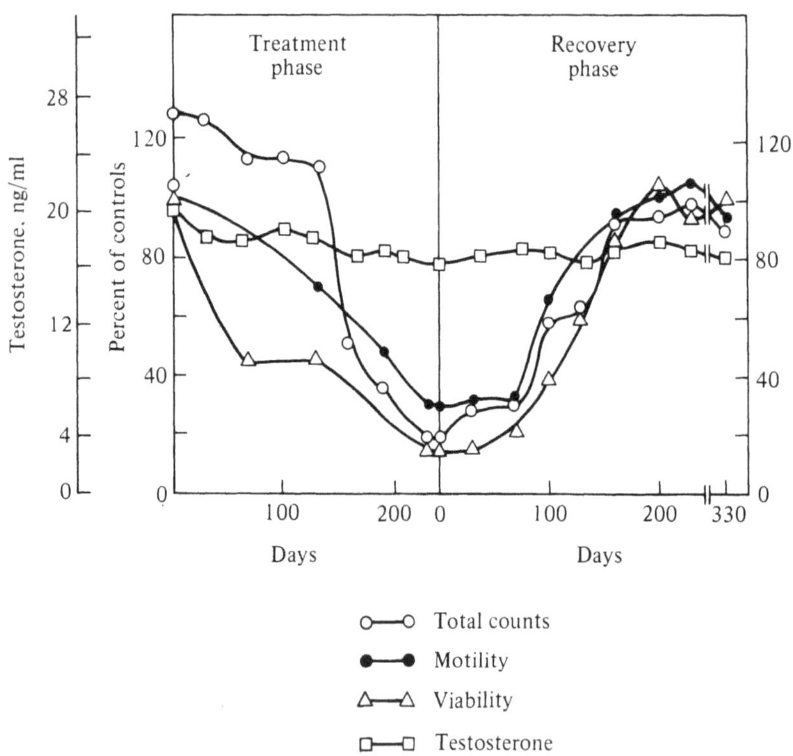

Fig. 23.1. Differential semen analysis and serum testosterone levels during treatment (FSH antibody) and recovery phase.

Table 23.2. Effect of FSH deprivation and restitution[a] on fertility of male bonnet monkeys.

	Antiserum treatment period				Antiserum withdrawal phase		
Group	Male (n)	Female used (n)	Pregnancies	Fertility[b] index (%)	Male (n)	Female used (n)	Return of fertility (days)
Controls	5	13	11	83.5	4	10	Fertile all through
Experimentals	4	18	0	0	4	19	138–194

[a] Restitution refers to antiserum withdrawal phase.
[b] Tested for ability to impregnate females between 95–223 days of treatment.

Current Status of the FSH Vaccine

The hormone oFSH, isolated in our laboratory from frozen sheep pituitaries and characterized by affinity chromatography for lack of LH contamination, has been used as the immunogen. Its efficacy has been tested thus far in more than 30 monkeys. In addition to our own tests, the efficacy of the vaccine was tested in two collaborating laboratories, one situated in New Orleans and the other in New Delhi. All three laboratories used the same protocol and adjuvant. The adjuvant used was the commercially available Alhydrogel, which, incidentally, is also approved for human usage. The protocol itself consisted of three to four primary immunizations given ten days apart and boosters given at intervals of 100 days. All immunized monkeys have responded positively by producing high titer antibodies capable of neutralizing hFSH. The latter was checked using an in vitro rat granulosa cell system standardized in the laboratory.

Sperm counts in ejaculates obtained periodically by electroejaculation start showing a sharp reduction by 150 days of immunization. This reduction ranges between acute oligospermia (10 million/ejaculate) to azoospermia, and the period during which azoospermia is maintained varies (Fig. 23.2). During this entire period of immunization, which has thus far covered 380 to 550 days, no significant change in serum testosterone has been observed. This is particularly advantageous as no testosterone supplementation was required to maintain libido.

Ten of the immunized monkeys have been mated (at a time when they started exhibiting acute oligospermia) with a total of 50 proven fertile females (average male:female ratio 1:5) during their most fertile period (days 9–14 of ovulatory cycles). None of the males were able to impregnate the females, thereby demonstrating that FSH immunization had rendered them infertile. This is highly significant when one considers that in our colony a minimum of 60% of the females become pregnant when exposed during a single ovulatory cycle to a fertile male. Thus the active immunization studies have essentially confirmed the results observed by us earlier using a passive immunization approach.

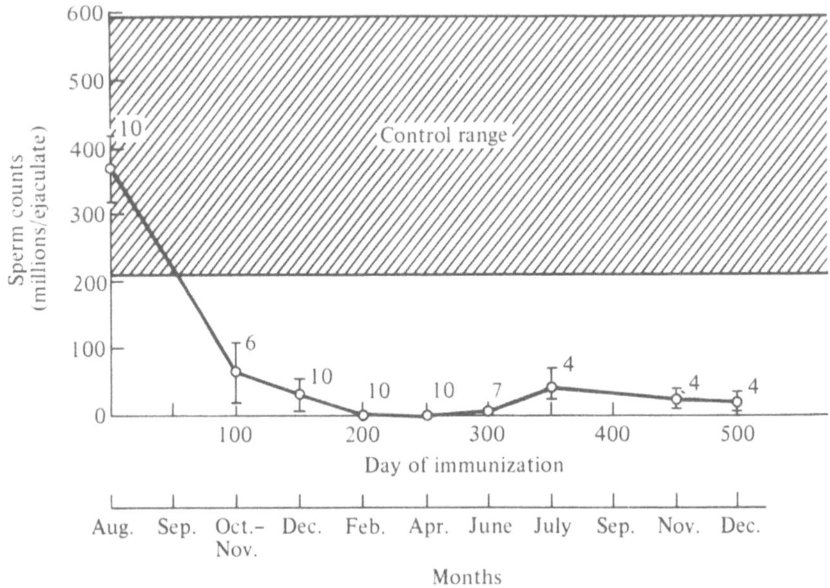

Fig. 23.2. Effect of active immunization of adult male bonnet monkeys with oFSH on total sperm counts.

Recently following withdrawal of booster injections and consequent reduction in antibody titers, the sperm counts of most monkeys have started increasing significantly and three out of the ten monkeys impregnated three females during a fresh mating study, thereby demonstrating that the antifertility effects of the FSH vaccine are reversible.

Discussion

We have succeeded in demonstrating the efficacy of FSH vaccine in rendering proven, adult male, fertile bonnet monkeys infertile, and this could be correlated with the production of reasonably high antibody titers capable of neutralizing primate FSH as well as the development of acute oligospermia or azoospermia. The effect appears reversible as stoppage of boosters leads to restoration of normal sperm counts and fertility.

One of the major considerations in the development of FSH vaccine is the long-term effect active immunization may have on (a) sperm production, particularly with respect to the quality and numbers of sperms produced, (b) recovery of testicular function following cessation of immunization, and (c) clinical well being, which includes absence of diseases due to immune complexes.

Srinath et al (10), in their 4½ years' study with four rhesus monkeys, observed that active immunization with oFSH resulted in production of high titers of neutralizing antibodies; this resulted in either azoospermia or oligospermia, but one to two monkeys did exhibit occasional high sperm counts. Although there was at the end of 4½ years an apparent gradual restoration of spermatogenesis, the diameter of seminiferous tubules remained decreased, and the germinal epithelium was depleted. Seasonal variations in testicular volume evident in the controls were not seen in the immunized monkeys, and the testicular volume continued to remain at the lowest level throughout the 4½ years of active immunization. In the absence of fertility testing, the inference that sperm counts are not "low enough" to induce infertility is at best subjective, and the conclusion that active immunization is not a feasible approach for male fertility regulation is questionable. Moudgal et al (6) followed bonnet monkeys actively immunized with FSH for upward of nine years and observed that (a) the sperm counts vary from azoospermic to oligospermic levels, (b) they remain infertile when mated in oligospermic state, and (c) the viability and motility of the sperms are markedly reduced compared with the controls.

Two basic questions to be considered in this context are (a) Is azoospermia prerequisite for fertility regulation? and (b) Is acute oligospermia produced under FSH deprivation compatible with a state of infertility? Contraceptive regimens (gestagen-androgen combinations) evaluated in the human male have so far produced varying degrees of oligospermia/necrospermia/azoospermia (11). Barfield et al (1) documented a few pregnancies when sperm counts were below 2×10^6/mL, although not all workers have observed this degree of unplanned pregnancies with the magnitude of oligospermia reported. In another study of about 300 volunteers, only seven pregnancies occurred over a six-year period; none of these pregnancies occurred when sperm counts were below 5×10^6/mL (8). Thus, azoospermia as a prerequisite for contraception, although is desirable, has not been unequivocally demonstrated to be mandatory. Infertility, in cases of oligospermia, has also been related to the poor quality of ejaculated sperms. Moudgal et al (6) showed that oligospermia induced by active immunization with FSH in bonnet monkeys results in infertility of all immunized monkeys; no pregnancies occurred in any of the 50 proven fertile females mated during ovulatory cycles with the ten immunized males (ratio of 1:5 females). This is perhaps due to the impairment of the quality of sperms produced in a microenvironment in the seminiferous epithelium in which Sertoli cell function is altered owing to FSH deprivation. Ejaculated sperms from actively immunized macaques show marked reduction in motility, hyaluronidase, and cAMP activity (H.G. Madhwa Raj, personal communication, July 1986). Thus, current studies conducted on a large sample of monkeys clearly indicate that oligospermia induced by active immunization with FSH is consistent with infertility. It is of particular interest that infertility could be achieved without compromising libido or resorting to testosterone supplementation.

It is encouraging to note that long-term active immunization has not resulted in deleterious effects on the health status of the monkeys [for more than nine years in bonnet monkeys (5) or for 4½ years in rhesus monkeys (7)]. Immune complexes were not detected in circulation, nor were precipitated immune complexes found in any of the organs/tissues (7). Although the nontoxic nature of the vaccine remains to be evaluated by a classic preclinical toxicology study, the data currently available are highly supportive of the suggestion that active immunization with FSH is a feasible approach for fertility regulation in man.

Acknowledgment. This project was aided by the Indian Council of Medical Research (ICMR), New Delhi.

References

1. Barfield A, Melo J, Coutinho E, et al: Pregnancies associated with sperm concentrations below 10 million/ml in clinical studies of a potential male contraceptive method, monthly depot medroxyprogesterone acetate and testosterone esters. *Contraception* 1979;20:121-127.
2. Bremner WJ, Matsumoto AM: Endocrine control of human spermatogenesis: possible mechanism for contraception, in Zatuchni GI, Goldsmith A, Spieler JM, Sciarra JJ (eds): *Male Contraception Advances and Future Prospects*. Philadelphia, Harper & Row, 1986, pp 81-88.
3. Goldberg E, Shelton JA: Immunologic properties of LDH-C_4 for contraceptive vaccine development, in Zatuchni GI, Goldsmith A, Spieler JM, Sciarra JJ (eds): *Male Contraception Advances and Future Prospects*. Philadelphia, Harper & Row, 1986, pp 435-446.
4. Murty GSRC, Sheela Rani CS, Moudgal NR, et al: Effect of passive immunization with specific antiserum to FSH on the spermatogenic process and fertility of adult male bonnet monkeys (*M. radiata*). *J Reprod Fertil* [*Suppl*] 1979;26:147-163.
5. Moudgal NR: A need for FSH in maintaining fertility of adult male subhuman primates. *Arch Androl* 1981;7:117-125.
6. Moudgal NR, Murthy GS, Rao AJ, et al: Development of oFSH as a vaccine for the male – a status report on the recent researches carried out using the bonnet monkey *M. radiata*, in Talwar GP (ed): *Proceedings of the International Symposium on 'Immunological Approaches to Contraception and Promotion of Fertility'.* New York, Plenum Press, 1986, pp 103-110.
7. Nieschlag E: Reasons for abandoning immunization against FSH as an approach to male fertility regulation, in Zatuchni GI, Goldsmith A, Spieler JM, Sciarra JJ (eds): *Male Contraception: Advances and Future Prospects*. Philadelphia, Harper & Row, 1986, pp 395-400.
8. Paulsen CA, Brenner WJ, Leonard JM: Male Contraception: Clinical Trials, in Mishell D (ed): *Advances in Fertility Research*. New York, Raven Press, 1982, pp 153-160.
9. Rao AJ, Moudgal NR: An immunochemical study of ovine pituitary follicle stimulating hormone (FSH). *Arch Biochem Biophys* 1970;138:189-198.
10. Srinath BR, Wickings EJ, Witting C, et al: Active immunization with follicle stimulating hormone for fertility control – a 4½ year study in male rhesus monkeys. Fertil Steril 1983;40:110-117.
11. World Health Organization: Annual Reports of the Special Programme of Research, Development and Research Training in Human Reproduction, Geneva, 1979, 1980.

Part VI
Gamete Antigens for Birth Control Vaccines

CHAPTER 24

Hormonal and Cellular Control of Seminiferous Tubular Function

C. Wayne Bardin

The seminiferous tubule is composed of three cell types: germ cells, myoid cells, and Sertoli cells. Of these, Sertoli cells comprise the epithelium of the tubule in which spermatogenesis occurs. These cells rest on a basal lamina, which, in turn, is surrounded by a layer of myoid cells. Sertoli cells are attached to one another by specialized junctional complexes near the basal portion of each cell. This arrangement is diagramed in Figure 24.1. The Sertoli cell junctional complexes limit the transport of fluid and molecules between the interstitial space of the testis and the seminiferous tubular lumen. The functional significance of this anatomic arrangement was first suggested by studies in which dyes, administered intravenously, failed to enter the tubular lumen. These and similar observations suggested that there is a blood-testis barrier similar to the barrier that exists between the vascular space and the brain.

In addition to Sertoli cells, spermatogonia are the only cellular elements of the tubule that are in contact with the lymphatic and blood vascular spaces. The remainder of germ cells are sequestered behind the blood-testis barrier in an unique environment where they are in intimate contact with Sertoli cell plasma membranes. Appreciation of this anatomic arrangement led to the conclusion that Sertoli cells are required to nourish germ cells as they move away from the basal lamina toward the avascular portion of the seminiferous epithelium next to the lumen. In addition to their role in germ cell maturation, Sertoli cells have a number of other important functions. Their junctional complexes provide electrical coupling between cells along the length of the seminiferous epithelium and maintenance of the blood-testis barrier. Sertoli cells also phagocytize damaged germ cells; produce proteins, including the hormone inhibin, that are secreted in the seminiferous tubular lumen and blood; metabolize steroid (i.e., produce estrogens from androgens); and serve as a major site of hormone action in the seminiferous tubule (Fig. 24.1).

In the 1960s assays for luteinizing hormone (LH), testosterone, and other androgens were developed, permitting investigators to define physiology and pathophysiology of Leydig cells in multiple species. Even though assays for follicle-stimulating hormone (FSH) were also developed at about the same time, an understanding of the function of the cells comprising the seminiferous tubules

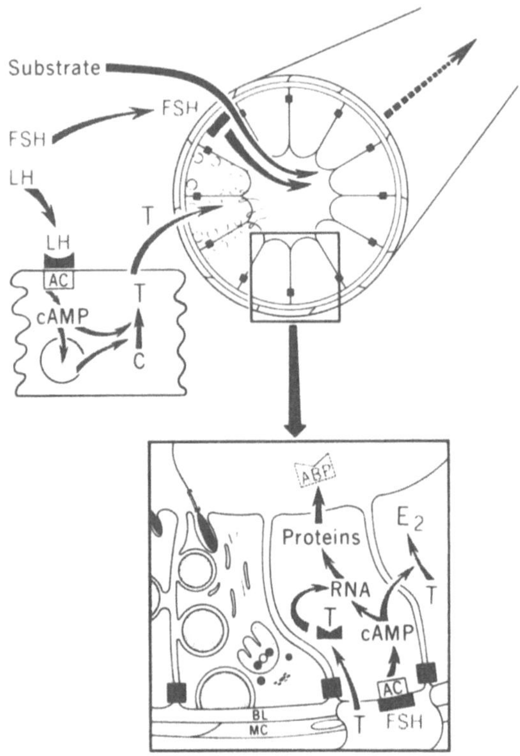

Fig. 24.1. The mechanism of action of pituitary hormones and testosterone on the seminiferous tubule. In the upper panel, LH acts by way of its Leydig cell membrane receptor to stimulate cAMP formation, which facilitates the conversion of cholesterol (C) to testosterone (T). Both testosterone and FSH act directly on Sertoli cells to stimulate the transport of various substances into the seminiferous tubular lumen. In the bottom panel, two Sertoli cells are shown. On the left, germ cells (stippled circles) develop, surrounded by Sertoli cell plasma membrane. On the right, testosterone and FSH stimulate Sertoli cell protein synthesis and androgen-binding protein (ABP). In addition FSH stimulates the conversion of testosterone to estradiol (E_2). Black squares, junctional complexes; AC, adenylate cyclase; BL, basal lamina; MC, myoid cells.

lagged behind that of Leydig cells. This was primarily due to the complexity of spermatogenesis, a process involving the interaction of multiple cells with several hormones over a relatively long period of time which could only be studied in vivo. In addition, the major product of the germinal epithelium (i.e., spermatozoa) could be quantified only in testicular biopsies by morphometric techniques or after they appeared in the ejaculate following passage through the epididymis. In view of the difficulty of studying the seminiferous tubular compartment, other products that could be measured more easily than spermatozoa were sought. It was realized that Sertoli cells were the major secretory element of the germinal epithelium and that their proteins might be used to study this portion of the testis. Accordingly, a number of Sertoli cell products were identified and used to expand our knowledge of seminiferous tubular physiology. This chapter discusses some of the secretory products that have been used to monitor the hormonal and cellular control of Sertoli cells and the seminiferous tubular epithelium. In addition, it reviews how the measurement of specific Sertoli cell proteins in blood and semen provides a method of assessing testicular physiology and pathophysiology in vivo.

The Secretory Products of Sertoli Cells

Androgen-Binding Protein

Androgen-binding protein (ABP) was one of the first of the Sertoli cell products to be identified. The importance of this protein relates to the fact that it has been extensively studied and is a prototype for other Sertoli cell secretory products (1,20). A variety of studies in the rat indicates that approximately 80% of this protein is secreted into the lumen of the seminiferous tubules, while the remainder is released into the blood, probably from the base of Sertoli cells (23). These early observations gave rise to the concept that seminiferous tubular physiology, like Leydig cell physiology, can be monitored by measuring secretory products. Accordingly, the structure and function of ABP has been studied in several species. This protein is similar in rat (rABP), rabbit (rbABP), and human (hABP) in that the native molecules are dimeric glycoproteins with molecular weights of 85 to 100,000. In the latter two species, the ABPs are remarkably similar to testosterone-estradiol binding globulin (TeBGs). The genes for rABP and hTeBG have been cloned and sequenced. There is a high degree of homology between these species, suggesting that these proteins in the same species will be similar, if not identical. The function of ABP is still disputed, but it is believed to be a transport protein for testosterone and other androgens in the male reproductive tract and fetal brain. Cell-surface receptors for ABP and TeBG have been identified in the epididymis and prostate, respectively. The binding sites are believed to be responsible for internalization of these proteins (7,28).

Testis-Specific Versus Serum Proteins

To identify other useful markers of testicular function, the secretory products of Sertoli cell cultures were examined. Many other testis-specific and serum proteins were identified (57). Unexpectedly, serum proteins, including transferrin and ceruloplasmin, were also demonstrated using a variety of immunological and chromatographic procedures (47,57). These observations suggested that Sertoli cells might be the origin of serum proteins known to exist in rete testis fluid. The major secretory products of the Sertoli cell thus far identified are summarized in Table 24.1. As will be noted below, many of these products have profound effects

Table 24.1. Testis-specific and plasma proteins that are produced by Sertoli cells.

Protein	Reference
Androgen-binding protein	Hansson et al (26)
	Bardin et al (1)
	Ritzén et al (43)
Inhibin	de Jong and Robertson (16)
Growth factors	Feig et al (18)
	Brown et al (5)
	Holmes et al (27)
Somatomedin	Hall et al (25)
Transferrin	Skinner and Griswold (47)
	Skinner et al (49)
Ceruloplasmin	Skinner and Griswold (48)
Retinol-binding protein	Huggenvik and Griswold (29)
	Carson et al (8)
H-Y antigen	Brunner et al (6)
Plasminogen activator	Lacroix et al (31)
	Vihko et al (55)
Clusterin	Fritz et al (21)
	Blaschuk et al (3)
	Tung and Fritz (54)
Cyclic proteins	Wright et al (58)
Dimeric acid glycoprotein	Sylvester et al (52)
Scm proteins	DePhilip and Kierszenbaum (17)
CMB proteins	Cheng et al (15)
Testibumin	Cheng and Bardin (13)

upon the function of cultured Sertoli cells. In addition, some of them may also influence the function of other cells in the testis and epididymis. Finally, each protein secreted by Sertoli cells can be a potentially useful marker for studying the seminiferous tubular compartment of the testes (37).

Testibumin

During a search for hormonally responsive products in media from Sertoli cell-enriched cultures, a protein that responds synergistically to FSH and testosterone was identified and designated CMB-1 (14). Structural and immunological analyses, including *N*-terminal sequence determination and immunoblots, indicate that CMB-1 is related to albumin and α-fetoprotein (AFP) (13). Furthermore, CMB-1 is highly concentrated in the testicular and epididymal compartments in adult rats. In view of its localization and its similarity to albumin and AFP, we postulate that this may be the albumin homologue of the adluminal compartment of the testis and have designated it testibumin. The fact that this protein is also found in the blood of adult female rats indicates that, unlike rABP, this protein is not synthesized exclusively in the testis.

Identification and Isolation of Other Sertoli Cell Proteins

As noted above, Sertoli cell products have been and will continue to be used as probes for studying Sertoli cells and their interactions with the multiple cell types in the adluminal and basal compartments. The isolation of these proteins has been difficult since, in many instances, the procedures used to identify them, including one- or two-dimensional gel electrophoresis, have not been useful in subsequent purification owing to their limited loading capacities. In addition, the use of radioactive amino acids to identify Sertoli cell proteins may confound a subsequent purification strategy since the intensity of labeling does not necessarily indicate which of the proteins in the medium are the major secretory products of Sertoli cells. This problem may be particularly pertinent in products having rapid turnover rates or a large number of the radiolabeled amino acid residues.

The recent application of newly developed high-pressure liquid chromatography (HPLC) columns to the rapid isolation of TeBGs from serum of human, rabbit, bull (11) and of ABP from human testes (12) encouraged us to take a similar approach for the identification and isolation of Sertoli cell secretory products. The first two steps of this procedure, anion-exchange HPLC and sodium dodecyl sulfate-polyacrylamide gel electrophoresis (SDS-PAGE), permit identification of more than 30 major acidic proteins, which have been termed CMB proteins (15). When culture media are prepared from Sertoli cells grown with testosterone, FSH, or both hormones, these proteins are classified as those responding to neither hormone, primarily to testosterone, primarily to FSH, or to both hormones (15). It is now possible to isolate many of these proteins from 1 to 2 L of medium using a variety of HPLC columns (13). It is of note that Sertoli cell-enriched cultures are likely to contain small amounts of peritubular cells and germ cells that can contribute directly or indirectly to the secreted proteins (50,51). Thus, proteins in the Sertoli cell media could be derived from peritubular and residual germ cells. In this regard, Sertoli cells do not secrete fibronectin or type I collagen into the medium when maintained for up to six days under serum-free conditions. By contrast, peritubular cells in culture secrete both of these proteins and when co-cultured with Sertoli cells, the secretion of both proteins is enhanced. Using a nonspecific fibronectin antiserum, no fibronectin-like macromolecules were present by Western blot test in the Sertoli cell-enriched cultures used for the isolation of CMB proteins. This suggests that most of the CMB proteins are of Sertoli cell origin. Nonetheless, the origins of CMB proteins must be established with more direct methods following purification, assay development, and sequence analysis.

The Sertoli Cell Hormone, Inhibin

Although many of the proteins and peptides produced by Sertoli cells have autocrine and paracrine effects on the testis, a few of these products are secreted into the blood and act as endocrine mediators on other tissues. The existence of water-soluble substances of gonadal origin that selectively suppress FSH secretion has been suspected for more than five decades (38). The quest to chemically characterize these molecules, referred to as inhibin, proceeded slowly until 1985 (9,16). In the past few years, several investigators have isolated inhibin-like molecules from a variety of sources. Primary structures of a 31-peptide and 94-peptide with inhibin-like activities were reported by Ramasharma et al (41) and Seidah et al (46). These peptides were isolated from human seminal plasma. Subsequently, two additional peptides of 52 and 92 amino acids were also reported (32). Studies using the synthetic 31-peptide demonstrated that it did not have the properties expected of inhibin since it did not suppress the plasma levels of FSH in castrated rats (44). Robertson and colleagues reported on the isolation of inhibin from bovine follicular fluid that had an apparent molecular weight of 56,000. This protein was characterized by two peptide chains of 14,000 and 46,000 connected by disulfide bonds (45). In serum this form of inhibin is converted to a 32,000 form without loss of biological activity. Inhibin from porcine follicular fluid was also isolated, but showed an apparent molecular weight of 32,000 (33,39). This peptide consisted of dissimilar polypeptide subunits of 18,000 (α-inhibin) and 14,000 (β-inhibin) linked by disulfide bonds (44). Our laboratory isolated a similar protein from ram rete testis fluid. The finding that the amino-terminal amino acid sequence of this material is almost identical to that isolated from porcine follicular fluid argues that ovarian and testicular inhibin are identical. All the preparations purified from ovarian and testicular fluids are several hundredfold more potent than and are not structurally related to any of the peptides isolated from seminal plasma. We conclude that the physiologically important inhibin is a 32,000 protein composed of nonidentical subunits.

Recently, the *N*-terminal amino acid sequence of α- and β-inhibin were determined, and the data used to identify cloned complementary DNAs correspond to the mRNA for proinhibin α and β (33–35). The complete amino acid sequences of these subunits were predicted from their nucleotide sequences, and it was found that β-inhibin shares a high homology with transforming growth factor β. It is of interest that molecules regulating reproduction and cell growth appear to be related. Recently a molecule was isolated from follicular fluid, which will stimulate FSH secretion, that is composed of two β-inhibin subunits.

The Humoral Control of Protein Secretion by Isolated Sertoli Cells

From studies in hypophysectomized animals, it is known that testes require FSH and testosterone for spermatogenesis. Several lines of evidence suggest that Sertoli cells are the site of action of both hormones. The establishment of Sertoli cell cultures permitted studies demonstrating that the hormonal control of Sertoli cell secretion is much more complex than once suspected from in vivo studies. Cultures of primary Sertoli cells along with established cell lines in serum-free medium provide an ideal system in which to study the effects of single and combined humoral agents under controlled conditions. These cultures consist of a single cell type whose environment can be controlled by adding known amounts of hormones, growth factors, attachment factors, vitamins, and binding proteins. Extensive studies on Sertoli cells in culture have indicated that these cells require insulin (Ins), transferrin (TF), and epidermal growth factor (EGF) for growth at low-cell densities in serum-free medium. In primary Sertoli cell cultures, Ins has the most marked effect on rABP secretion and a lesser effect on cell survival. TF and EGF alone have only minor effects on rABP secretion, but the three together are better than Ins alone. For this reason, supplementation with Ins, TF, and EGF (3H) is used as the basal condition for studies of Sertoli cell secretion. The actions of all other hormones are augmented by the presence of these three factors (42).

FSH is known to stimulate Sertoli cells in the presence of 3H. These cells can also be stimulated by representatives from several steroid classes including progesterone (P), hydrocortisone (HC), and testosterone (T).

Fat soluble vitamins E (α-tocoperol) and A [retinoic acid (RA) and retinol (R-ol)] are important for Sertoli cell function, since deficiencies of these vitamins lead to sterility in male animals. Both of the vitamins stimulate growth and secretion in Sertoli cells of the TM_4 line and primary Sertoli cells. From these studies, a hormone-supplemented medium that is considered optimal for secretion of the Sertoli cell product, rABP, was devised. This includes 3H + FSH, T, HC, P, and vitamin E (8H).

These observations indicate that multiple humoral agents, in addition to testosterone and FSH, stimulate growth and protein secretion of Sertoli cells. These include other hormones, growth factors, and binding proteins. It is significant that many of the factors that stimulate these cells are also made by them. This suggests that some Sertoli cell products may have an autocrine function. In addition, it is possible that several of these proteins may have a paracrine function on other cellular elements of the seminiferous tubular epithelium. These conclusions are consistent with the postulate that Sertoli secretory products contribute to the unique environment in which spermatogenesis occurs.

The Cellular Control of Sertoli Cells

When isolated Sertoli cells are placed in culture, their function changes dramatically even in the presence of optimal concentrations of hormones, vitamins, transport proteins, and growth factors. One reason for this is that Sertoli cell function is modified by the presence of other cells and the extracellular matrix of the testis.

Myoid Cell-Sertoli Cell Interaction

Surrounding the seminiferous tubule outside the basal lamina is a thin layer of cells. Since these cells are responsible for the contractility of the seminiferous tubule, they are called myoid cells (Fig. 24.1). Recent studies using cultured

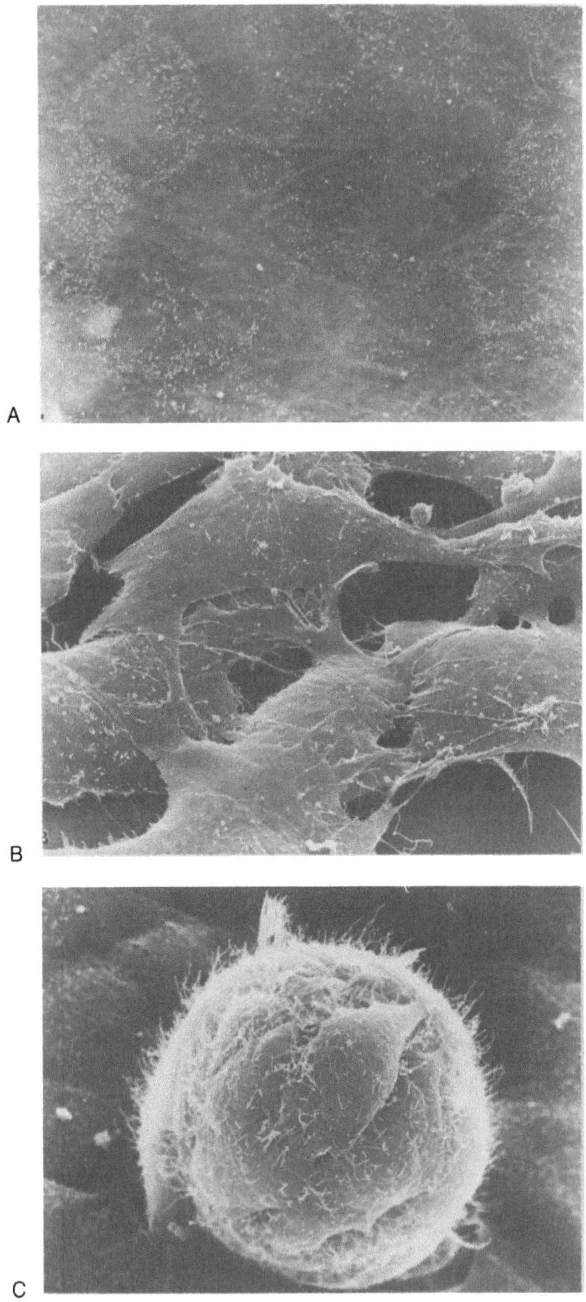

Fig. 24.2. Scanning electron micrographs of myoid and Sertoli cells in vitro. **A**: Myoid cells; **B**: Sertoli cells; **C**: Sertoli cells plated on myoid cells form balls in vitro. See text for details.

myoid cells show that they also have profound effects on Sertoli cells. When myoid and Sertoli cells are cultured separately, they both grow as monolayers (Fig. 24.2A,B). However, when primary or established Sertoli cell lines (TM_4 and TR-ST) are plated on top of a monolayer composed of myoid cells, they round up and form tubular or ball-like structures. Myoid cells then grow over the balls. This dramatic myoid-directed "differentiation" of Sertoli cells is dependent upon unique plasma membrane antigens, since other cells do not form "balls" on myoid cells, nor do Sertoli cells when plated on other cells types (Fig. 24.2C). Ball formation is therefore believed to be an in vitro model for tubular differentiation. In addition to producing dramatic changes in the way Sertoli cells associate with each other, myoid cells also increase the basal and hormone-stimulated secretion rates of rABP and other Sertoli cell proteins (36,37).

Another function of myoid cells is production of the extracellular matrix, which is believed to form a portion of the basal lamina. The matrix produced by myoid cells can substitute for many of the effects that this cell type has on primary Sertoli cells. For example, myoid matrix increased the basal and FSH-stimulated secretion of rABP and other Sertoli cell proteins (36).

Germ Cell-Sertoli Cell Interaction

The impetus for studying the interaction between germ and Sertoli cells is based on the observation that the spermatogenic cells are not randomly distributed in the seminiferous epithelium, but rather are strictly arranged in distinct cell associations (40). These cell associations can be identified according to cytological characteristics of young spermatids. The spermatogenic cycle is defined as the succession of different stages at a fixed point in the seminiferous tubule as a function of time, while the wave of the seminiferous epithelium is the succession of cell associations *along the length* of the seminiferous tubule. One complete wave is limited by two identical stages, between which all other stages occur. For each species the cycle has a constant duration and may be viewed as a biological clock.

The wave of the seminiferous epithelium can be visualized in many species by transillumination of fresh unstained tubules. The individual stages are distinguishable by their differential light absorption. This variation along the length of the tubule is due to the condensation of the nuclei of late spermatids and to their relation with Sertoli cells. This is shown in Figure 24.3 for the rat. Each stage is associated with several morphological and biochemical events important in spermatogenesis. For example, stage 8 (the place of greatest light and dark transition in Fig. 24.3) is the point at which mature spermatozoa are released into the seminiferous tubular lumen. Stage 13 is the end of meiotic prophase where the primary spermatocytes enter reduction division in stage 14. It should be noted that the staging and organization of individual stages along the tubule vary between species (40).

The influence of different germ cell associates on Sertoli cell function can be studied by isolating tubular segments from defined stages of the spermatogenic

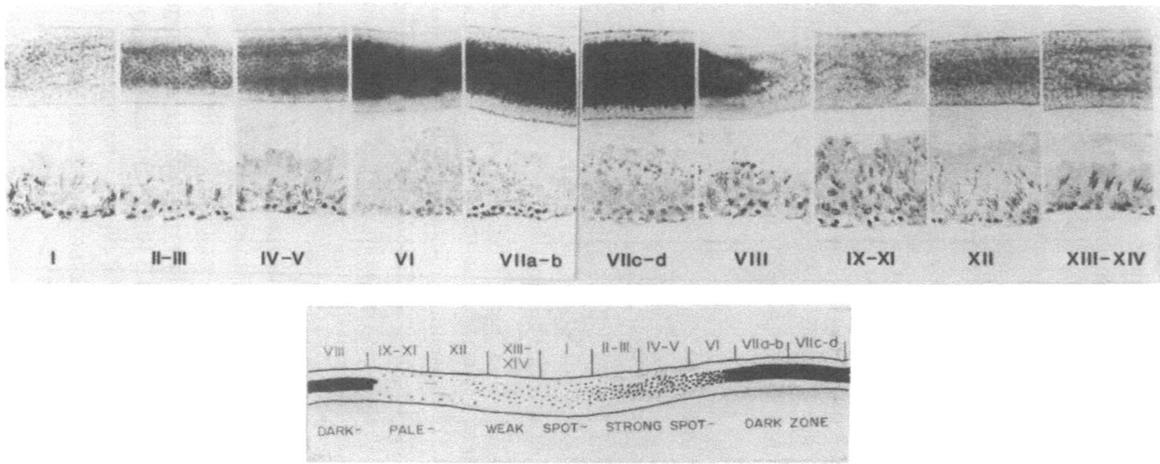

Fig. 24.3. The transillumination pattern of the freshly isolated rat seminiferous tubulus related to the stages of the cycle of the seminiferous epithelium (top). Detailed depiction of the epithelial stages. After photography in freshly isolated unstained condition, the tubular segments were fixed and processed for histology. It can be recognized that the transillumination pattern of the seminiferous tubulus is dependent on the location of the step 13–19 spermatids in the seminiferous epithelium: the spermatid bundles of low density are recognized as weak spots (stages XIII-I) and those of high density and deep location in Sertoli cells as dark spots (stages II-V). At late steps of maturation, the spermatids form a continuous layer at the top of the seminiferous epithelium that give rise to the dark absorption pattern at stages VII and VIII. The spermiation is seen as an abrupt stop of the dark absorption at stage VIII. Transillumination pictures × 100; histological pictures, × 400 (bottom). Schematic drawing of the main absorption zones of the transilluminated seminiferous tubulus: pale, weak spot, strong spot, and dark zones, and their relationship to the epithelial stages. The vertical lines indicate the cutting point during the transillumination-assisted microdissection procedure. (From Parvinen, reference 40)

cycle under the dissecting microscope. When tubular segments are incubated in vitro, Sertoli cell proteins are secreted into the medium where they can be measured by immunoassay. Such studies indicate that Sertoli cell products are secreted cyclically as a function of the stages of the seminiferous epithelial cycle (58). Furthermore, the amount of products secreted at any given stage is increased if peritubular myoid cells are present (37). Analysis of these secretory products by two-dimensional gel electrophoresis indicates that different patterns of proteins are secreted at different stages of the cycle. That is, many proteins are secreted cyclically, but peak secretion occurs at different stages. By contrast, few of the nonsecreted cellular proteins demonstrate a definite cycle.

The findings indicate that Sertoli cells secrete many proteins in a cyclical manner. The cycle of secretion for each protein is linked to the cycle of the seminiferous epithelium. Thus, the germ cells appear to interact with Sertoli cells in such a way as to modulate the pattern of secretion of some proteins. Myoid cells, in turn, seem to influence the magnitude of Sertoli cell secretion at all stages of the cycle.

Leydig Cell-Sertoli Cell Interactions

It has long been known that steroids from Leydig cells stimulate Sertoli cell growth and protein secretion. Additional studies suggest that other products of the Leydig cell can have effects on Sertoli cells, and perhaps additional cells, in the seminiferous epithelium. Recently β-endorphin was found in extracts of rat testes. β-Endorphin in testis raised the possibility that it could arise from POMC and that Proopiomelanocortin (POMC)-derived products may have some function in the male reproductive tract. Immunohistochemical and chemical studies demonstrated that β-endorphin and des-acetyl-α-melanocyte-stimulating hormone (MSH) are present in Leydig cells. The observation that POMC mRNA is present in Leydig cells is consistent with the hypothesis that POMC-derived peptides are made in this cell type (2,10,30,53).

The finding of POMC-derived peptides in the testis raises the question of their function in this organ. The concentrations of these peptides in the testis are much lower than those in brain areas where they may function as neurotransmitters. Recent in vivo and in vitro studies suggest that POMC-derived peptides may have a variety of autocrine and paracrine functions in the testis. α-MSH, β-MSH, γ-MSH, and ACTH stimulate Sertoli cell growth, cAMP secretion, and FSH-induced changes in the function of this cell type. Interestingly, no difference in the potencies of α-MSH and des-acetyl α-MSH could be detected on Sertoli cells; this is strikingly different from brain and melanocytes in which the potencies of these two latter peptides are markedly different (4). In contrast to the POMC-derived peptides with MSH activity, β-endorphin and/or other testicular opiates inhibit Sertoli cell division in fetal and neonatal animals. In addition, opiates also inhibit Sertoli cell secretion of proteins, and receptors for this class of peptides are present on this cell type (22).

In addition to peptides of Leydig cell origin influencing Sertoli cells, the reverse has also been shown. There is some evidence to suggest that LHRH can regulate Leydig cell activity. The secretion of luteinizing hormone-releasing hormone (LHRH), or a closely related peptide, has been demonstrated by Sertoli cells. In addition, Leydig cell size appears to be regulated by the stage of the spermatogenic cycle of adjacent tubules.

Sertoli Cell Products as Markers of Testicular Function In Vivo

The Bidirectional Secretion of rABP and Other Proteins

As noted previously in the rat, rABP may be useful for studying testicular physiology and pathophysiology in vivo. The major portion of this protein is secreted into the lumen of the seminiferous tubule and is transported via the rete testis and efferent ductules to the epididymis, where it is concentrated in the caput and slowly degraded. Since testicular proteins are not stored by Sertoli cells but are secreted immediately after synthesis, the concentration of rABP in the epididymis is much higher than in the testis. As a consequence, epididymal content of this protein has been proposed as an index of the amount of rABP secreted into the seminiferous lumen. The finding of rABP in the blood raises the interesting question as to how a specific Sertoli cell protein could enter this compartment. A variety of studies suggest that rABP is not released into the blood from the lumen of the seminiferous tubule, the rete testis, or the epididymis. If this is true, then it is possible that a fraction of rABP is released from the basal aspect of Sertoli cells into the interstitial space of the testis where it is taken up by lymphatics and blood. This concept implies that Sertoli cells secrete rABP in a bipolar fashion into both the luminal and basal compartments of the testis. Additional studies indicate that transferrin and testibumin are also secreted from both the apex and the base of Sertoli cells (23). Kinetic analysis suggests that in the normal rat at least 80% of rABP is secreted into the tubular lumen and transported to the epididymis and 20% is secreted into the blood. It should be noted at this point that all studies to date on the bidirectional secretion of proteins by the testes are based on studies in the rat. Assays for specific Sertoli cell proteins have yet to be established in other species. Until these procedures are available, the broad application of the bidirectional model remains uncertain.

Secretion of rABP in Testicular Disease

H^{re} Gene. Animals heterozygous for the H^{re} gene ($H^{re}/+$) develop seminiferous tubular failure. The testicular morphology and hormonal profiles of these animals resemble those of humans with a Sertoli cell-only syndrome. Some studies suggest that $H^{re}/+$ animals have an abnormality of Sertoli cell function that is possibly secondary to germ cell depletion (24). This postulate is supported

by studies showing transport of rABP to the epididymis to be only 3% of normal as contrasted to release into the blood, which was 70% to 80% normal. The marked reduction in epididymal rABP relative to that in blood cannot be explained by an increased turnover rate of rABP in the epididymis or by a decreased clearance of this protein from the blood. Virtually identical patterns of rABP secretion are also present in rats with germ cell depletion secondary to other gene defects or prenatal x-ray treatment. These observations suggest that germ cell depletion per se results in decreased rABP synthesis and that secretion into the lumen of the seminiferous tubule is reduced much more than that into the blood. In fact, some animals with marked abnormalities of seminiferous tubular function have near normal rABP levels in the blood. We conclude from these observations that some testicular disorders will result in an unbalanced or disordered pattern of rABP secretion.

Transient Hypogonadotropic Hypogonadism. There are several models of abnormal Sertoli cell development. One of these is the rat with transient hypogonadal hypogonadism produced by a single passive immunization with LHRH antiserum (LHRH-AS) during the neonatal period (56). Even though the serum gonadotropins and testosterone levels return to normal after several weeks, testicular weights are about 60% of controls at all ages from days 10 to 90. In spite of the reduction in total organ weight, tubular diameter and the seminiferous epithelium were indistinguishable from normal. In addition, both serum and epididymal rABP contents were reduced commensurate with the reduction of testicular weight, which is in contrast to rats with the H^{re} gene. These results indicate that neutralization of LHRH at a crucial period in development impairs early testicular growth and may cause "permanent" reduction in testicular weight. The total volume of the seminiferous tubules in adult life is reduced, even though gonadotropin concentrations are normal. These observations in the rat may be comparable to those in patients with a hereditary defect in LHRH secretion (hypogonadotropic hypogonadism, Kallmann's syndrome). If these patients are not exposed to gonadotropins during puberty, the testes are small and may never achieve adult size no matter how long gonadotropin treatment is continued as adults. An analogous situation has also been seen in mice with this disorder.

The combination of direct and indirect studies suggests that Sertoli cell secretion in the rat is polarized to two compartments (seminiferous tubule and blood/lymph) in much the same way as hepatocytes are polarized to the blood and bile ducts. The ability to measure specific Sertoli cell products in blood and semen may, therefore, be used to examine the physiology and pathophysiology of the seminiferous tubule. As is the case in assessing hepatic function, it will probably be necessary to use a battery of specific markers to study Sertoli cell function in vivo. Thus, other products that can be measured in blood and/or semen must be identified. The recent observation that a major portion of transferrin in human semen is of testicular origin suggests that this protein can also be used to examine the function of the tubular compartment in vivo.

References

1. Bardin CW, Musto N, Gunsalus G, et al: Extracellular androgen binding proteins. *Ann Rev Physiol* 1981;43:189–198.
2. Bardin CW, Shaha C, Mather J, et al: Identification and possible function of pro-opiomelanocortin-derived peptides in the testis. *Ann NY Acad Sci* 1984;438:346–364.
3. Blaschuk OW, Fritz IB: Isoelectric forms of clusterin isolated from ram rete testis fluid and from secretions of primary cultures of ram and rat Sertoli cell enriched preparations. *Can J Biochem Cell Biol* 1984;62:456–461.
4. Boitani C, Mather JP, Bardin CW: Stimulation of cAMP production in rat Sertoli cells by α-MSH and des-acetyl α-MSH. *Endocrinology* 1986;118:1513–1518.
5. Brown KD, Blakeley DM, Henville A, et al: Rete testis fluid contains a growth factor for cultured fibroblasts. *Biochem Biophys Res Commun* 1982;105:391–397.
6. Brunner M, Moreira-Filho CA, Wachtel G, et al: Secretion of H-Y antigen. *Cell* 1984;37:615–620.
7. Byers SW, Musto NA, Dym M: Culture of ciliated and nonciliated cells from rat ductuli efferentes. *J Androl* 1985;6:271–278.
8. Carson DD, Rosenberg LI, Blaner WS, et al: Synthesis and secretion of a novel binding protein for retinol by a cell line derived from Sertoli cells. *J Biol Chem* 1984; 209:3117–3123.
9. Channing CP, Gordon WL, Liu VK, et al: Minireview: Physiology and biochemistry of ovarian inhibin. *Biol Med* 1985;178:339–361.
10. Chen C-L, Mather JP, Morris PL, et al: Expression of pro-opiomelanocortin-like gene in the testis and epididymis. *Proc Natl Acad Sci USA* 1984;81:5672–5675.
11. Cheng CY, Bardin CW, Nagendranath N, et al: Purification of testosterone-estradiol-binding globulins from mammalian sera by anion-exchange high-performance liquid chromatography. *Int J Androl* 1985;8:1–12.
12. Cheng CY, Musto NA, Gunsalus GL, et al: The role of the carbohydrate moiety on the size heterogeneity and immunologic determinants of human testosterone-estradiol binding globulin. *J Steroid Biochem* 1985;22:127–134.
13. Cheng CY, Bardin CW: Rat testicular testibumin is a glycoprotein responsive to FSH and testosterone that shares immunodeterminants with albumin. *Biochemistry* 1986;25:5276–5288.
14. Cheng CY, Grima J, Lee WM, et al: The distribution of rat testibumin in the male reproductive tract. *Biol Reprod* 1987 (in press).
15. Cheng CY, Mather JP, Byer AL, et al: Identification of hormonally responsive proteins in primary Sertoli cell culture medium by high performance liquid chromatography. *Endocrinology* 1986;118:480–488.
16. de Jong FH, Robertson DM: Inhibin: 1985 Update on action and purification. *Mol Cell Endocrinol* 1985;42:95–103.
17. DePhilip RM, Kierszenbaum AL: Hormonal regulation of protein synthesis, secretion, and phosphorylation in cultured rat Sertoli cells. *Proc Natl Acad Sci USA* 1982;79:6551–6555.
18. Feig LA, Bellve AR, Erickson NH, et al: Sertoli cells contain a mitogenic poly peptide. *Proc Natl Acad Sci USA* 1980;77:4774–4778.
19. Feig LA, Klagsbrun M, Bellve AR: Mitogenic polypeptide of the mammalian seminiferous epithelium: Biochemical characterization and partial purification. *J Cell Biol* 1983;97:1435–1443.

20. French FS, Ritzén EM: A high affinity androgen binding protein (ABP) in rat testis: Evidence for secretion into efferent duct fluid and absorption by epididymis. *Endocrinology* 1973;93:88–95.
21. Fritz IB, Burdzy K, Setchell B, et al: Ram rete testis fluid contains a protein clusterin which influences cell-cell interactions in-vitro. *Biol Reprod* 1983;28:1173–1188.
22. Gerendai I, Shaha C, Gunsalus GL, et al: The effects of opiod receptor antagonists suggest that testicular opiates regulates Sertoli and Leydig cell function in the neonatal rat. *Endocrinology* 1986;118:2039–2044.
23. Gunsalus GL, Musto NA, Bardin CW: Bidirectional release of a Sertoli cell product, androgen binding protein, into the blood and seminiferous tubule, in Steinberger A, Steinberger E (eds): *Testicular Development, Structure and Function*. New York, Raven Press, 1980, pp 291–297.
24. Gunsalus GL, Musto NA, Becker RR, et al: The bidirectional secretion of androgen binding protein in rats with germ cell depletion secondary to the H^{re} gene and prenatal x-ray. *Endocrinology* 1987 (in press).
25. Hall K, Ritzén EM, Johnsonbaugh RE, et al: Secretion of somatomedin-like compound from Sertoli cells *in vitro*, in Spencer EM (ed): *Insulin-Like Growth Factors/Somatomedins*. New York, Walter de Gruyter, 1983, pp 611–614.
26. Hansson V, Ritzén EM, French FS, et al: Androgen transport and receptor mechanisms in testis and epididymis, in Hamilton DW, Greep RO (eds): *Handbook of Physiology, Section 7: Endocrinology, Volume V: Male Reproductive System*. Washington DC, American Physiological Society, 1975, pp 173–201.
27. Holmes SD, Spotts G, Smith RG: Rat Sertoli cells secrete a growth factor that blocks epidermal growth factor (EGF) binding to its receptor. *J Biol Chem* 1986;261: 4076–4080.
28. Hryb DJ, Khan SM, Rosner R: Testosterone-estradiol-binding globulin binds to human prostatic cell membranes. *Biochem Biophys Res Commun* 1985;128:432–440.
29. Huggenvik, J, Griswold MD: Retinol binding protein in rat testicular cells. *J Reprod Fertil* 1981;61:403–408.
30. Krieger DT, Margioris AN: Liotta AS, et al: Pro-opiomelanocortin (POMC)-derived peptides in the rodent male reproductive tract, in Delitala G, Motta M, Serio M (eds): *Opioid Modulation of Endocrine Function*. New York, Raven Press, 1984, pp 223–235.
31. Lacroix M, Smith FE, Fritz IB: Secretion of plasminogen activator by Sertoli cell enriched cultures. *Mol Cell Endocr* 1977;9:227–236.
32. Li CH, Hammonds RG, Ramasharma K, et al: Human seminal α-inhibins: Isolation, characterization, and structure. *Proc Natl Acad Sci USA* 1985;82:4041–4044.
33. Ling N, Ying S-Y, Ueno N, et al: Isolation and partial characterization of a M_r 32,000 protein with inhibin activity from porcine follicular fluid. *Proc Natl Acad Sci USA* 1985;82:7217–7221.
34. Mason AJ, Hayflick JS, Ling N, et al: Complementary DNA sequences of ovarian follicular fluid inhibin show precursor structure and homology with transforming growth factor-β. *Nature* 1985;318:659–663.
35. Mason AJ, Niall HD, Seeburg PH: Structure of two human ovarian inhibins. *Biochem Biophys Res Commun* 1986;135:957–964.
36. Mather JP, Zhuang L-Z, Perez-Infante V, et al: Culture of testicular cells in hormone-supplemented serum-free medium, in Bardin CW, Sherins RJ (eds): *The Cell Biology of the Testis*. New York, New York Academy of Sciences, 1982, vol 383, pp 44–68.

37. Mather JP, Gunsalus GL, Musto NA, et al: The hormonal and cellular control of Sertoli cell secretion. *J Steroid Biochem* 1983;19:41-51.
38. McCullagh DD: Dual endocrine activity of the testis. *Science* 1932;76:19-20.
39. Miyamoto K, Hasegawa Y, Fukuda M, et al: Isolation of porcine follicular fluid inhibin of 32K daltons. *Biochem Biophys Res Commun* 1985;128:396-403.
40. Parvinen M: Regulation of the seminiferous epithelium. *Endocr Rev* 1982;3:404-417.
41. Ramasharma K, Sairam MR, Seidah NG, et al: Isolation, structure and synthesis of a human seminal plasma peptide with inhibin-like activity. *Science* 1984;223: 1199-1202.
42. Rich KA, Bardin CW, Gunsalus GL, et al: Age-dependent pattern of androgen binding protein (ABP) secretion from rat Sertoli cells in primary culture. *Endocrinology* 1983;113:2284-2293.
43. Ritzén EM, Hansson V, French FS: The Sertoli cell, in Burger H, de Kretser D (eds): *The Testis*. New York, Raven Press, 1981, pp 171-194.
44. Rivier J, McClintock R, Vaughan J, et al: Partial purification of inhibin from ovine rete testis fluid, in Zatuchni G, Goldsmith A, Sciarra J (eds): *Male Contraception: Advances and Future Prospects*. Philadelphia, Harper & Row, 1986, pp 401-407.
45. Robertson DM, Foulds LM, Leversha L, et al: Isolation of inhibin from bovine follicular fluid. *Biochem Biophys Res Commun* 1985;126:220-226.
46. Seidah NG, Ramasharma K, Sairam MR, et al: Partial amino acid sequence of a human seminal plasma peptide with inhibin-like activity. *FEBS Lett* 1984;167: 98-102.
47. Skinner MK, Griswold MD: Sertoli cells synthesize and secrete transferrin-like protein. *J Biol Chem* 1980;255:9523-9525.
48. Skinner MK, Griswold MD: Sertoli cells synthesize and secrete a ceruloplasmin-like protein. Biol Reprod 1983;28:1225-1230.
49. Skinner MK, Cosand WL, Griswold MD: Purification and characterization of testicular transferrin secreted by rat Sertoli cells. *Biochem J* 1984;218:313-320.
50. Skinner MK, Fritz IB: Androgen stimulation of Sertoli cell function is enhanced by peritubular cells. *Mol Cell Endocrinol* 1985;40:115-122.
51. Skinner MK, Fritz IB: Testicular peritubular cells secrete a protein under androgen control that modulates Sertoli cell functions. *Proc Natl Acad Sci USA* 1985;82: 114-118.
52. Sylvester SR, Skinner MK, Griswold MD: A sulfated glycoprotein synthesized by Sertoli cells and by epididymal cells is a component of the sperm membrane. *Biol Reprod* 1984;31:1087-1102.
53. Tsong SD, Phillips DM, Halmi N, et al: β-Endorphin is present in the male reproductive tract of five species. *Biol Reprod* 1982;27:755-764.
54. Tung PS, Fritz IB: Immunolocalization of clusterin in the ram testis, rete testis, and excurrent ducts. *Biol Reprod* 1985;33:177-186.
55. Vihko KK, Suominen JJO, Parvinen M: Cellular regulation of plasminogen activator secretion during spermatogenesis. *Biol Reprod* 1984;31:383-390.
56. Vogel DL, Gunsalus GL, Bercu BB, et al: Sertoli cell maturation is impaired by neonatal passive immunization with antiserum to luteinizing hormone-releasing hormone. *Endocrinology* 1983;112:1115-1121.
57. Wright WW, Musto NA, Mather JP, et al: Sertoli cells secrete both testis-specific and serum proteins. *Proc Natl Acad Sci USA* 1981;78:7565-7569.
58. Wright WW, Parvinen M, Musto NA, et al: Identification of stage-specific proteins synthesized by rat seminiferous tubules. *Biol Reprod* 1983;29:257-270.

CHAPTER 25

LDH-C_4 as the Model Sperm Antigen for a Contraceptive Vaccine

Erwin Goldberg

Development of an immunological contraceptive as an alternative birth control method requires an antigen that is gamete-specific, readily available, safe, and effective. With respect to these requirements, lactic dehydrogenase isoenzyme C_4 (LDH-C_4), found exclusively on sperm and in testis, is a promising candidate for a vaccine. This antigen has been extensively analyzed by immunological and biochemical methods. It has been found to reduce fertility when administered systematically to females in a number of animal models, including mice, rabbits, and baboons. Local immunization into the reproductive tract of female mice to produce IgA at the site of fertilization also results in marked fertility reduction. Rather than attempting to isolate large quantities of LDH-C_4 for a vaccine, antibody binding sites on the LDH-C_4 molecule are being studied to determine whether they can act as synthetic antigens to replace intact LDH-C_4. One such synthetic peptide representing amino acids 5-15 of mouse LDH-C_4 was attached to a macromolecular carrier, diphtheria toxoid, and used to immunize female baboons. These animals made antibody that bound to the native LDH-C_4 molecule. Mating resulted in a 70% reduction in their fertility over that of nonimmune controls. This contraceptive effect was reversed when immunizations were discontinued. There were no apparent immunopathological side effects of the vaccine. These experiments confirm the feasibility of using a synthetic antigenic determinant of sperm-specific LDH-C_4 as a vaccine to control fertility. Furthermore, specificity, availability, effectiveness, reversibility, and safety requirements for a successful immunological contraceptive technology can be satisfied by this approach.

Recent advances in the construction of synthetic vaccines (2,13) have encouraged and reinforced our efforts to predict and define antigenic sites of LDH-C_4. The application of today's new techniques of immunochemistry to development of a contraceptive vaccine based on this sperm-specific enzyme shows great promise. Immunosuppression of fertility by LDH-C_4 has already been demonstrated. The fact that LDH-C_4 represents the most completely characterized sperm antigen allows us to use it as a model system for studying sperm antigens in general and to pursue development of this specific sperm protein toward a human contraceptive technology.

A major obstacle to progress in immunocontraception by sperm antigens is the limited availability of purified material that can be isolated for further characterization. LDH-C_4, on the other hand, is readily isolated from the testes of mice and other animals and can be purified routinely in relative abundance (29). Consequently, it has been possible to evaluate the biochemical and immunological properties of this enzyme in terms of its potential as a fertility controlling vaccine.

Biochemical characterization has revealed that the LDH-C_4 molecule is composed of four identical subunits (7). The three-dimensional structure of these subunits has been determined by x-ray crystallography (19). Complete primary sequences are also available for both mouse and rat LDH-C_4 (20). The LDH-C subunit is relatively homologous structurally to the LDH-A_4 and LDH-B_4 somatic isozymes (15) and they can form heterotetramers in vitro (7). However, there are major changes in the amino acid sequence of LDH-C_4 (4,14). Such primary sequence substitutions and ensuing modifications in its three-dimensional structure amount for the total lack of immunological cross-reactivity among LDH-C_4 and its somatic isozymes (5,16). Thus, the immune response to LDH-C_4 is absolutely sperm-specific.

Previous studies designed to evaluate the immune response to LDH-C_4 have shown that parenteral immunization of males and females with mouse LDH-C_4 in combination with an adjuvant results in serum antibody production (5). These antibodies will react directly with LDH-C_4 on the surface of sperm (3). Alvarez and Storey (1) determined biochemically that approximately 10% of LDH-C_4 activity is associated with the sperm plasma membrane and immunofluorescence has localized LDH-C_4 to the principal piece of the sperm tail (7). Binding of anti-LDH-C_4 to sperm results in agglutination and cytotoxicity in the presence of complement (8). Enzymatic activity is also inhibited by antibody binding (5), which suggests that sperm energy metabolism and, consequently, motility might be immunologically impaired.

Systemic immunization with LDH-C_4 provokes a cell-mediated immune response in addition to the humoral response. Kille et al (10) reported a delayed-type hypersensitivity in mice injected with LDH-C_4 in Freund's complete adjuvant. Shelton and Goldberg (23) detected cell-mediated cytotoxicity in LDH-C_4 immune mice.

From these observations, we may conclude that LDH-C_4 injection will result in immunosuppression of fertility in females and that a variety of immunological mechanisms can contribute to this phenomenon. In support of this conclusion are the observations that following LDH-C_4 immunization, fertility is suppressed by 40% to 60% in mice (17), by 70% in rabbits (6), and by 63% in baboons (9). It is instructive to note that the correlation between circulating antibody titers and protection against pregnancy is imperfect. In the studies with female baboons, some animals with high levels of circulating antibodies to LDH-C_4 did become pregnant while others with lower titers were protected (9). These observations suggest that we still must define the appropriate parameter to predict immunological infertility.

On the other hand, there are numerous clinical observations of immunological infertility presumably associated with serum antisperm antibodies. In fact, in one study (24) antibodies to LDH-C_4 were elevated in patients identified as having infertility problems. Thus, at our current level of knowledge, we can only be sure that the presence of circulating antibodies to LDH-C_4 suggests protection from conception, and further analyses are required.

It is likely that these sperm-specific antibodies function to block fertilization in the female reproductive tract. Antibodies to LDH-C_4 may be detected in cervical mucus, uterine fluids, and oviducal secretions (11) and impede sperm transport to the site of fertilization (12). Therefore, stimulation of a local immune response may increase the level of protection against pregnancy by LDH-C_4 antibodies. Although the systemic response consists primarily of IgG, which presumably enters the female reproductive tract as a transudate of serum (11), the major class of antibody bathing the reproductive tract is secretory IgA (sIgA), which is present locally in reproductive tract secretions (27). IgA is produced by plasma cells migrating from Peyer's patches to draining lymph nodes and then to the mucosal lining of the reproductive tract. The secretory component is attached to dimeric IgA as it migrates through mucosal epithelium. Thus, an immunization designed to stimulate this mucosal immune response may ensure higher levels of antibody at the site of fertilization than those resulting from systemic immunization. We have immunized female mice with LDH-C_4 directly into the uterus. These animals produced secretory IgA specific for the enzyme in their uterine fluids. When mated, only 20% of the mice immunized with LDH-C_4 became pregnant as opposed to 70% of nonimmune controls (21,25). Local immunization is therefore an effective means of suppressing fertility. We are currently exploring the proper dose of immunogen and the possibilities of increasing the local response by indirect stimulation of the gut-associated lymphatic tissues (GALT) (22), which would take advantage of the commonality of the mucosal immune system (18). The ability to administer a vaccine orally to stimulate the GALT and, indirectly, the mucosal lining of the reproductive tract is an attractive one. In addition, the short duration of the memory response for sIgA should afford ready reversibility of the antifertility effects (28).

As noted above, our strategy to develop a contraceptive vaccine is based on provoking an immune response to LDH-C_4 with a synthetic peptide that bears an antigenic determinant of the native protein. We have approached the problem of developing a synthetic antigen by identifying the sites on the LDH-C_4 molecule where antibodies bind. Once these epitopes have been mapped, they can be chemically synthesized, attached to a macromolecular carrier, and used for immunization. Antibodies made to these synthetic peptide fragments are then tested for their ability to bind native LDH-C_4. To date, eight such sites have been identified (Table 25.1) (30). Two tryptic peptides comprising sequences 5-15, 211-220, and a third sequence 97-110 have been chemically synthesized, attached to either diphtheria toxoid (DT) or to bovine serum albumin (BSA) and inoculated into rabbits (31). The 97-110 peptide cannot be isolated from a tryptic digest of LDH-C_4, but was chosen as a possible antigenic determinant because it

Table 25.1. Antigenic peptides of the LDH-C subunits.

Region	Structure and location
1	5 10 14a 14b Glu-Gln-Leu-Ile-Gln-Asn-Leu-Val-Pro-Glu-Asp-Lys
2	41 50 55 Gly-Leu-Ala-Asp-Glu-Leu-Ala-Leu-Val-Asp-Ala-Asp-Asp-Thr-Asp-Lys
3	211 220 Ser-Leu-Asn-Pro-Ala-Ile-Gly-Thr-Asp-Lys
4	221 230 Asn-Lys-Gln-His-Trp-Lys-Asn-Val-His-Lys
5	231 240 243 Gln-Val-Val-Glu-Gly-Gly-Tyr-Glu-Val-Leu-Asp-Met-Lys
6	283 290 Glu-Glu-Val-Phe-Leu-Ser-Ile-Pro-Cys-Val-Leu-Gly-Glu-Ser-Gly 300 303 Ile-Thr-Asp-Phe-Val-Lys
7	304 310 316 Val-Asn-Met-Thr-Ala-Glu-Glu-Glu-Gly-Leu-Leu-Lys-Lys
8	317 320 330 Ser-Val-Asp-Thr-Leu-Trp-Asn-Met-Gln-Lys-Asn-Leu-Glu-Leu

comprises the coenzyme binding site of LDH-C_4, would be accessible to antibody, and shows extensive sequence modification from the other isozymic forms. Rabbits were immunized with these three peptides. Although there was considerable variability among individual rabbits, certain patterns were observed. The rabbits immunized with peptide 5-15 responded best, regardless of the carrier. Peptide 97-110 was the poorest antigen, and 211-220 elicited a median response among the three peptides, with some differences in response appearing to be due to the carrier. BSA seemed to be slightly more effective than DT with the 211-220 peptide.

Based on these data, DT-MC 5-15 was chosen for further immunization and fertility studies in the female baboon. Fourteen female baboons were immunized with DT-MC 5-15 in squalene-arlacel A plus CGP 11,637, a muramyl dipeptide (MDP) adjuvant (provided by Ciba-Geigy Limited, Basel, Switzerland) (26). Fourteen control animals were immunized with the carrier plus adjuvant alone. All experimental animals produced antibodies to MC5-15 that also reacted with LDH-C_4. No antibodies to either MC 5-15 or to LDH-C_4 were detectable in control animals. Subsequently, fertility was tested by mating each animal at least four times following immunizations or until pregnant. The result was a 71% reduction in the fertility of the DT-MC 5-15 immunized baboons when compared with the control animals. Furthermore, when immunizations were discontinued, matings of experimental animals resulted in pregnancy. Thus, the immunological effect on fertility was reversible.

While these results demonstrate the promise that this approach holds, fertility reduction must be increased to a level acceptable for human contraceptive practice. Possibly, another peptide or combination of peptides might be more immunogenic. This avenue is currently being explored.

Recently, we have obtained quantitative data on the relative immunogenicity of some of the synthetic peptides as follows: 5-15>304-316>97-110>49-58 (Hogrefe, 1987, unpublished observations). Computer graphics analyses have been employed to refine the molecular structure of LDH-C and to determine the relative accessibility to antibody of these antigenic peptides on the surface of the native protein (Hogrefe et al, submitted). Peptide 5-15 has a relatively higher averaged surface accessibility. Therefore, this antigenic region of LDH-C_4 appears best suited for further study. Refinement of this sequence relative to human LDH-C_4 and perhaps as part of a surface simulation determinant may yield the synthetic vaccine construct to more effectively block fertilization at a level acceptable by human populations for contraceptive use.

It is clear that the availability of a defined and well-characterized sperm antigen such as LDH-C_4 is necessary to manipulate the afferent and efferent arms of the immune response in order to develop a contraceptive vaccine. Our studies with this system clearly demonstrate that inhibition of fertility by immunization with a fragment of this sperm-specific enzyme can be achieved. Data obtained thus far provide encouragement that vaccine technology will offer an alternative form of contraception.

Acknowledgments. This work was supported by National Institutes of Health grant HD-05863 and by the Program for Applied Research on Fertility Regulation, Northwestern University, under a cooperative agreement with the Agency for International Development (AID/DPE-0586-A-00-1003). The views of the author do not necessarily reflect AID policy.

References

1. Alvarez JG, Storey BT: Assessment of cell damage caused by spontaneous lipid peroxidation in rabbit sperm. *Biol Reprod* 1984;30:323–331.
2. Arnon R: in Schechter AN, Dean A, Goldberger RF (eds): Impact of Protein Chemistry on the Biomedical Sciences. Orlando, Florida, Academic Press, 1984, p 187.
3. Beyler SA, Wheat TE, Goldberg E: Binding of antibodies against antigenic domains of murine lactate dehydrogenase-C_4 to human and mouse spermatozoa. *Biol Reprod* 1985;32:1201–1210.
4. Eventoff W, Rossmann MG, Taylor SS, et al: Structural adaptations of lactate dehydrogenase isozymes. *Proc Natl Acad Sci USA* 1977;74:2677–2681.
5. Goldberg E: Immunochemical specificity of lactate dehydrogenase-X. *Proc Natl Acad Sci USA* 1971;68:349–352.
6. Goldberg E: Infertility in female rabbits immunized with lactate dehydrogenase X. *Science* 1973;186:273–278.

7. Goldberg E: Isozymes in testes and spermatozoa, in Ratazzi MS, Scandalios JG, Whitt GS (eds): *Isozymes: Current Topics in Biological and Medical Research*. New York, Alan R Liss Inc, 1977, vol 1, pp 79–124.
8. Goldberg E, Gonzales-Prevatt V, Wheat TE: Immunosuppression of fertility in females by injection of sperm-specific LDH-C_4 (LDH-X): Prospects for development of a contraceptive vaccine, in Semm K, Mettler L (eds): *Human Reproduction*, Proceedings of the IIIrd World Congress. Amsterdam, Excerpta Medica, 1981, pp 360–363.
9. Goldberg E, Wheat TE, Powell JE, et al: Reduction of fertility in female baboons immunized with lactate dehydrogenase C_4. *Fertil Steril* 1981;35:214–217.
10. Kille JW, Wheat TE, Mitchell G, et al: Strain differences in the immune response of mice to homologous sperm-specific lactate dehydrogenase (LDH-C_4). *J Exp Zool* 1978;204:259–266.
11. Kille JW, Goldberg E: Female reproductive tract immunoglobulin responses to a purified sperm-specific antigen (LDH-C_4). *Biol Reprod* 1979;20:863–871.
12. Kille JW, Goldberg E: Inhibition of oviducal sperm transport in rabbits immunized against sperm-specific lactate dehydrogenase (LDH-C_4). *J Reprod Immunol* 1980; 2:15–21.
13. Lerner RA: Tapping the immunological repertoire to produce antibodies of predetermined specificity. *Nature* 1982;299:592–596.
14. Li SS-L, Feldmann RJ, Okabe M, et al: Molecular features and immunological properties of lactate dehydrogenase C_4 isozymes from mouse and rat testes. *J Biol Chem* 1983;258:7017–7032.
15. Li SS-L, Fitch WM, Pan Y-CE, et al: Evolutionary relationships of vertebrate lactate dehydrogenase isozymes, A_4 (muscle), B_4 (heart), and C_4 (testis). *J Biol Chem* 1983;258:7029–7032.
16. Liang Z-G, Shelton JA, Goldberg E: Non-crossreactivity of antibodies to murine LDH-C_4 with LDH-A_4 and LDH-B_4. *J Exp Zool* 1986;240:377–384.
17. Lerum JE, Goldberg E: Immunological impairment of pregnancy in mice by lactate dehydrogenase-X. *Biol Reprod* 1974;11:108–115.
18. Montgomery PC, Lemaitre-Coelho IM, Vaer J-P: A common mucosal immune system. Antibody expression in secretions following gastro-intestinal stimulation. *Immunol Commun* 1980;9:705–713.
19. Musick WDL, Rossmann MG: The structure of mouse testicular lactate dehydrogenase isoenzyme C_4 at 2.9A resolution. *J Biol Chem* 1979;254:7611–7620.
20. Pan, Y-CE, Sharief FS, Okabe M, et al: Amino acid sequence studies on lactate dehydrogenase C_4 isozymes from mouse and rat testes. *J Biol Chem* 1983;258: 7005–7016.
21. Shelton JA, Abe M, Goldberg E: Cellular and local immunity to sperm-specific lactate dehydrogenae-C_4. *J Reprod Immunol* [Suppl] 1983;7:26.
22. Shelton JA, Alexander NJ, Goldberg E: Local secretory immunity to sperm-specific lactate dehydrogenase C_4 in the reproductive tract of the female rhesus macaque. *Biol Reprod* [Suppl 1] 1984;30:74.
23. Shelton JA, Goldberg E: Induction of cell-mediated cytotoxic immunity to sperm-specific lactate dehydrogenase-C_4 in SJL/J female mice. *Biol Reprod* 1985;32: 556–560.
24. Shelton JA, Goldberg E: Serum antibodies to LDH-C_4. *J Reprod Immunol* 1985;8:321–327.

25. Shelton JA, Goldberg E: Local reproductive tract immunity to sperm-specific lactate dehydrogenase C_4. *Biol Reprod* 1986;35:873–876.
26. Stevens VC, Cinader B, Powell JE, et al: Preparation and formulation of a human chorionic gonadotropin anti-fertility vaccine: selection of adjuvant and vehicle. *Am J Reprod Immunol* 1981;1:315–321.
27. Tomasi TB: The secretory immune system, in Dhindsa DS, Schumacher GFB (eds): *Immunological Aspects of Infertility and Fertility Regulation*. New York, Elsevier/North Holland, 1980, pp 23–31.
28. Tomasi TB, Bienenstock J: Secretory immunoglobulins, in Dixon FJ Jr, Kunkel HG (eds): *Advances in Immunology*. New York, Academic Press, 1968, vol 9, pp 1–96.
29. Wheat TE, Goldberg E: Isolation of the sperm-specific lactate dehydrogenase from mouse, rabbit and human testes and human spermatozoa, in Boettcher B (ed): *Immunological Influences on Human Fertility*. Sydney, Australia, Harcourt Brace Jovanich, 1977, pp 221–227.
30. Wheat TE, Goldberg E: Antigenic domains of the sperm-specific lactate dehydrogenase C_4 isozyme. *Molec Immunol* 1985;22:643–649.
31. Wheat TE, Shelton JA, Gonzales-Prevat V, et al: The antigenicity of synthetic peptide fragments of LDH-C_4. *Molec Immunol* 1985;22:1195–1199.

CHAPTER 26

Should Cell-Mediated Immunity be Considered in the Implementation of Antifertility Vaccines?

Deborah J. Anderson and Joseph A. Hill

Most studies of infertility due to immunological mechanisms have focused on antibody responses to sperm and gestational tissue antigens. Limited experimental data exist to indicate that reproductive tissue antigens also stimulate cell-mediated immune responses and that such responses are associated with infertility (3,18,26,32,34). Cell-mediated immunity (CMI) as a cause of infertility has been difficult to address because in vivo and in vitro assays to detect specific CMI responses until lately have provided imprecise information. However, recent technological advances have enabled the purification, characterization, and mass production by genetic engineering of several of the mediators of CMI responses. Our laboratory has begun systematic investigation of the effects of soluble products of activated macrophages and lymphocytes on reproductive tissue viability and function. Our studies to date, which are reviewed here, indicate that these factors may have dramatic effects on fertility. This information, combined with recent data on accessibility of female and male reproductive tracts to immune effectors, indicates that strategies to enhance local cell-mediated immunological responses should be considered for improving the efficacy of antifertility vaccines.

Soluble Mediators of Immune Responses

It is now appreciated that antibodies comprise but a subset of the soluble mediators of immune responses. Classical immune responses are initiated by foreign antigens, which are processed by macrophages and presented to thymic-derived (T) lymphocytes. Antigen-activated macrophages also produce soluble factors, termed monokines, which induce lymphocytes to differentiate and produce soluble products, termed lymphokines, which further amplify and also regulate immune responses. This subject has recently been reviewed (6,42).

Cell-mediated immunity is achieved through cell-cell interactions [i.e., direct cellular lysis mediated by cytotoxic T cells, natural killer cells or antibody-dependent cellular cytotoxicity (ADCC) mechanisms], or indirectly by soluble products produced by macrophages and lymphocytes. The activities of monokines and lymphokines characterized to date are complex and can be classified into at least four broad categories:

1. growth factors (e.g., interleukin-2, B cell growth factor)
2. differentiation factors (e.g., colony-stimulating factor, B cell differentiation factor)
3. inducers of DNA expression (α- and γ-interferon, interleukin-1, tumor necrosis factor α)
4. cytotoxic factors (tumor necrosis factors α and β, γ-interferon, interleukin 1)

The effects of lymphokines and monokines are not restricted to cells of the immune system. Examples of effects on other cell types include (a) tumor necrosis factor, interleukin 1, and γ-interferon are toxic to neoplastic and virally infected cells (37,47,48), (b) interferons and tumor necrosis factor induce or enhance the expression of major histocompatibility antigens on a variety of cell types including epithelial, endothelial, and neoplastic cells (7,17,40,46), (c) interleukin-1 is a growth factor for fibroblasts (20), and (d) α-interferon affects cell motility (44).

Evidence for Accessibility of Reproductive Tissues to Immune Effectors

It has recently become apparent that both male and female reproductive tissues are immunologically dynamic environments. Numerous immunoglobulin-secreting plasma cells reside in the human cervix and function as an immunological barrier to microbes and in the case of some infertile women, to the passage of sperm into the uterus (15,16,31,33). The uterus is not an immunologically privileged site; numerous studies have documented that leukocytes can be attracted to the uterus and that effective immune responses can occur in this location (2,9,30). Histologic examination of the endometrial lining of the uterus reveals numerous lymphocytes, especially in the late luteal phase, and in decidua (12,28,36,51). Leukocytes are also residents of the vagina, oviduct, and peritoneal cavity (21,22,38).

The seminiferous tubules of the human testis are sequestered from the immune system by the Sertoli cell blood-testis barrier; however, numerous cells of lymphocytic and monocytic morphology and phenotype have been located in other areas of the male reproductive tract. The rete testis and caput epididymis are regions of pronounced leukocytic infiltration, and normal human semen contains an average of 1×10^6 viable leukocytes [reviewed in Alexander and Anderson (1)]. We have recently applied immunohistologic techniques to identify lymphocyte subpopulations and macrophages in normal human male reproductive tissues and semen. Studies currently underway in the laboratory indicate that semen from some infertile men contain elevated levels of certain immunologically relevant leukocyte subpopulations, notably activated macrophages and plasma cells [Anderson et al, in preparation; (50)].

Spermatozoa and gestational tissues express antigens that can stimulate auto- and allogeneic immune responses. The concept that male and female reproductive tissues are immunologically sequestered has given way to a new concept that these tissues are accessible to the immune system and are protected by active local immunoregulatory mechanisms mediated by suppressor lymphocytes and immunosuppressive factors produced by reproductive tissues themselves (1,14). It is now thought that infertility and recurrent spontaneous abortion may be mediated in some individuals by activated lymphocytes that arise because of a breakdown in such natural immunoprotective mechanisms. A better understanding of the immune mediators and their effects in reproductive tissues is critical to an understanding of immunological mechanisms that may cause infertility.

Evidence That Products of Activated Leukocytes Affect Sperm Motility

It has been reported that many infertile men have increased numbers of leukocytes in their semen (19,35,43). Furthermore, high numbers of seminal leukocytes have been associated with decreased sperm motility (10,13,43) and impairment of zona-free hamster egg penetration (10,29). Most studies have implicated polymorphonuclear leukocytes (PMNs) in these infertility effects. However, lymphocytes and macrophages are also present in these samples. We have recently studied the effects of soluble products of activated lymphocytes and macrophages on human sperm motility (24). Supernatants from Ficoll-hypaque-purified (PMN depleted) activated peripheral blood leukocyte cultures significantly decreased sperm motility. In studies of effects of individual lymphokines and monokines on sperm motility, significant antimotility effects were observed when spermatozoa were incubated with the lymphokine γ-interferon, and the monokine, tumor necrosis factor α. Based on these data we hypothesized that soluble products of activated lymphocytes and macrophages could account for poor semen parameters and subfertility observed in some men with high seminal leukocyte counts, or in women with activated leukocytes residing in the reproductive tract. It may therefore be advantageous to explore means to locally activate lymphocytes and macrophages to produce such factors in effecting sperm-antigen-based contraceptive vaccines.

Evidence That Products of Activated Leukocytes Are Toxic to the Preimplantation Embryo and Fetal Membranes

Products of polymorphonuclear leukocytes have been implicated in preimplantation pregnancy loss (2,39,41), but few if any studies to date have systematically addressed the possible effects of activated lymphocytes and macrophages on embryo survival and development. The preimplantation embryo, like testicular germ cells and spermatozoa, lacks expression of class I major histocompatibility complex (MHC) antigens (4,5,49) and therefore would not be recognized as a cellular target by activated cytotoxic T cells. However, we have recently found that soluble products of mitogen- and mixed-lymphocyte-culture-activated lymphocytes and macrophages affect mouse embryonic development in vitro. When individual lymphokines and monokines were tested for effects on mouse blastocyst formation in vitro, it was found that γ-interferon, colony-stimulating factor, and B cell growth factor significantly arrested embryonic development at presumed physiologic doses (25). We hypothesized that infertility could result from local production of such embryotoxic factors by activated lymphocytes and macrophages in response to stimulation by microorganisms or reproductive tissue antigens in the female reproductive tract.

We have also recently explored the effects of lymphokines and monokines on viability, growth, and MHC antigen expression in trophoblast (choriocarcinoma) cell lines. We observed that γ-interferon enhanced the expression of class I MHC transplantation antigens in one of the cell lines with a normally low level of expression of these transplantation antigens (7). It has since been confirmed that γ-interferon induces or enhances the expression of class I MHC antigens in normal human amnion (27) and placental trophoblast (23). This evidence suggests a possible mechanism of cytotoxic T cell-mediated abortion in women with uterine immune activity (and local γ-interferon production), possibly as a result of chronic infection or sensitization to gestational tissues.

We have also recently tested the effects of purified lymphokines and monokines on choriocarcinoma (trophoblast) cell growth and viability in vitro. γ-Interferon and tumor necrosis factor-α significantly inhibited cell proliferation as measured by ^{3}H-thymidine uptake and colony formation assays (11).

Conclusions

Evidence now exists that both human male and female reproductive tracts are accessible to cellular and molecular mediators of immune responses. Furthermore, recent data indicate that spermatozoa, the preimplantation embryo, and trophoblast cells are adversely affected by soluble mediators of cell-mediated immune responses. Efficacy of antifertility vaccines would likely be enhanced by the use of adjuvant-antigen combinations that elicit local cellular as well as humor immune responses.

Acknowledgments. The authors wish to thank the other members of the Fearing Laboratory for contributing to this research: H. Faris, D. Meehan, Drs. R. Berkowitz, F. Haimovici, C. Xu, H. Wolff, and S. Emery-Taylor. This project is supported by PHS grant CA-42738 from the National Institutes of Health, Bethesda, MD, and by the Fearing Laboratory Endowment, Boston, MA.

References

1. Alexander NJ, Anderson DJ: Immunology of semen. *Fertil Steril* 1987;47:192–205.
2. Anderson DJ, Alexander NJ: Induction of uterine leukocytosis and its effect on pregnancy in rats. *Biol Reprod* 1979;21:1143–1152.
3. Anderson DJ, Alexander NJ, Fulgham DL, et al: Immunity to tumor-associated antigens in vasectomized men. *J Natl Cancer Inst* 1982;69:551–559.
4. Anderson DJ, Bach DL, Yunis EL, et al: Major histocompatibility antigens are not expressed on human epididymal sperm. *J Immunol* 1982;129:452–454.
5. Anderson DJ, Narayan P, DeWolf WC: Major histocompatibility antigens are not detectable on post-meiotic human testicular germ cells. *J Immunol* 1984;133:1962–1965.
6. Anderson DJ: Immunogenetic considerations in the development of contraceptive vaccines based on testicular or sperm antigens and anti-sperm immune responses, in Zatuchni GI (ed): *Male Contraception, Recent Advances and Future Prospects.* Philadelphia, Harper & Row/Lippincott, 1985, pp 419–428.
7. Anderson DJ, Berkowitz RS: γ-interferon enhances expression of Class I MHC antigens in the weakly HLA⁺ human choriocarcinoma cell line BeWo, but does not induce MHC expression in the HLA-choriocarcinoma cell line Jar. *J Immunol* 1985;135:2498–2501.
8. Anderson DJ, Hill JA: CD4 (T4⁺) lymphocytes in human male reproductive tissues and semen: implications for the venereal transmission of AIDS. *Fertil Steril* (in press).
9. Beer AE, Billingham RE: Host responses to intra-uterine tissue, cellular and fetal allografts. *J Reprod Fertil*, 1974; [Suppl 21]:59–88.
10. Berger RE, Karp LE, Williamson RA, et al: The relationship of pyospermia and seminal fluid bacteriology to sperm function as reflected in the sperm penetration assay. *Fertil Steril* 1982;37:557–564.
11. Berkowitz RS, Hill JA, Kurtz CB, et al: Influence of lymphokines and monokines on the growth of Jeg 3 human gestational choriocarcinoma (CCA) cells *in vitro*. *Am J Obstet Gynecol* (in press).
12. Bulmer JM, Sunderland CA: Immunohistological characterization of lymphoid cell populations in the early human placental bed. *Immunology* 1984;52:349–356.
13. Caldamone AA, Emilson LBV, Al-Juburi A, et al: Prostatitis: prostatic secretion dysfunction affecting fertility. *Fertil Steril* 1980;34:602–603.
14. Clark DA, Slapsys RM, Croy BA, et al: Local active suppression by suppressor cells in the decidua: a review. *Am J Reprod Immunol* 1984;5:78–83.
15. Clarke GN, Stojanoff A, Cauchi MN, et al: Detection of anti-spermatozoal antibodies of IgA class in cervical mucus. *Am J Reprod Immunol* 1984;5:61.
16. Clarke GN: Detection of antispermatozoal antibodies of IgG, IgA and IgM immunoglobulin classes in cervical mucus. *Am J Reprod Immunol* 1984;5:195–201.
17. Collins T, Lapierre LA, Fiers W, et al: Recombinant tumor necrosis factor increases m-RNA levels and surface expression of HLA-A,B antigens in vascular endothelial cells and dermal fibroblasts *in vitro*. *Proc Natl Acad Sci USA* 1986;83:446–454.
18. Dorsman BG, Tumbh-Oeri AG, Roberts TK: Detection of cell-mediated immunity to spermatozoa in mice and man by the leukocyte adherence-inhibition test. *J Reprod Fertil* 1978;53:277–286.
19. Eliasson R: Clinical examination of infertile men, in Hafez ESE (ed): *Human Semen and Fertility Regulation in Men*. St Louis, CV Mosby, 1976, pp 321–331.

20. Freundlich B, Bomalaski JS, Neilson E, et al: Regulation of fibroblast proliferation and collagen synthesis by cytokines. *Immunol Today* 1986;7:303–307.
21. Haas GG, Beer AE: Immunologic influences on reproductive biology: Sperm gametogenesis and maturation in the male and female genital tracts. *Fertil Steril* 1986; 46:753–766.
22. Haney AF, Muscato JJ, Weinberg JB: Peritoneal fluid cell populations in infertility patients. *Fertil Steril* 1981;35:696–709.
23. Head JR: Expression of MHC antigens on murine trophoblast and their modulation by interferon. *J Immunol* in press.
24. Hill JA, Haimovici F, Politch JA, et al: Effects of soluble products of activated lymphocytes and macrophages (lymphokines and monokines) on human sperm motion parameters. *Fertil Steril* 1987;47:460–465.
25. Hill JA, Haimovici F, Anderson DJ: Products of activated lymphocytes and macrophages inhibit mouse embryo development *in vitro. J Immunol* (in press).
26. Hill JA, Haimovici F, Schiff I, et al: Trophoblast antigens induce production of toxic factors by leukocytes from a subgroup of women with unexplained recurrent abortion: Further evidence for a cellular immune mechanism in spontaneous abortion, Abstract, *Proc Soc Gynecol Invest* 1987.
27. Hunt JS, Wood GW: Gamma interferon induces Class I HLA and B_2 microglobulin expression by human amnion cells. *J Immunol* 1986;136:364–367.
28. Kabawat SE, Mostoufi-adeh M, Driscoll SG, et al: Implantation site in normal pregnancy. A study with monoclonal antibodies. *Am J Pathol* 1985;118:76–92.
29. Maruyama DK, Hale RW, Rogers BJ: Effects of white blood cells on the *in vitro* penetration of zona-free hamster eggs by human spermatozoa. *J Androl* 1985;6: 127–135.
30. McAnulty PA, Morton DB: The immune response of the genital tract of the female rabbit following systemic and local immunization. *J Clin Lab Immunol* 1978;1:255–267.
31. McDermott MR, Bienenstock J: Evidence for a common muscosal immunological system. I. Migration of B immunoblasts into intestinal, respiratory and genital tissues. *J Immunol* 1979;122:1892–1909.
32. McShane PM, Schiff I, Trentham D: Cellular immunity to sperm in infertile women. *JAMA* 1985;253:3555–3558.
33. Murdocj AJM, Buckley CH, Fox H: Hormonal control of the secretory immune system of the human uterine cervix. *J Reprod Immunol* 1982;4:23–31.
34. Nagarkatti RS, Rao SS: Cell-mediated immunity to homologous spermatozoa following vasectomy in the human male. *Clin Exp Immunol* 1976;26:239–249.
35. Nahoum CRD: Medical treatment: inflammation and infection, in Bain J, Schill WB, Schwarztein L (eds): *Treatment of Male Infertilty.* New York, Springer-Verlag, 1982, pp 5–32.
36. Noyes RW, Hertig AT, Rock J: Dating the endometrial biopsy. *Fertil Steril* 1950;1:3–17.
37. Onozaki K, Matsushima K, Kleinerman ES, et al: Role of interleukin 1 in promoting human monocyte mediated tumor cytotoxicity. *J Immunol* 1985;135:314–327.
38. Pandya IJ, Cohen J: The leukocytic reaction of the human uterine cervix to spermatozoa. *Fertil Steril* 1985;43:417–436.
39. Parr EL, Shirley RL: Embryotoxicity of leukocyte extracts and its relationship to intrauterine contraception in humans. *Fertil Steril* 1976;27:1067–1077.

40. Pober JA, Gimbrone MA, Cotran RS, et al: Ia expression by vascular endothelium is inducible by activated T cell and by human interferon. *J Exp Med* 1983;157: 1339–1349.
41. Smith DM, EL Sahwi S, Wilson N, et al: Effect of polymorphonuclear leukocytes on the development of mouse embryos cultured from the two-cell stage to blastocysts. *Biol Reprod* 1971;4:74–83.
42. Smith KA: Lymphokine regulation of T and B cell function, in Paul WE (ed): *Fundamental Immunology*. New York, Raven Press, 1984, pp 559–576.
43. Sokol S, Jacobson CV, Derrick FC: Use of methenamine hippurate in male infertility. *Urology* 1975;6:59–62.
44. Tamm I, Pfeffer LM, Wang E, et al: Interferon-induced changes in cell motility and structure. *Lymphokines* 1984;9:37.
45. Taylor-Emery S, Battaile A, Anderson D: An immunohistological study of lymphoid cells in the human male reproductive tract. *J Reprod Immunol* 1986; [Suppl]:71.
46. Todd I, Pujol-Borrell R, Bottazzo GF, et al: Epithelial MHC class II subregion expression in autoimmunity. *Immunol Today* 1986;7:6.
47. Trinchieri G, Perussia B: Immune interferon: A pleiotropic lymphokine with multiple effects. *Immunol Today* 1985;6:131–134.
48. Urban JL, Shepard HM, Rothstein JL, et al: Tumor necrosis factor: a potent effector molecule for tumor cell killing by activated macrophages. *Proc Natl Acad Sci USA* 1986;83:5233–5237.
49. Webb CG, Gall WE, Edelman GM: Synthesis and distribution of H-2 antigens in preimplantation embryos. *J Exp Med* 1977;146:923–929.
50. Wolff H, Faris HMP, Hill JA, et al: Characterization and quantitation of white blood cell subsets in human semen, Abstract, *J Androl*, 1987; Suppl.
51. Xu C, Hill JA, Anderson DJ: Identification of T lymphocyte subpopulations in normal and abnormal human endometrial biopsies, Abstract, *Proc Soc Gynec Invest* 1987.

CHAPTER 27

Contraceptive Potential of Secretory Immunity

Nancy J. Alexander

Inhaled and ingested antigens continually are in contact with all orifices, including the respiratory and gastrointestinal tract. Mucus-producing cells lining these tracts serve as a first-defense line, a physical barrier, against an onslaught of environmental antigens. An important aspect of this defense is the local immune response, which can be triggered by antigen stimulation independently of a systemic response. Portions of the female reproductive tract are part of this external mucosal system. Cervical mucus, a complex mixture of mucin and other macromolecules, provides an important component of this barrier.

There is considerable interest in the development of a contraceptive vaccine. Any immune method for fertility control involving antigens of spermatozoa will involve producing antibodies in the genital tract. Since stimulation of local immunity is an important consideration, this chapter will serve as a review and a consideration of the possibilities of this approach.

Local Immune Systems

The mucosal immune system (mucosa-associated lymphoid tissue, MALT) is present at many sites throughout the body, although it is perhaps best developed in the gut (gut-associated lymphoid tissue, GALT), and lungs (bronchiole-associated lymphoid tissue, BALT). Stimulation of these tissues results in repopulation of these and other mucosal surfaces with IgA-producing cells (5).

The GALT is composed of Peyer's patches and diffuse cells in the lamina propria of the small and large intestines. Special epithelial cells (M cells) sample the environment and pass potentially antigenic material to immunocompetent cells. These cells have numerous cytoplasmic vesicles but no lysosomes. The M cells engulf luminal antigens by pinocytosis and endocytosis. Thus, M cells transport and process particulate and macromolecules from proteins, viruses, bacteria, and bacterial cell wall components.

The Peyer's patch itself is divided into three regions: a dome covered by the M cells, a lymphoid follicle that contains a germinal center of replenishing cells, and a parafollicular area rich in B cells (Fig. 27.1). The Peyer's patches contain all

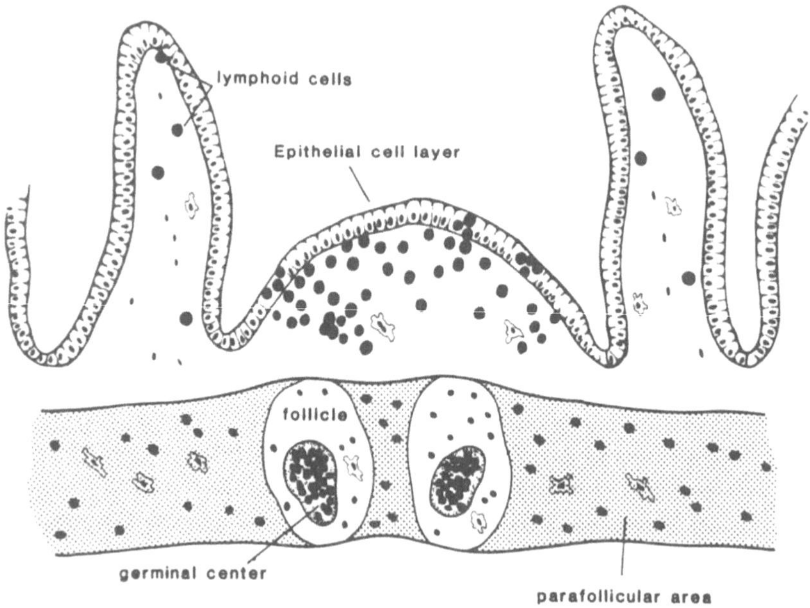

Fig. 27.1. Diagram of Peyer's patch. Microfolds covered with M cells that sample luminal antigens overlay lymphoid follicles and the surrounding parafollicular area.

major lymphoreticular cell types including B lymphocytes, T lymphocytes, macrophages, and I-A$^+$ dendritic cells. Although Lyt-2$^+$ (suppressor T cells) are present, Lyt-1$^+$ (helper T cells) are more common. Whether antigen administration results in tolerance or immunization is little understood (4). The environment plays an important role in the secretory immune response; there are no mucosal antibody-producing cells in the intestines of germ-free animals and newborns. Constant exposure often results in tolerance but serum factors are also involved. An intravenous injection of antibody that is of the IgG_1 class will result in local immune system tolerance. From 85% to 92% of all the B cells bear surface IgM (7). Isotype switching from IgM to IgA results in the high frequency of IgA committed B cells (16). Cultures of Peyer's patch cells from mice orally primed with a T-dependent antigen produce in vitro a secondary immune response that is largely of the IgA isotype (11).

After the M cells present antigen to the germinal follicles, helper T cells contribute to lymphoid cell maturation. IgA-bearing lymphoblasts that migrate to the mesenteric lymph nodes and enter into the circulation are formed (Fig. 27.2). Generally the lymphoblasts return to the same tissue where they were produced. When Peyer's patches are transferred into an irradiated recipient, although other mucosal areas may become seeded, the majority migrate to the lamina propria and become IgA plasma cells.

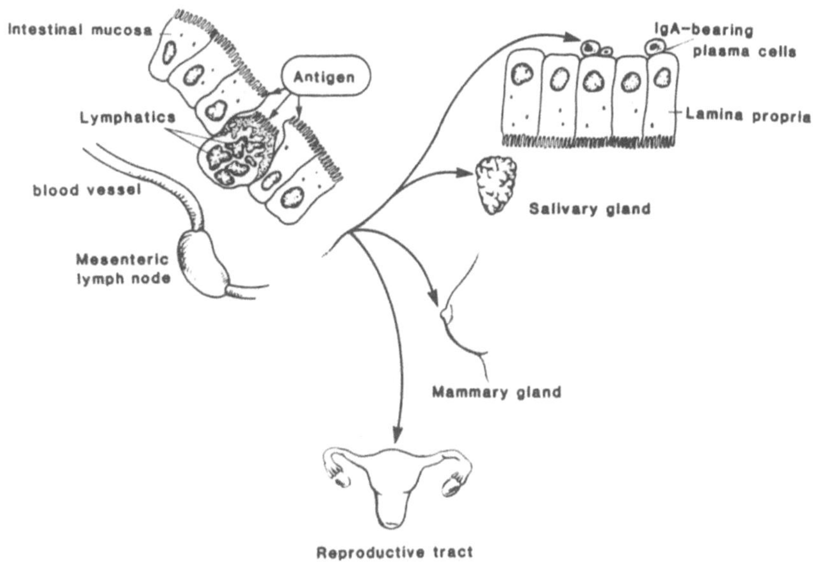

Fig. 27.2. Development of an IgA response. Exposure of the Peyer's patches of the ilium to antigen stimulates lymphocytes underlying the M cells within the patches. These lymphoblasts migrate to the mesenteric lymph nodes where they further mature, enter into the systemic circulation, and subsequently home to various mucosal epithelia.

Evidence that the female reproductive tract may become stimulated from local antigen presentation comes from the report that rudimentary lymphoid aggregates analogous to Peyer's patches have been found in the endocervix, particularly in pregnant and multiparous subjects (6).

IgA

There are two subclasses of IgA in man, namely IgA_1 and IgA_2 (Fig. 27.3). The most notable difference between them is the hinge region; the IgA_1 monomer is covalently linked between the heavy and light chains whereas the IgA_2 monomer is not. In IgA_1 the hinge includes a duplicate 8 amino acid proline sequence with 5-linked oligosaccharides, whereas in IgA_2 the region is a shorter sequence of proline residues (29). Disulfide bonding binds the dimer. The two monomers are joined covalently by a peptide J chain. The secretory component is a unique component that changes IgA so that it is resistant to proteolytic digestion, providing optimum function in a hostile environment. This secretory component binds only to the polymeric form of IgA and interacts noncovalently with the two monomer subunits, forming a unit with a molecular weight of 390,000. IgA does not fix complement by either the classical or alternate pathway. Thus IgA can be

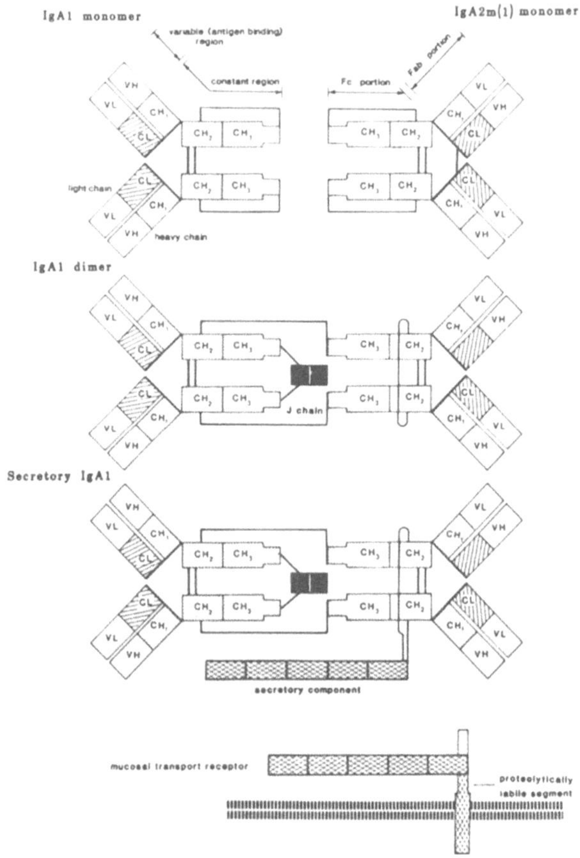

Fig. 27.3. Depiction of the structure of monomeric, dimeric, and secretory IgA.

in contact with the mucosal surface without causing damaging reactions. On the other hand, IgA can initiate antibody-dependent cellular cytotoxicity. Furthermore, dimeric IgA has an agglutinating capacity several times that of serum monomeric IgA.

Production of IgA

After the lymphoblast cells home to the lamina propria, they mature to plasma cells. As these cells secrete IgA, it is transferred to epithelial cells where it combines with secretory piece synthesized by the epithelial cells. This complex of secretory IgA (sIgA) is transported across the epithelial cells in vesicles and released at the apical surface (2,19) (Fig. 27.4). Plasma cells of the IgG type, although not in large numbers, have been observed in the vaginal mucosa (3). Free secretory piece must be produced by the vaginal epithelium, since it is found

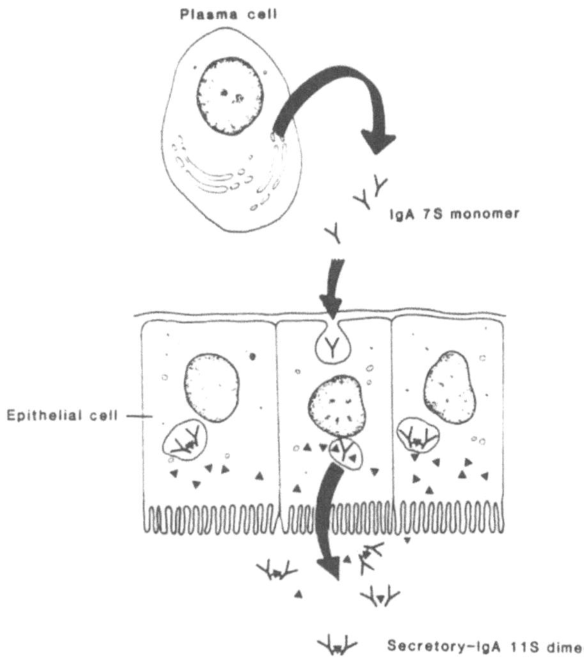

Fig. 27.4. Production of secretory IgA. Monomeric IgA is secreted by matured lymphoblasts that become plasma cells. The IgA is taken up by the epithelial cells and combines with secretory piece produced by the epithelial cells. This complex is released from vesicles at the apical surface.

in the vagina (21). Further support for an active local immune system comes from the finding of an in vitro synthesis of IgA and secretory piece from vaginal explants from the mucocutaneous regions (12).

It seems likely that the cervical mucosa has a more active role in local immunity than the vagina. IgA- and IgG-containing cells are found in the lamina propria of the human endocervix (9), and secretory piece is found in the glandular epithelium (14). Specific antibodies (both IgA and IgG) to *Candida albicans* (8,30), *Poliomyelitis* virus (20), and *Neisseria gonorrhoeae* (22) have been described.

Reproductive Tract

Little is known about the secretory immune system of the male and the precise amounts of immunoglobulin that traverse in the normal animal. Although more studies have been conducted on the female, results here, too, are unclear. The cervical mucus contains significant amounts of IgG, IgA, and IgM; the ratio of IgA to IgG is higher than that found in serum (17). Immunoglobulin levels vary with the menstrual cycle. Reproductive hormones modulate the secretions of immunoglobulins throughout the tract. Schumacher (24–26) showed that IgA levels in cervical secretions are low in the estrogen phase and rise in the progesterone phase. Tauber et al (28), on the other hand, found that periovulatory levels of both IgG and IgA were higher in both cervix and uterus than amounts collected from the follicular and luteal phases (Table 27.1). Levels from the Fallopian tubes were highest in the follicular phase. The highest levels of diffusable IgA were present in the cervical mucosa. The total amount of diffusable complement activity was higher in the internal and external os than in other parts of the reproductive tract. Furthermore, the C3 component was highest in the cervical canal, internal os, and first part of the uterine cavity. Nonetheless, the fact that only C3 is present in measurable quantities brings into question the role of sperm-immobilizing antibodies and infertility. Thus, different organs respond independently to changes in hormone levels, producing different amounts of diffusable proteins. Apparently estrogen stimulates the local immune system of the cervix and uterus but either little local immunity or a different response exists for the Fallopian tubes.

Murdoch et al (18) suggested that the amount of sIgA is dependent on the migration of IgA immunoblasts as well as on the production of secretory components in the epithelial cells. If the control mechanisms of these two cells differ, then a complex regulatory system may be involved. The suggestion that has been made (18) that progesterone causes increased plasma cell localization in the cervix whereas estrogen causes a decrease.

Wira and Sandoe (32) found that in the rat IgA secretion is stimulated by estrogen and depressed by progesterone. However, Parr and Parr (23) showed that the numbers of uterine plasma cells in mice increased with pregnancy, a time of high progesterone. The human cervix, in contrast to the mouse cervix, contains well-developed glands and many plasma cells. Cervical glands may be a major source of immunoglobulins in cervical-vaginal secretions (10).

When considering immunoglobulins and complement levels, it must be stressed that vaginal secretions cannot be assumed to be produced by the vagina,

Table 27.1. Cyclic variation in immunoglobulins from genital tract mucosa.[a]

		F	P	L
Cervix	IgG	3.88 ± 0.25	4.92 ± 0.32	2.41 ± 0.23
	IgA	0.33 ± 0.05	0.63 ± 0.06	0.39 ± 0.06
Uterus	IgG	2.41 ± 0.18	4.66 ± 0.43	3.24 ± 0.22
	IgA	0.25 ± 0.03	0.49 ± 0.05	0.33 ± 0.01
Fallopian tubes	IgG	3.78 ± 0.21	1.66 ± 0.21	2.70 ± 0.16
	IgA	0.37 ± 0.07	0.12 ± 0.02	0.17 ± 0.02

[a] From Tauber et al (28).
F, follicular; P, periovulatory; L, luteal.

but are instead a mixture of secretory products from the Fallopian tubes, endometrium, and cervix. Immunoglobulin in cervical mucus is not produced solely from the cervix. If the studies by Wira and Tauber are generally applicable in women, then the chance of a vaccine with a local immune component is much more enticing.

Vaccine Considerations

Site of Administration

The comparison of various studies is difficult because different animal models were used and because of the many differences in experimental protocol, including sites of administration and use of adjuvants. When McAnulty and Morton (15) administered horseradish peroxidase to rabbits, they found that intrauterine administration caused no measurable IgA or IgG production, either systematically or locally. By far the best response was achieved with vaginal immunization, which resulted in local production of IgA and IgG, as well as systemic antibody levels.

In a study done in mice (27), intrauterine immunization with lactate dehydrogenase isoenzyme C_4 (LDH-C_4) caused an immune response and a lower pregnancy rate than in control animals. These researchers used a systemic (footpad) injection with Freund's complete adjuvant and a week later implanted an agarose cylinder containing LDH-C_4 in the left uterine horn. IgG levels rose in those receiving the intrauterine insemination but not in those who did not. Similar studies in mice by our group in collaboration with Goldberg also suggest that local stimulation enhances a systemic priming immunization, causing an increase in local and systemic responses. Since the animals that develop a systemic response are not necessarily the same as those that develop a local one, local production of IgG is implicated.

In retrospect, perhaps the mouse is a poor experimental model for studies of local immunity. Less than 1% of an intrauterine antigen dose reaches the immune system (13). Furthermore, antigen uptake from the uterus is very slow. Histologic examination of tissues after intrauterine innoculation with India ink revealed that material was retained in the uterine cavity even after 14 days (13). Suppression, similar to that in an allogeneic pregnancy rather than immune activation, may occur.

Whether stimulation of GALT alone can reduce fecundity with no local stimulation is an interesting question, both from the aspect of understanding homing of immunocompetent cells and the development of an oral contraceptive vaccine. Mice fed epididymal sperm by gastric intubation had reduced fecundity, especially when immunized while they were in proestrus (1). High IgA antisperm antibody levels were observed in the treated females 20 days after intragastric immunization.

Considerations for the Future

More studies must be done to better understand the mode of administration (oral only, systemic followed by gastric, use of a local stimulation to assist homing). Further studies comparing the responses in different species will allow better study designs and the efficacious use of primates. Whether adjuvants are necessary for a response in all individuals (individual variation even within a strain is common) and whether local or systemic administration of adjuvants is more appropriate is a consideration.

Finally, whether IgG or IgA is more important when considering a local immune response is not clear. An IgA response in the reproductive tract may not be important. In the gut, secretions containing IgA bind immune complexes, which are subsequently removed with mucous flow. Removal of complexed sperm from the reproductive tract might reduce fertility but a more complete immunologic approach will be necessary. Secretory IgA does not activate the complement cascade nor does it opsonize the appropriate antigens either with or without complement. Unless the binding of IgA to the sperm surface can prevent normal function, then IgA antibodies to sperm may have little effect. Furthermore, there is little evidence for an anamnestic response; little immunologic memory can be expected.

Systemic immunization may be more suitable than a local approach. Circulating antibodies to sperm would have access to all regions of the reproductive tract. The main antibody, IgG, can mediate cytolytic reactions. Furthermore, we suggest that IgG may be produced locally, providing additional immune defense. Since antibodies to the sperm surface can prevent fertilization, further studies on enhancement of local immunity are warranted and important.

Acknowledgment. The work described in this chapter, Publication No. 1497 of the Oregon Regional Primate Research Center, was supported by National Institutes of Health grant RR-00163. Support [PARFR-355] for this project was also provided by the Program for Applied Research on Fertility Regulation, Northwestern University, under a Cooperative Agreement with the US Agency for International Development (A.I.D.) (DPE–0546-A-00-1003-00). The views expressed by the author do not necessarily reflect the views of AID.

References

1. Allardyce RA: Effect of ingested sperm on fecundity in the rat. *J Exp Med* 1984; 159:1548–1553.
2. Allen WD, Smith CG, Porter P: Localization of intracellular immunoglobulin A in porcine intestinal mucosa using enzyme-labelled antibody. *Immunology* 1973;25: 55–70.
3. Behrman SJ, Lieberman ME: Biosynthesis of immunoglobulins by the human cervix, in Blandau RJ, Moghissi K (eds): *The Biology of the Cervix*. Chicago, University of Chicago Press, 1973, pp 235–249.

4. Bhogal BS, Karkhanis YD, Bell MK, et al: Production of auto-anti-idiotypic antibody during the normal immune response. XII. Enhanced auto-anti-idiotypic antibody production as a mechanism for apparent B-cell tolerance in rabbits after feeding antigen. *Cell Immunol* 1986;101:93–104.
5. Bienenstock J, Befus AD: Mucosal immunology. *Immunology* 1980;41:249–270.
6. Chipperfield EJ, Evans BA: The influence of local infection on immunoglobulin formation in the human endocervix. *Clin Exp Immunol* 1972;11:219–233.
7. Frangakis MV, Koopman WJ, Kiyono H, et al: An enzymatic method for preparation of dissociated murine Peyer's patch cells enriched for macrophages. *J Immunol Methods* 1982;48:33–44.
8. Govers J, Girard JP: Some immunological properties of human cervical and vaginal secretions. *Gynecol Invest* 1972;3:184–194.
9. Hulka JP, Omran KF: The uterine cervix as a potential local antibody secretor. *Am J Obstet Gynecol* 1969;104:440–442.
10. Jalanti R, Isliker H: Immunoglobulins in human cervico-vaginal secretions. *Int Arch Allergy Appl Immunol* 1977;53:402–408.
11. Kiyono H, McGhee JR, Wannemuehler MJ, et al: *In vitro* immune response to a T cell-dependent antigen by cultures of disassociated murine Peyer's patch. *Proc Natl Acad Sci USA* 1982;79:596–600.
12. Lai a Fat RFM, Suurmond D, Furth R van: In vitro synthesis of immunoglobulins, secretory component and complement in normal and pathological skin and the adjacent mucous membranes. *Clin Exp Immunol* 1973;14:377–395.
13. Lande IJM: Systemic immunity developing from intrauterine antigen exposure in the nonpregnant rat. *J Reprod Immunol* 1986;9:57–66.
14. Lippes J, Ogra S, Tomasi TB Jr, et al: Immunohistological localization of γG, γA, γM, secretory piece and lactoferrin in the human female genital tract. *Contraception* 1970;1:163–183.
15. McAnulty PA, Morton DB: The immune response of the genital tract of the female rabbit following systemic and local immunization. *J Clin Lab Immunol* 1978;1: 255–260.
16. McGhee JR, Kiyono H, Alley CD: Gut bacterial endotoxin: Influence on gut-associated lymphoreticular tissue and host immune function. *Surv Immunol Res* 1984;3:241–252.
17. Moghissi KS, Neuhaus OW: Composition and properties of human cervical mucus. II. Immunoelectrophoretic studies of the proteins. *Am J Obstet Gynecol* 1962;83: 149–155.
18. Murdoch AJM, Buckley CH, Fox H: Hormonal control of the secretory immune system of the human uterine cervix. *J Reprod Immunol* 1982;4:23–30.
19. Nagura H, Nakane PK, Brown WR: Translocation of dimeric IgA through neoplastic colon cells *in vitro*. *J Immunol* 1979;123:2359–2368.
20. Ogra PL, Ogra SS: Local antibody response to poliovaccine in the human female genital tract. *J Immunol* 1973;110:1307–1311.
21. Ogra PL, Yamanaka T, Losonsky GA: Local immunologic defenses in the genital tract, in Gleicher N (ed): *Reproductive Immunology*. New York, Alan R Liss Inc, 1981, pp 381–394.
22. O'Reilly RJ, Lee L, Welch BG: Secretory IgA antibody responses to *Neisseria gonorrhoeae* in the genital secretions of infected females. *J Infect Dis* 1976;133:113–125.
23. Parr MB, Parr EL: Immunohistochemical localization of immunoglobulins A, G and M in the mouse female genital tract. *J Reprod Fertil* 1985;74:361–370.

24. Schumacher GFB: Soluble proteins in cervical secretions, in Sherman AI (ed): *Pathways to Conception. The Role of the Cervix and the Oviduct in Reproduction*. Springfield, Illinois, Charles C Thomas, 1971, pp 168–187.
25. Schumacher GFB: Soluble proteins of human cervical mucus, in Elstein M, Moghissi KS, Borth R (eds): *Cervical Mucus in Human Reproduction–W.H.O. Coloquium*. Copenhagen, Scriptor, 1972, pp 93–113.
26. Schumacher GFB: Soluble proteins in cervical mucus, in Blandau RJ, Moghissi K (eds): *The Biology of the Cervix*. Chicago, University of Chicago Press, 1973, pp 201–233.
27. Shelton JA, Goldberg E: Local reproductive tract immunity to sperm-specific lactate dehydrogenase-C_4 in SJL/J female mice. *Biol Reprod* 1985;35:873–876.
28. Tauber PF, Wettich W, Nohlen M, et al: Diffusable proteins of the mucosa of the human cervix, uterus, and fallopian tubes: Distribution and variations during the menstrual cycle. *Am J Obstet Gynecol* 1985;151:1115–1125.
29. Underdown BJ, Schiff JM: Immunoglobulin A: Strategic defense initiative at the mucosal surface. *Annu Rev Immunol* 1986;4:389–417.
30. Waldman RH, Cruz JM, Rowe DS: Intravaginal immunization of humans with *Candida albicans*. *J Immunol* 1972;109:662–664.
31. Wira CR, Sandoe CP: Sex steroid hormone regulation of IgA and IgG in rat uterine secretions. *Nature* 1977;268:534–536.
32. Wira CR, Sandoe CP: Hormonal regulation of immunoglobulins: influence of estradiol on immunoglobulins A and G in the rat uterus. *Endocrinology* 1980;106: 1020–1026.

CHAPTER 28

Establishment of Human-Mouse Hybridoma to Human Seminal Plasma Antigen and Its Contribution to a Contraceptive Vaccine

Shinzo Isojima, Minoru Shigeta, Kinu Kameda, and Koji Koyama

Characteristics of Sperm-Immobilizing Factor in Sterile Women

In 1964 Franklin and Dukes reported that women who were sterile owing to unknown causes possessed a high incidence of sperm agglutinating antibodies (SA-Abs) in their sera (2,3). We, however, recognized the occurrence of spontaneous sperm agglutination in many cases on examining patients' sera. To avoid false positive results due to nonspecific sperm agglutination, we instituted a sperm-immobilizing reaction to detect the antibodies, because this reaction required a complement and was more specific in showing the nature of the antibody. Since 1965 the sperm immobilization test (SI-test) has been applied to examine antisperm antibodies in sterile women, and the sperm immobilizing factor was found in 12% of women sterile owing to unknown causes (9). This factor was proved to require a complement to immobilize spermatozoa, and it was fractionated in IgG and IgM sediments, absorbed specifically with washed human spermatozoa, but not with other human tissues. Fifty percent activity of this factor was dissociated from the absorbed spermatozoa by heating at 60°C for 20 minutes or by glycine-HCl (pH 2.5) buffer. Therefore it was proved that this factor is likely to have antibody characteristics. When the patients' sera were examined for antibody activities by using both SA-test and SI-test, the positive correlation of sterility to positive antibodies was much greater in SI-Abs than SA-Abs, as shown in Table 28.1.

Human-Sperm-Specific Antigens as Candidates for Immunological Contraceptive Vaccine

Sperm antigens may play an important role in the preparation of an immunological contraceptive vaccine, based on the fact that many women, without any physical problems, are sterile owing to antisperm antibodies. Therefore, if we could purify the antigens corresponding to antisperm antibodies in patients' sera, the artificial induction of sterility might be possible, judging from our experimental results.

Table 28.1. Percentage of positive results by sperm immobilization (SI) test (Isojima's method) and sperm agglutination (SA) test (Franklin-Dukes' method).

Types of patients	SI-test	SA-test
Sterility		
Unknown cause	37/297 (12.4%)	33/82 (40.2%)
Known cause	6/648 (0.9%)	34/138 (24.6%)
Incomplete survey	8/585 (1.4%)	17/56 (30.4%)
Control group		
Pregnant women	1/202 (0.5%)	45/100 (45.0%)
Unmarried women	0/91 (0%)	5/56 (8.9%)

The presence of two kinds of sperm antigens are well known; one is sperm membrane antigens and the other is human seminal plasma (HSP)-specific antigens such as sperm-coating antigens. From our studies it was learned that antibodies to HSP could immobilize and agglutinate spermatozoa; thus, HSP antigens are the most convenient components for purification because they are already solubilized and there is no need to extract them from cell membrane. HSP antigens were analyzed by immunoelectrophoresis, using rabbit anti-HSP that was obtained by immunizing rabbits with human azoospermic semen. After absorption of antiserum with human serum, kidney and liver tissues, six HSP-specific antigens (Ag) were found by immunoelectrophoresis (Nos. 3, 4, 6, 7, 9, 10 HSP-Ag) (10). Among them, HSP-Ag No. 3 and No. 7 were found to be lactoferrin (6) and ferrisplan (13), respectively. Both antigens contained iron in the molecule and shared antigenicity with human milk. Therefore they are not really HSP-specific and would not be suitable for making a contraceptive vaccine. To purify four other HSP-specific antigens (HSP-Ag Nos. 4, 6, 9, 10), physicochemical and biochemical procedures were used, but it was found difficult to get a single component by polyacrylamide gel electrophoresis (PAGE) analysis. For this reason, immunoaffinity chromatography was attempted to purify the HSP-Ag by using monoclonal antibody (MCA) to HSP.

Establishment of Hybridomas to Secrete MCAs Against HSP

In 1980 we first succeeded in establishing MAC to HSP-Ag (16) and could purify the Ag corresponding to the MCA (13). Later, several investigators reported the establishment of hybridomas to human spermatozoa (8,15). Our procedures are briefly described.

Rat-Mouse and Mouse-Mouse Hybridomas Producing MCAs to HSP-Antigen

Fischer rats were immunized with human azoospermic semen emulsified with complete Freund's adjuvant. The spleen cells from immunized mice were fused with mouse myeloma cells (P3U1) in the presence of PEG-1000. After HAT selection and limiting dilution, the hybridoma cell line that secreted strong SI-Ab was cloned twice, establishing the hybridoma 1C4. The MCA 1C4 possessed strong SI and SA activities and was revealed as IgM. It reacted with human spermatozoa and HSP, but not with other human tissues, except human milk protein. The localization of corresponding antigen to this MCA was postnuclear cap, which was assessed by fluorescent staining.

Unexpectedly, this MCA showed a precipitin line to HSP by immunodiffusion test and was proved to cross-react with human milk protein. This did not react to lactoferrin but reacted to ferrisplan (No. 7 HSP-Ag). To purify the corresponding antigen to MCA 1C4, the supernatant of the cell culture medium was precipitated by 40% saturation with ammonium sulfate, and the γ-globulin fraction was obtained. The MCA 1C4 was bound to CNB*r*-activated Sepharose 4B, and HSP protein was passed through the column. After extensive washing, the bound HSP-Ag corresponding to MCA 1C4 was eluted with a glycine-HCl buffer (pH 2.5), and the dissociated HSP component was purified 196 times after reimmuno-affinity chromatography. The purified 1C4 Ag lost the ability to make a precipitin line to MCA 1C4 by immunodiffusion test, but showed a single band by PAGE. After the addition of sodium dodecyl sulfate (SDS), the molecule was split to a molecular weight of 15 kDa, even though it was present in seminal plasma as a macromolecule of more than 100 kDa before the addition of SDS. The corresponding Ag to MCA 1C4 was purified as a single component possessing a 15 kDa molecule, but it can not be utilized as a contraceptive vaccine because its antigenicity is shared with human milk protein.

MCA 2C6 Specific to HSP-Ag

In seeking to obtain an MCA specifically reacting to HSP specific Ag, trials to make other hybridomas to HSP were carried out, and a hybridoma that secreted strong SI- and SA-MCA was successfully cloned. This hybridoma was named 2C6 and the MCA was designated MCA 2C6. Spleen cells from immunized BALB/C mice with HSP were fused with P3U1 mouse myeloma cells, and the limiting dilution procedures were used for cloning. MCA 2C6 was IgM (κ) and specific to HSP-Ag. It did not cross-react with human milk protein or other human tissue antigens. MCA 2C6 possessed high titers of SI- and SA-activities and also showed a blocking effect on human in vitro fertilization. Therefore, it seemed to be a good candidate for the preparation of a contraceptive vaccine. The MCA stained the acrosome region of human spermatozoa. The purification procedures are shown in Figure 28.1.

After gel filtration of HSP by Sephacryl S-300, evaluating the inhibition of reaction between MCA 2C6 and HSP-Ag showed that the fraction of major peak indicated 3.2-fold increased antigenic activity per protein. In the second step, ion exchange chromatography by DEAE-Sephacel (Pharmacia) was used and the purity per protein was increased 1.8 times; the immunoaffinity chromatography bound MCA 2C6 was used for further purification of HSP-Ag. The fraction eluted by glycine-HCl buffer (pH 2.5) revealed 4.8 times more antigenic activity

Pooled human azoospermic semen

Salted out with 100% $(NH_4)_2SO_4$

Precipate

Sephacryl S-300

First fraction (A1)

DEAE-Sephacel

Main fraction (A2)

Immunoaffinity chromatography
bound MCA 2C6
Eluted with glycine HCl buff. pH 2.5

Bound fraction (A3)

Fig. 28.1. Procedures for purification of HSP-Ag corresponding to MCA 2C6.

per protein, but the purity did not go up on re-chromatography. The overall increase of purity was 27.8 times per protein (Table 28.2). The purification efficiency was very low compared with the case of MCA 1C4.

By using partially purified HSP-Ag corresponding to MCA 2C6, the character of the antigenic molecule was analyzed. The blocking effect of this partially purified 2C6-Ag on the enzyme-linked immunosorbent assay (ELISA) reaction between coated human sperm antigen and MCA 2C6 was used for analysis. The blocking activity of the 2C6-Ag on micro-SI-test by MCA 2C6 was also used to analyze the change of antigenicities caused by various treatments. By treating 2C6-Ag with periodic acid in 50 mmol/L at 4°C overnight, the antigenicity was completely destroyed. It also lost antigenic activity by mixed glycosidase in 0.1 mg/mL at 37°C for 16 hours, by pronase in 0.1 mg/mL at 37°C for 16 hours, and by trypsin in 0.1 mg/mL at 37°C for 16 hours, but not by lipase in 0.1 mg/mL at 37° for 16 hours. The summarized results are shown in Table 28.3. From the results of treating 2C6-Ag with various procedures, the antigenic molecule of 2C6-Ag seems to be glycoprotein. Therefore, further purification will not be easy, even though it may be an ideal antigen for producing a contraceptive vaccine.

Table 28.2. Purification efficiency by each procedure for purifying 2C6-Ag.

Purification procedure	Increased antigen activities per protein	Purification index
Gel filtration	3.2 times	3.2
Ion exchange chromatography	1.8 times	5.8
Immunoaffinity chromatography	4.8 times	27.8

Table 28.3. Summaries of the experiments treating 2C6-Ag with periodic acid and various enzymes.

Treatment with	Antigen activities of 2C6-Ag after treatment
Periodic acid	Decreased
Mixed glycosidase	Decreased
Pronase	Decreased
Trypsin	Decreased
Lipase	Not changed

HSP-Specific MCA 2E5

As with the other MCA, we obtained MCA 2E5, which was HSP-specific and its subclass was $IgG_3(\kappa)$. It expressed strong SI and SA activities but did not have any blocking effect on human in vitro fertilization. The reaction between spermatozoa coated on wells of ELISA plate and MCA 2E5 was examined, before and after treating sperm with various enzymes and periodic acid. The results showed that the corresponding antigen epitope to MCA 2E5 must be present on carbohydrate moieties because it was destroyed by periodic acid treatment (10 mmol/L, room temperature, 10 minutes), mixed glycosidase (1 mg/mL, 37°C, 1 hour), and neuraminidase (0.2 U/mL, 37°C, 30 minutes), but not by pepsin (200 μg/mL, 37°C, overnight), trypsin (100 μg/mL, 37°C, 20 minutes), α-chymotrypsin (100 μg/mL, 37°C, 20 minutes), and lipase (200 μg/mL, 37°C, overnight). The summarized results are shown in Table 28.4.

Human-Mouse Hybridoma Producing MCAs to Human-Sperm-Coating-Antigen

We checked for inhibition of reactions between human spermatozoa and MCAs 1C4, 2C6, and 2E5 by patients' sera possessing SI-Abs, but unexpectedly no serum inhibited this reaction. The possibilities seemed to be that: (a) The anti-

Table 28.4. Summaries of the experiments treating 2E5-Ag with periodic acid and various enzymes.

Treatment with	Antigen activities of 2E5-Ag after treatment
Periodic acid	Decreased
Mixed glycosidase	Decreased
Neuraminidase	Decreased
Pepsin	Not changed
Trypsin	Not changed
α-Chymotrypsin	Not changed
Lipase	Not changed

genic epitopes of human sperm, which induced antibody production, are xenogenic for the mouse, and there is a difference in recognized antigenic epitopes isotypic in humans and xenogenic in mice. (b) Mouse MCA (IgM) to HSP might have stronger avidity to sperm-coating antigen than human antibody. Therefore it became essential to establish human-mouse hybridomas that produce the same anti-HSP antibodies as are raised in sterile women in order to study these antigenic epitopes.

Establishment of Human-Mouse Hybridomas Producing MCAs Against HSP Ag (14)

Peripheral blood lymphocytes (PBL) (6–8 $\times$ 10^6) from patients with high titers of circulating SI-Ab (11) and a history of infertility for more than six years were isolated by Ficoll-Conray density gradient centrifugation. The lymphocytes obtained were cultured for five to six days with washed spermatozoa of the patient's husband (3–4 $\times$ 10^6) in RPMI-1640 medium, with pokeweed mitogen (PWM) supplemented with 10% fetal calf serum (FCS). The stimulated PBL were fused with NS-1 or SP2/0 mouse myeloma cells in the presence of PEG-1000 (fusion ratio 1:1) using a slight modification of the method of Butler et al (1). The fused cells were distributed in a 96-well Corning plate at a concentration of 44 $\times$ 10^3 cells per well. Culture medium was changed to HAT medium after 24 hours. The supernatants of wells with growing hybrid cells were tested 21 days after the cell fusion, for the production of human immunoglobulin (Ig) by a sandwich ELISA method with sensitivity in the range 10 ng to 100 μg/mL. Growing hybrid cells were observed in 214 of 632 wells in three independent fusion experiments, and 21 of the 214 hybridomas growing produced human Igs. All hybridomas positive for production of human Ig were subsequently given the SI-test (12) and micro-SA-test (4).

Three hybridomas produced human SA-Abs. The three positive hybridomas were distinguished from each other in the level of Ig production and in specific reactions to sperm and stability in culture (Table 28.5). Hybridoma H3-1B12 cells were found to produce antibody having both SI and SA activities, whereas the other two hybridomas (H2-3H1 and H3-1B5) produced Abs with SA activity only. The agglutination titer of H3-1B5 supernatant was 1:16, as determined by micro-SA-test, and the minimum concentration for sperm agglutination was 6.25 μg/mL as Ig. Although cloning by limiting dilution was carried out immediately after detection of Ig production, hybridoma H2-3H1 and H3-1B12 cells lost Ab secretion ability after 40 days. The hybridoma H3-1B5 cells, however, preserved human Ig production after two consecutive clonings and were stable for more than 90 days in culture but stopped Ab production thereafter. The H3-1B5 cells secreted IgM(κ) Ab, but stopped producing Ab after three months, because human chromosomes in their cells might easily segregate by well-known phenomenon in heterohybridomas. The supernatant of hybridoma H3-1B5 mass culture was fractionated, and the γ-globulin fraction (3 mg/mL) obtained was biotinylated with N-hydroxysuccinimido (NHD)-biotin, as described by

Table 28.5. Reactivity and stability of human-mouse hybridomas producing antihuman sperm antibodies.

Designation of heterohybridomas	Ig class	Production of human immuno-globulins (mL)	Reactivity against human sperm		Stability
			Micro-SIT	Agglutination	
H2-3H1	NT	100 ng	–	+	Unstable, about 40 days
H3-1B12	NT	10 μg	±	+	Unstable, about 40 days
H3-1B5	IgM(κ)	100 μg	–	+	Stable, more than 90 days

Micro-SIT, microsperm immobilization test; NT, not tested.

Guesdon et al (5), and used for fluorescent staining by using fluorescein isothiocynate (FITC)-avidin conjugate as the second layer. Various human organs, such as liver, kidney, spleen, testis, adrenal cortex, stomach, lymphnode, epididymis, prostate, seminal vesicle, pancreas, thyroid gland, blood vessel, and intestine, were not stained by MCA H3-1B5, but the ejaculated human sperm showed positive staining.

Herr et al (7) reported the establishment of human-mouse heterohybridoma secreting human Ig, which possessed binding activity to human sperm, from lymphocytes of a vasectomized man, but did not show any biological activity of their MCA on human spermatozoa. Therefore, MCA H3-1B5 may be the first biologically active MCA obtained from human-mouse hybridoma that was established from a patient. The summarized results of the course in establishing human-mouse heterohybridoma are shown in Figure 28.2.

Establishment of Human-Hybridoma Secreting MCA With Extremely High Titers of SI and SA Activities

The three human-mouse heterohybridomas that secreted antisperm MCAs were successfully established in our laboratory, but the SI and SA activities of these MCAs were weak. These hybridomas eventually could not survive more than 90 days. Therefore we tried to obtain a stable hybridoma that secretes high titers of SI- and SA-Abs and finally succeeded. We took for cell fusion PBLs from a 39-year-old nulliparous woman possessing SI- and SA-Abs, and with a history of infertility for more than 15 years. Routine medical examinations did not show any reasonable cause for her infertility except for the presence of high titers of antisperm antibodies such as SI- and SA-Abs in her serum. The antibody titers of her serum were 45.0 SI_{50} (11) for SI-Ab, and 1:32 dilutions for SA-Ab (4). The PBLs isolated (1.0×10^7) were mixed with washed spermatozoa (5×10^6) from a healthy donor in a medium (RPMI 1640) containing PWM (final concentration 1:100) and 10% FCS. The mixture was incubated at 37°C for five days, and the stimulated lymphocytes were washed twice and suspended in the RPMI

Heparinized peripheral blood lymphocytes (PBL) — Semen

centrifuged on Ficoll-Conray — washed in RPMI 1640

PBL — Spermatozoa (SP)

mixed (PBL:SP=2:1)

cultured in 10% FCS-RPMI 1640 c̄ 1% PWM for 5 days

Stimulated PBL — NS-1 or SP 2/0

mixed (PBL:NS-1=1:1)

Cell fusion

fused by 42.5% PEG-1000-RPMI 1640 c̄ 15% DMSO
HAT selection in 15% FCS–RPMI 1640 c̄ HAT
Cloning by limiting dilution
Screening of antibody production by ELISA and micri-SIT

Hybridomas

Fig. 28.2. Procedure for establishment of human-mouse heterohybridomas.

1640 medium. The stimulated lymphocytes (5.3×10^6) were fused with the same number of mouse myeloma cells (NS-1) in the presence of PEG-1000, and 100 μL of the fused cells were distributed to each of 96 wells and cultured. Culture medium was changed to a HAT selection medium 24 hours later, and the supernatant of each well growing hybrid cells was tested on the 12th, 19th, and 23rd day after cell fusion for the production of human Igs by a sandwich ELISA method.

On the 23rd day after cell fusion, hybrid cell growth was observed in 58 of 96 wells. Human Ig production was detected in the culture supernatants from 22 of 58 wells with cell growing, but the antibody indicating antisperm activities was detected in only one well. After three clonings, the strongest antibody-producing clone was selected.

One of the colonies that showed clonal growing was selected for further experiments and designated as hybridoma H6-3C4. The established hybridoma H6-3C4 cells were transferred into a 24-well plastic plate (Corning-25820) and cultured. After cell growth, the cell culture was expanded to tissue culture flasks (Corning-25100 and 25110, USA). The supernatant obtained from culture flasks contained approximately 1.5 μg Ig/mL as human IgM and strong SI- and SA-antibody activities. It had SI-Ab titers of 5,000 SI_{50} units detected by the quantitative SI-test and gave a positive SA-test as high as 1:1,600 dilutions.

Hybridoma H6-3C4 cells continued to grow for more than 11 months without any loss of IgM production, indicating the same strong level of SI and SA activities as at the beginning (Table 28.6). Ig subclass was IgM(λ). Chromosome

Table 28.6. Follow-up studies of IgM concentrations and sperm-immobilizing and agglutinating antibody activities in culture supernatants of hybridoma H6-3C4.

		Antibody activity	
		SI	SA
Days after cell fusion	IgM (μg/mL)	(SI_{50} units)	(dilutions)
63	1.5	5,000	1:1,600
120	ND	8,400	ND
189	2.1	6,900	1:3,200
249	2.0	5,900	1.3,200

SI, sperm immobilization; SA, sperm agglutination.
ND, not determined.

Table 28.7. Summary of characteristic properties of H6-3C4 MCA.

Immunoglobulin class	Human IgM(λ)
Biological activity	
Sperm immobilization	+ (SI_{50} unit: 5,000)
Sperm agglutination	+ (1:1,600 dilutions)
Tissue specificity (reactivity to human specimens)	
Ejaculated spermatozoa	+
Seminal plasma	+
Milk proteins	−
Serum proteins	−
Testis	−
Epididymis	not conclusive
Seminal vesicle	+
Prostate	±
Liver	−
Spleen	−
Brain	−
Red blood cell	−
White blood cell	−
Species specificity (reactivity to spermatozoa)	
Human	+
Boar	±
Goat	−
Rabbit	−
Hamster	−
Rat	−

+ strongly positive, ± weakly positive, − negative.

analysis showed that hybridoma H6-3C4 cells contained two types of chromosomes, an acrocentric type from mice and a metacentric type from humans, and at least seven human chromosomes, including chromosome numbers 5, 9, 11, 12, 14, 17, and 22, which were identified by Q-band staining. The nature of MCA H6-3C4 was examined by antibody absorption assay, as shown in Table 28.7. The mAb was absorbed only with human spermatozoa but not with leukocytes or erthrocytes. The MCA was absorbed with human azoospermic semen but not with human milk, serum, liver, kidney, brain, or spleen tissues. The absorption capabilities for the MCA were also present in seminal vesicles and also weakly in the prostate. The absorption of MCA by human sperm and HSP were species specific and those of bulls, goats, and rabbits did not have any absorption capabilities. Boar spermatozoa had minor cross-reactivity. The competitive binding inhibition of ^{125}I-labeled MCA H6-3C4 to human sperm by patient's serum showed that the patient's serum, whose lymphocytes were donated to make this hybridoma, clearly inhibited antibody binding (Fig. 28.3).

It was proved that the antigen of HSP corresponding to MCA H6-3C4 was in the 670 kDa fraction of macromolecules after gel filtration of HSP by Sephacryl

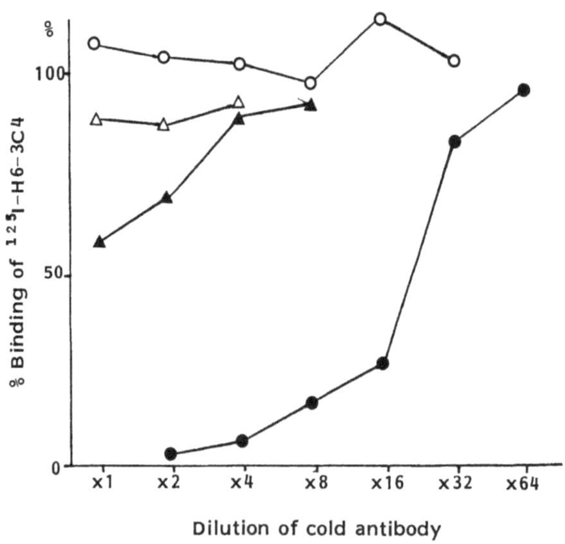

To wells coated with human spermatozoa, ^{125}I-labeled MCA and serially diluted test specimens were added simultaneously for competitive binding to human spermatozoa

●–● Culture supernatant of H6-3C4 hybridoma
▲–▲ Serum of patient who donated lymphocytes for cell fusion
△–△ Control of human serum without SI antibidies
○–○ Control culture supernatant form a human-mouse hybridoma (1C2) that secretes human IgM (λ) with unknown specificity.

Fig. 28.3. Competitive binding inhibition of ^{125}I labeled MCA H6-3C4 to human spermatozoa by patient's serum.

S-300. It was not split by trypsin digestion, but was divided into 15 to 25 kDa molecules after SDS-PAGE in reduced condition.

As the reduced condition itself did not affect molecular size, the detergent seems to have some role of dissociating macromolecule to about 20 kDa fragment. The 20 kDa HSP molecule corresponding to MCA H6-3C4, which was transblotted on nitrocellulose membrane after SDS-PAGE, reacted with ^{125}I-IgM of MCA H6-3C4. Therefore the antigen epitope corresponding to MCA is in the macromolecule around 670 kDa in HSP and it may be split to 20 kDa molecule after SDS-PAGE.

Contraceptive Vaccine From HSP

HSP-Ags are already solubilized and there is no need to extract them from cell membranes like sperm membrane antigens. SI- and SA-Abs found in sterile women which were raised against HSP-Ags, were not harmful to the women's health, except that they caused sterility. Therefore, HSP-Ags would be ideal immunogens if they could be purified as a single component. Immunoaffinity chromatography may be one of the best procedures for purifying antigen corresponding to a certain antibody and obtaining a large amount of human MCA to HSP-specific antigen is required for this purpose. From our current studies, human MCA H6-3C4 may be a very useful tool for making a contraceptive vaccine because it is an antisperm MCA that must be the same Ab as in the serum of a sterile woman. Hybridoma H6-3C4 still secretes human MCA with antibody titers as high as at the beginning of cell culture, even more than 11 months later. First, mass production of this MCA must be studied to purify the antigen, and, second, the permanent maintenance of this hybridoma cell line must be studied. Other problems in making a contraceptive vaccine are the antigenicity of purified HSP-Ags because the immune responses of women to a certain purified HSP-Ag may possibly be different for each owing to genetic problems. For this reason, a mixture of several purified HSP-Ags corresponding to antisperm antibodies should be considered.

Acknowledgments. We are grateful for technical assistance from Mrs. N. Fujiwara. This research was supported by grants from the Japanese Ministry of Education and the Japanese Foundation of Private Schools for Promoting Science.

References

1. Butler JH, Lane CH, Fanci AC: Delineation of optimal conditions for producing mouse-human heterohybridomas from human peripheral blood B cells of immunized subjects. *J Immunol* 1983;130:165–168.
2. Franklin RR, Dukes CD: Antispermatozoal antibody and unexplained infertility. *Am J Obstet Gynecol* 1964;89:6–9.

3. Franklin RR, Dukes CD: Further studies on sperm agglutinating antibody and unexplained infertility. *JAMA* 1964;190:682–683.
4. Frieberg J: A simple and sensitive micro-method for demonstration of sperm-agglutinating activity in serum from infertile men and women. *Acta Obstet Gynecol Scand* [Suppl] 1974;36:21–29.
5. Guesdon JL, Ternynck T, Avrameas S: The use of Avidin-Biotin interaction of immunoenzymatic techniques. *J Histochem Cytochem* 1979;27:1131–1139.
6. Hekman A, Rümke P: The antigens of human seminal plasma; with special reference to lactoferrin as a spermatozoa coating antigen. *Fertil Steril* 1969;20:312–323.
7. Herr JC, Fowler JE Jr, Howards SS, et al: Human antisperm monoclonal antibodies constructed postvasectomy. *Biol Reprod* 1985;32:695–711.
8. Isahakia M, Alexander NJ: Interspecies cross-reactivity of monoclonal antibodies directed against human sperm antigens. *Biol Reprod* 1984;30:1015–1026.
9. Isojima S, Li TS, Ashitaka Y: Immunologic analysis of sperm immobilizing factor found in sera of women with unexplained sterility. *Am J Obstet Gynecol* 1968;101: 677–683.
10. Isojima S, Koyama K, Tsuchiya K: The effect on fertility in women of circulating antibodies against human spermatozoa. *J Reprod Fertil [Suppl]* 1974;21:125–150.
11. Isojima S, Koyama K: Quantitative estimation of sperm immobilizing antibody in the sera of women with sterility of unknown etiology: the 50% sperm immobilization unit (SI_{50}). *Excerpta Medica, Intl Cong Series* 1976;370:10–15.
12. Isojima S, Koyama K: Microtechnique of sperm immobilization test, in Bratanov K, Vulchanov VH, Dikov V, Georgieva R, Somler B (eds): Immunology of Reproduction. Sofia, Bulgarian Academy of Sciences, 1979, pp 215–219.
13. Isojima S, Koyama K, Fujiwara N: Purification of human seminal plasma No. 7 antigen by immunoaffinity chromatography on bound monoclonal antibody. *Clin Exp Immunol* 1982;49:449–456.
14. Kyurkchiev S, Shigeta M, Koyama K, et al: A human-mouse hybridoma producing monoclonal antibody against human sperm coating antigen. *Immunology* 1986;57: 489–492.
15. Lee C-YG, Huang Y-S, Huang C-H, et al: Monoclonal antibodies to human sperm antigens. *J Reprod Immunol* 1982;4:173–181.
16. Shigeta M, Watanabe T, Matsuyama S, et al: Sperm immobilizing monoclonal antibody to human seminal plasma antigens. *Clin Exp Immunol* 1980;42:458–462.

CHAPTER 29

An Immunoreactive Human Sperm Antigen in Rat Spermatogenic Cells

Tatsuji Haneji and S.S. Koide

Introduction

Antisperm antibodies have been demonstrated in sera and cervical fluids of infertile subjects (3,5,7,10). The condition is complicated by the findings that there is a marked variability in the antibody titers and that multiple antibodies interacting with different sperm proteins are produced. Hence, a cause and effect relationship between antisperm antibodies and infertility has not been established. There appears to be a consensus that sperm agglutinating and/or immobilizing antibodies are responsible for the infertile condition (5,9,10).

To prove that the production of antisperm antibodies is a cause of infertility, the following postulates need to be fulfilled:

1. Demonstrate significant titer of antisperm antibodies in sera and/or cervical fluids obtained from infertile subjects of undetermined etiology;
2. Isolate and characterize the antisperm antibodies and the interacting antigens;
3. Immunization of a suitable experimental animal with the antigen results in infertility;
4. Active immunization of human subjects with the antigen induces an infertile condition.

An experimental animal model simulating the clinical syndrome of "immunological infertility" needs to be developed. To induce infertility in an experimental animal by administrating a human sperm protein will be an important step in our quest for an antifertility vaccine. This task is feasible in view of the findings that sera from infertile women interact with sperm of various species (6) and that monoclonal antibodies (MCA) raised against human sperm or those of other species show interspecies cross-reactivity (2,4,8,13). We initially cultivated a hybridoma cell line that secretes MCA inducing agglutination of human sperm (19). This MCA, designated as YWK-I, did not cross-react with sperm of other species tested. Subsequently, we have prepared and cultivated a second hybridoma cell line, producing MCA that induces sperm agglutination (20). This chapter describes our experimental results with this MCA, designated as YWK-II.

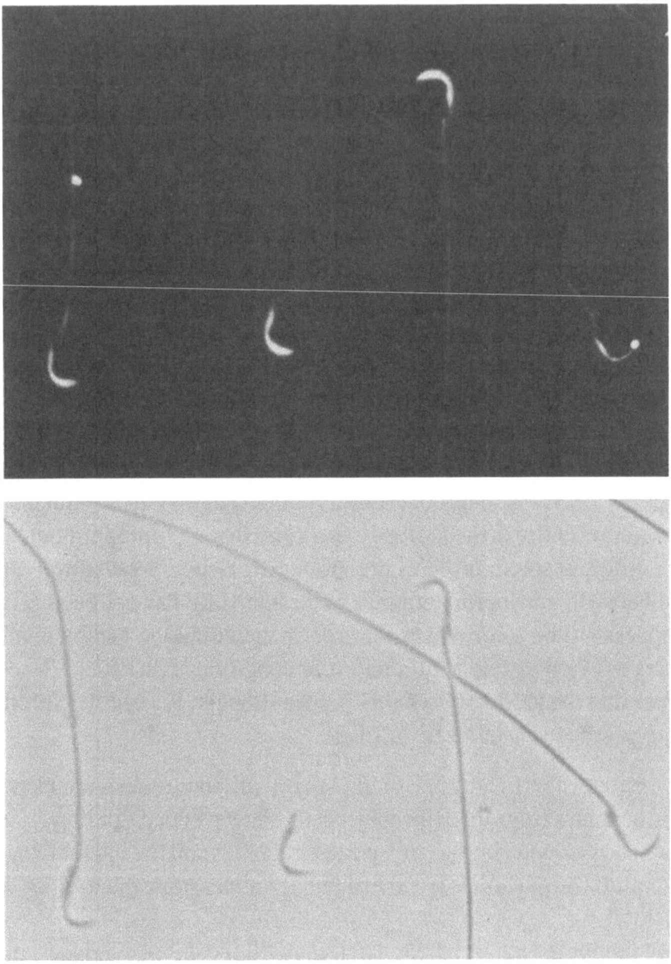

Fig. 29.1. Indirect immunofluorescent staining of rat spermatozoa from the caput epididymis with the YWK-II monoclonal antibody. Fluorescence (**A**) and phase contrast (**B**) microphotographs. Note the intense fluorescence staining over the surface of the head of the spermatozoa except for the postacrosomal region. (× 800)

Experimental

Localization of the YWK-II Antibody in Rat Spermatozoa, Testis, and Epididymis

The YWK-II ascitic fluid was used as the primary antibody, and its location on the surface of rat spermatozoa was determined by an indirect immunofluorescence method. Intense staining was found over the sperm head, while minimal to no staining occurred over the midpiece and tail (Fig. 29.1). In contrast, the antibody localized to the equatorial region of the head of human sperm (20).

With paraffin sections of rat testis, intense staining occurred around the luminal cells of the seminiferous tubules (Fig. 29.2). The heads of the elongated spermatids and spermatozoa of the luminal region stained intensely. Spermatogonia, leptotene and zygotene spermatocytes in the basal layers and pachytene spermatocytes and round spermatids in the luminal region also showed weak to moderate but significant staining (Fig. 29.2). Spermatozoa located in the lumen of the epididymis and the epididymal epithelium stained intensely (Fig. 29.3).

Localization of YWK-II Antibody With Cultured Rat Testicular Cells

Sertoli cell-enriched cultures with germ cells were prepared from 25-day-old rat testis according to Dorrington and Fritz (1). The peritubular myoid cells were cultured by the method of Tung and Fritz (18). Germ cells were obtained by gentle pipetting or by hypotonic shock treatment with distilled water from those attached to Sertoli cells (21). Sertoli cell-germ cell co-cultures and peritubular myoid cell cultures were incubated with YWK-II MCA and stained with FITC-conjugated IgG fraction of goat antimouse IgM. Germ cells, principally sperm-

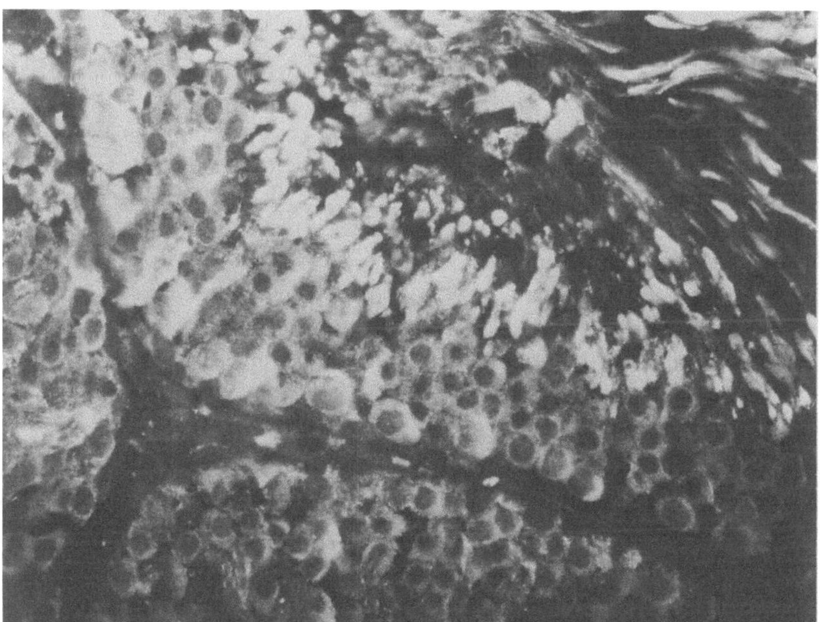

Fig. 29.2. Indirect immunofluorescent staining of adult rat testis with the YWK-II monoclonal antibody. Note the intense fluorescence of the spermatozoa located in the lumen of the seminiferous tubule. Basal layers of the tubules are also stained. (× 600)

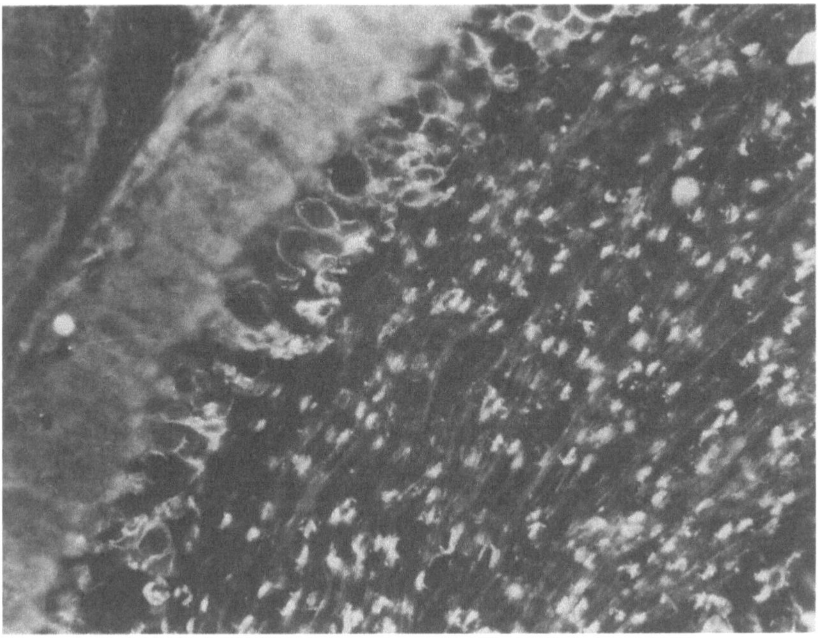

Fig. 29.3. Indirect immunofluorescent staining of adult rat epididymis with the YWK-II monoclonal antibody. Note that the spermatozoa in the lumen and the epithelial cells of the epididymis stained intensely. (× 600)

atogonia and early spermatocytes, attached to the Sertoli cell monolayer stained strongly (Fig. 29.4). Sertoli cells and peritubular myoid cells consistently remained unstable (Fig. 29.4). To validate further that the YWK-II MCA interacts with the germ cells, germ cells were isolated from rat testis by collagenase/dispase treatment or cultured with Sertoli cells as described above. Germ cells, immediately after isolation, showed weak immunofluorescence staining that increased with the duration of co-culture. Germ cells, isolated from the Sertoli cell monolayers after culturing for six days, showed intense staining (Fig. 29.4).

Estimation of M_r of the YWK-II Antigen

Extract of sperm proteins was prepared with sperm collected from the epididymis of adult Wistar-strain rats. Cytosol and membrane fractions from adult rat testes were prepared by homogenization and centrifugation. The proteins were analyzed by sodium dodecyl sulfate (SDS) polyacrylamide gel electrophoresis according to Laemmli (11) and by the immunoblot procedure of Towbin et al (16). The immunoreactivity was determined by incubation of the nitrocellulose sheets with YWK-II ascitic fluid or purified antibody (diluted 1:100) followed by biotin-conjugated $F(ab')_2$ fragment goat antimouse IgM (diluted 1:2,000), and

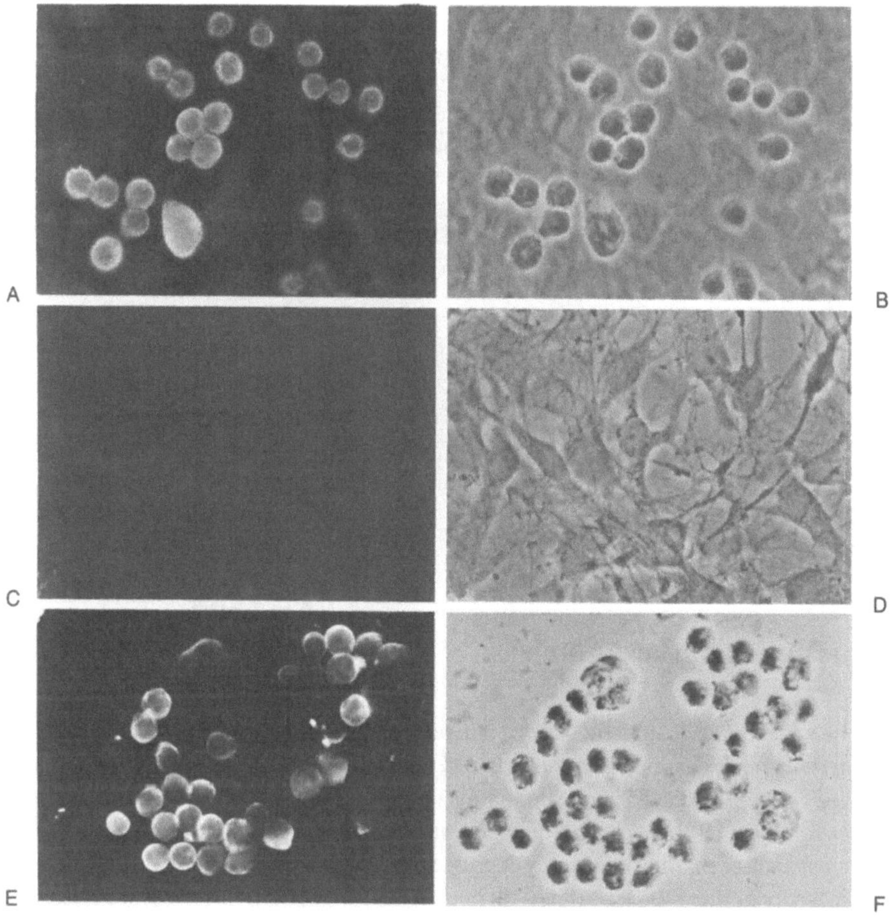

Fig. 29.4. Indirect immunofluorescent staining of cultured rat testicular cells with the YWK-II monoclonal antibody. Fluorescent staining (left) and phase contrast microphotographs (right). **A**,**B**: Sertoli cells and germ cells, respectively, in co-culture; **C**,**D**: peritubular myoid cells; **E**,**F**: germ cells isolated from Sertoli germ cell co-cultures by hypotonic treatment. (× 400)

by peroxidase conjugated avidin (diluted 1:2,000). Reaction was developed in 4-chloro-1-naphthol and hydrogen peroxide.

Immunoblotting of the membrane proteins of the testis showed a major positive band of 88 kDa and a minor band of 115 kDa (Fig. 29.5). The cytosol proteins did not stain. Immunoblotting of epididymal sperm proteins showed staining of a single band of an estimated M_r of 115 kDa (Fig. 29.5). The finding of a single 115 kDa antigen in the epididymal sperm and a major 88 kDa in the testis suggests variation in the expression of the gene coding for the YWK-II antigenic variants during spermiogenesis.

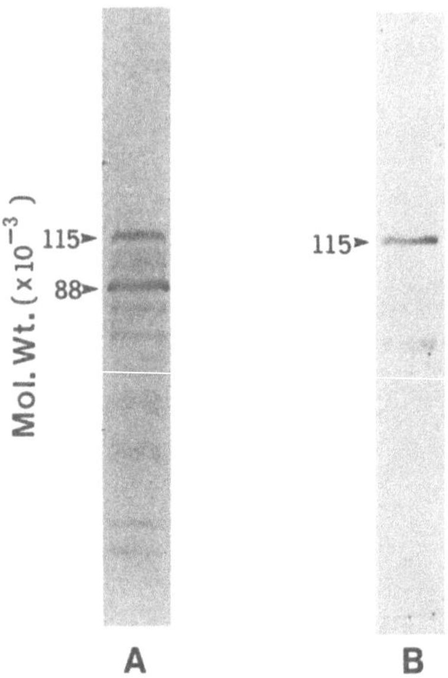

Fig. 29.5. Identification of the YWK-II antigen in adult testicular membrane fractions and in proteins of epididymal spermatozoa. **A**: Immunoblot of testicular membrane proteins separated on a 12.5% SDS-polyacrylamide gel and probed with the YWK-II monoclonal antibody. **B**: Immunoblot of spermatozoa antigens separated on a 10% SDS-polyacrylamide gel and probed with the YWK-II monoclonal antibody.

Germ Cell Surface Antigens

The appearance of specific surface antigens on pachytene spermatocytes during spermiogenesis has been demonstrated serologically (12,14,18) while other antigens are expressed on round spermatids (15,17). YWK-II antigen is localized in germ cells at all stages of spermatogenesis, including spermatogonia, spermatocytes, and spermatids. Thus it is expressed early in spermatogenesis. The cross-reactivity of the YWK-II MCA with human and rat sperm is not unusual because MCAs raised against human sperm proteins have been shown to interact with sperm and spermatogenic cells of various mammalian species (8,13).

Significance of the YWK-II Antigen

Since the YWK-II antigen is localized to the sperm and germ cells of rat testis, it may serve as a useful marker of rat germ cells. By Western immunoblot method two variants of the YWK-II antigen were found, a 115 kDa in the epididymal sperm and an 88 kDa in the testis membrane faction. This finding suggests that a switch in the production of the YWK-II variants may have occurred during spermiogenesis. The presence of the antigen in the epididymal epithelial cells suggests that it may be a sperm-coating protein and may participate in sperm maturation. The rat would be a useful experimental animal model to observe the antifertility effect of immunization with the YWK-II protein.

Summary

A mouse-fused spleen cell line, raised against human sperm proteins, was cultivated and designated as YWK-II. The monoclonal antibody (MCA) induced agglutination of human sperm and was used to locate the antigen on rat sperm and testicular cells by an indirect immunofluorescent method. The YWK-II MCA localized over the acrosomal region of the head of rat sperm while the postacrosomal region of head and tail remained unstained. Staining of germ cells occurred at every developmental stage while Sertoli cells and peritubular myoid cells were not stained. With co-cultures of Sertoli cells and germ cells, staining was restricted to germ cells.

To identify the YWK-II antigen, rat sperm and testis proteins were analyzed by sodium dodecyl sulfate polyacrylamide gel electrophoresis and by an immunoblot technique. Staining occurred with a single sperm protein of estimated M_r of 115 kDa and with two testis proteins of estimated M_r of 88 and 115 kDa.

It can be concluded that rat germ cells and sperm contain antigens that interact with the YWK-II MCA. These antigens may play a role in the differentiation of spermatozoa.

Acknowledgment. The study was supported by a grant from the Mellon Foundation.

References

1. Dorrington J, Fritz IB: Cellular localization of 5α-reductase and 3α-hydroxysteroid dehydrogenase in the seminiferous tubules of the rat testis. *Endocrinology* 1975;96: 879–889.
2. Feuchter FA, Vernon RB, Eddy EM: Analysis of the sperm surface with monoclonal antibodies: Topographically restricted antigens appearing in the epididymis. *Biol Reprod* 1981;24:1099–1110.
3. Franklin RR, Dukes CD: Antispermatozoa antibody and unexplained infertility. *Acta Obstet Gynecol Scand* 1964;89:6–9.
4. Gaunt SJ: A 28K-dalton cell surface autoantigen of spermatogenesis: Characterization using a monoclonal antibody. *Dev Biol* 1982;89:92–100.
5. Haas GG Jr, Cines DB, Schreiber AD: Immunologic infertility: Identification of patients with antisperm antibody. *N Engl J Med* 1980;303:722–727.
6. Hanson KB: Immunofluorescent studies on mammalian spermatozoa: Antigenic relationship between human, bull and rabbit spermatozoa. *J Reprod Fertil* 1972;29: 19–28.
7. Hargreave TB: Incidence of sperm agglutinating and immobilizing sperm antibodies in infertile couples. *Int J Fertil* 1982;27:90–94.
8. Isahakia M, Alexander NJ: Interspecies cross-reactivity of monoclonal antibodies detected against human sperm antigens. *Biol Reprod* 1984;30:1015–1026.

9. Isojima S, Koyama K, Tsuchiya K: The effect of fertility in women on circulating antibodies against human spermatozoa. *J Reprod Fertil* [Suppl] 1974;21:125–150.
10. Jones WR: Immunologic infertility: Fact or fiction? *Fertil Steril* 1980;33:577–586.
11. Laemmli UK: Cleavage of structural proteins during the assembly of the head of bacteriophage T_4. *Nature* 1970;227:680–685.
12. Millette CF, Bellve AR: Temporal expression of membrane antigens during mouse spermatogenesis. *J Cell Biol* 1977;74:86–97.
13. Naz RK, Alexander NJ, Isahakia M, et al: Monoclonal antibody to a human germ cell membrane glycoprotein that inhibits fertilization. *Science* 1984;225:342–344.
14. O'Rand MG, Romrell LJ: Appearance of cell surface auto- and isoantigens during spermatogenesis in the rabbit. *Dev Biol* 1977;55:347–358.
15. O'Rand MG, Romrell LJ: Appearance of regional surface autoantigens during spermatogenesis: Comparison of anti-testis and anti-sperm autoantisera. *Dev Biol* 1980;75:431–441.
16. Towbin H, Staehlin T, Gordin J: Electrophoretic transfer of proteins from polyacrylamide gels to nitrocellulose sheets: Procedure and some applications. *Proc Natl Acad Sci USA* 1979;76:4350–4354.
17. Tung KSK, Han LPB, Evan AP: Differentiation autoantigen of testicular cells and spermatozoa in the guinea pig. *Dev Biol* 1979;68:224–238.
18. Tung PS, Fritz IB: Isolation and culture of testicular cells, in Hafez ESE (ed): *Techniques of Human Andrology.* Amsterdam, North-Holland, 1977, pp 125–146.
19. Yan CY, Wang LF, Sato E, et al: Monoclonal antibody inducing human sperm agglutination. *Am J Reprod Immunol* 1983;4:111–115.
20. Yan CY, Wang LF, Mitsudo SM, et al: Characterization of an antisperm monoclonal antibody inducing human sperm agglutination, in Talwar P (ed): *Immunological Approaches to Contraception and Promotion of Fertility.* New York, Plenum Press, 1986, pp 231–240.
21. Ziparo E, Germia R, Russo MA, et al: Surface interaction in vitro between Sertoli cells and germ cells at different stages of spermatogenesis. *Am J Anat* 1980;159: 385–388.

CHAPTER 30

Role of the Fertilization Antigen (FA-1) in Immunoregulation of Fertility and Involuntary Infertility in Humans

Rajesh K. Naz

Introduction

Since 1899, when Landsteiner and Metchnikoff independently demonstrated that injection of sperm from heterospecies can produce antibody response in animals (22,31), sperm have been of great interest as the target antigen for the control of fertility. Now it has been successfully documented that deliberate immunization of animals of various species with preparations of isologous sperm or mature testis results in infertility. The cause of infertility after isoimmunization has been attributed to either fertilization failure or preimplantation embryo mortality, or both, as reported in mice, rabbits, guinea pigs, and cattle (7,18,20,21,28).

Effects of Antisperm Antibodies on Fertility in Humans

Sperm have both auto- as well as isoantigenic potentials in humans, as is evident from the data on vasectomy and involuntary infertility. About 50% to 70% of men form antisperm antibodies after vasectomy (56), and up to 30% of infertility is associated with the presence of antisperm antibodies in at least one partner in infertile couples (29). The evidence that these antisperm antibodies are the causative factor of infertility is indicated by the following data:

1. The restoration of fertility has been recorded in the majority of (but not all) men after vasovasostomies, and the nonrestoration of fertility in these men has been attributed to the presence of circulating antisperm antibodies (11).
2. Successful conceptions in immunoinfertile couples treated with immunosuppressive agents have been attributed to the drop in antibody titers, indicating the correlation of antibodies with infertility (1,14).
3. Active immunization of men with male germ cells has been reported to cause aspermatogenic orchitis (25). Active immunization (iso-) of women with human spermatozoa has been reported to induce infertility associated with the production of antisperm antibodies (2,5,6).
4. Clear evidence that antisperm antibodies can block fertilization has been reported using in vitro fertilization (IVF) techniques. Tsukui et al (53) showed

that antisperm antibody present in serum of infertile women, when added to the insemination medium used in IVF, blocked human sperm penetration through human zona pellucida. When these antibodies were absorbed with washed spermatozoa, the blocking effects disappeared. This cumulative evidence supports the contention that sperm are immunogenic (auto- and iso-) in humans and that the antibodies reacting with them can induce infertility when present in situ at a point and time when fertilization occurs.

Cross-Reactivity of Sperm Cells With Somatic Cells

Although the whole sperm can produce antibodies that can induce infertility, they cannot be employed for the development of a vaccine per se. There are many antigens of the male germ cells that are likely to be shared with other somatic tissues. Besides the presence of various subsurface "internal" antigens common to somatic and germ cells, there are various proteins on sperm surface that are likely to be shared with somatic cell plasma membranes. Sperm have antigens that can be shared with antigens on brain (9), liver, kidney (4), erythrocytes [sperm can carry blood group antigens (19)], and lymphocytes (27). Sperm antigens can also react with several soluble proteins, such as lactoferrin, present in body fluids, e.g., milk, serum, and saliva (26). A few of these cross-reacting antigens have been delineated. A nervous system antigen (NS-6) has been reported to be present on the cell surface of brain cells, kidney cells, and spermatozoa (4). Another testis antigen was found to be related to the brain D2 adhesion protein (17). Many of the sperm antigens are shared with developing embryo (30) and oncofetal antigens. F-9 antigen is present on sperm, embryo, and teratocarcinoma (16).

Theoretically, the immunization with whole sperm antigens may elicit various immunopathologic effects in the cross-reacting tissues, leading to autoimmune diseases and to embryotoxic effects, and increase the chances of spontaneous neoplasia (54). However, there are only a few reports [e.g., Freund et al (9)] of the actual disease occurring in the cross-reacting tissue after active immunization with the whole sperm. In most of these studies on cross-reactions, in vitro techniques were used. If antibodies to any antigen bind with a tissue other than the target tissue in vitro, their effects in vivo still have to be elucidated. Although the sperm antigens have been reported to cross-react with numerous other tissue antigens, the majority of infertile patients and vasectomized men who are loaded with antisperm antibodies in their circulation are rarely, if ever, found to have concomitant autoimmune disease of the cross-reacting organs. Baskin (2) did not find any consitutional symptoms, including the duration and character of menses over a period of two years, in 20 fertile women immunized with whole semen, although these women developed antisperm antibodies leading to infertility in all of them. Moreover, binding of antibodies to a tissue does not necessarily mean that these antibodies will cause a disease or an effect. Although many of the antisperm monoclonal antibodies bind to sperm surface, they do not affect fertilization in vitro or in vivo (37,43). Thus, the cross-reaction needs to be carefully analyzed using procedures that are relevant to the biological system.

Utility of Sperm Enzymes as "Antigens" for Immunoregulation of Fertility

Only those antigens that have been carefully analyzed for sperm specificity should be employed for the development of an antisperm contraceptive vaccine. The utility of any antigen for the contraceptive vaccine is contingent on its tissue specificity, its involvement in fertilization and fertility, and its homogeneity (35). A few sperm-specific enzymes, which are purified to homogeneity, have been tried for immunoregulation of fertility.

Acrosin. An acrosomal enzyme, acrosin has proteolytic properties and is required for penetration of zona pellucida (58). Active immunization with acrosin did not show a reduction in fertility of female rabbits (51) and female sheep (33).

Hyaluridonase. Another acrosomal enzyme, hyaluridonase has the ability to disperse the cells of the cumulus oophorus. It is organ and species specific. It has been tried for isoimmunization to control fertility in females. Metz (32) reported the reduction in fertility in female rabbits by active and passive immunization. However, Morton and McAnulty (33) did not find a reduction in fertility of female sheep actively immunized with highly purified ram hyaluronidase, although the animals did have high titers of circulating antibodies to hyaluronidase.

Sorbitol Dehydrogenase. This enzyme is present in the testes of both vertebrates and invertebrates. It has been extracted from the guinea pig testes (3). It is present in mitochondria and cytoplasm of germ cells and has a molecular weight of 100,000 daltons. No one, to our knowledge, has tried to investigate the effects of active immunization with sorbitol dehydrogenase on the fertility of female animals.

Lactate Dehydrogenase-x (LDH-C_4). The testes-specific LDH-C_4 has been isolated and characterized from mouse testes and has a molecular weight of 140,000 daltons. Although organ specific, it is not species specific. Active immunization with murine LDH-C_4 has been reported to cause a reduction in fertility of female rabbits, female mice, and female baboons (12).

In conclusion, active immunization with none of the sperm enzymes except LDH-C_4 caused a significant reduction in fertility. Using LDH-C_4, there is always a reduction (up to 50%) in fertility rather than a complete block, even when potentially toxic and nonpermissible adjuvants such as Freund's complete adjuvant (FCA) are employed for immunization. LDH-C_4 is not the antigen of high immunogenicity in spermatozoa (55) and is rarely, if ever, involved in involuntary immunoinfertility in humans (45). Thus, there is a need to purify additional sperm-specific antigens that are involved in the fertilization process and that can elicit an immune response sufficient to intercept fertility.

Monoclonal Antibodies as Probes to Delineate Target Antigens

The advent of hybridoma technology has made it possible to identify sperm-specific antigens that are involved in fertility and infertility. There are reports of studies on monoclonal antibodies (MCAs) against the sperm of mice (8,46,49), rabbits (44), guinea pigs (34), and humans (23,57). Some of these sperm-specific MCAs are involved in fertilization and fertility. Saling et al (48) described an MCA that blocks fertilization in mice. This antibody of IgM subclass binds to mouse spermhead and prevents fertilization at the level of sperm and egg plasma membrane interaction. Saling and Lakoski (47) described another MCA that inhibits sperm penetration through the zona pellucida in mice. Lee et al (24) described a series of MCAs that prevent IVF in mice at various levels of sperm-zona interactions. O'Rand and Irons (44) described MCAs developed against rabbit sperm autoantigens that inhibit IVF in rabbits.

In most of these studies on MCAs, the target antigens against which the MCAs are directed have been delineated using Western blot procedure. The majority of these MCAs are species specific and do not react with human sperm. Those MCAs that cross-react with human spermatozoa need further investigation. Being of high specificity and low affinity, the MCAs are invaluable tools in the isolation of target antigens. The target antigens need to be isolated in homogeneous form and characterized. The characterized antigens should be tested for their effects on fertility in actively immunized animals with which they cross-react. The polyclonal antisera raised against these antigens need to be reinvestigated for their tissue specificity.

Sperm-Specific Antigens Isolated Using Selected MCAs and Their Effects on Fertility

The work on some of the bioeffective MCAs that we developed has progressed one step ahead. We have purified and characterized two antigens, the germ-cell antigen (GA-1) and the fertilization antigen (FA-1), from sperm membranes using sperm-specific MCAs that blocked fertility. The effects of active immunization with these characterized antigens have also been investigated.

Germ-Cell Antigen (GA-1)

Out of several hybridoma lines, two MCAs produced against rabbit sperm (8C4.1 and 8C10.5) inhibited fertility when female rabbits were artificially inseminated with sperm treated with these MCAs. These MCAs did not show any effect on the fertilization rates and inhibited fertility at the postfertilization stage (43). Both

Table 30.1. Characteristics of the germ-cell antigen (GA-1) isolated from rabbit testis.

Apparent molecular weight	63 kDa on 10% gel and on reduction form two subunits composed of numerous spots in 50–70 kDa range in 2D-gel
Amino acid analysis	High percentage of glycine
Carbohydrate analysis	21% carbohydrate
Topographic localization	Plasma membrane antigen localized on the acrosomal region of the sperm and is also present in the intrastructure of tail
Ontogenesis	Synthesized in testis and first develops on spermatocytes
Phylogenetic relationship	Cross-reacting antigen on mouse, rabbit, bull, monkey, and human
Immunogenicity	Immunogenic in mice and rabbits

these monoclones bound to the same antigens of 63,000 in the Western blot involving rabbit testes. Using MCA 8C10.5 in an affinity chromatography, the target antigen, GA-1, was isolated and characterized (42). Some of the characteristics of GA-1 are summarized in Table 30.1.

Female rabbits and mice were actively immunized against GA-1 of 63 kDa molecular mass isolated from rabbit male germ cells. There was a significant reduction in fertility of rabbits actively immunized with GA-1 when compared with controls as seen by the percentage of nine-day implants/corpora lutea ratio (Table 30.2). In mice there was again a significant ($P < .01$) reduction in fertility as seen by mean seven- to nine-day implants per mated mouse actively immunized with GA-1. The antisera from these actively immunized animals were negative for sperm agglutinating and immobilizing antibodies. In the Western blot enzyme-immunobinding procedure (38), the antisera showed specific binding to a single protein of 63 kDa. The incidence of fertilization of eggs recovered from rabbits inseminated with anti-GA-1 antibodies-treated sperm was not significantly different from control rabbits. The percentage of fertilized eggs obtained from rabbits inseminated with anti-GA-1 antibodies-treated sperm that reached the blastocyst stage upon in vitro incubation was, however, significantly

Table 30.2. Fertility at 9 days postinsemination of dose immunized against germ-cell antigen (GA-1) isolated from rabbit testis.

Treatment	Rabbits (No.)	Viable implants/CL (No.)	Viable implants/CL (%)	Modal ELISA titers
Rab testis extract (DOC-RT)	3	13/30	43.3[a]	100
GA-1	6	20 + 1 deg/55	36.3[a]	6400
Adjuvant control	3	24/28	85.7	<25

[a] Mean differs significantly from control mean ($P < .05$).

CL, corpora lutea; ELISA, enzyme-linked immunosorbent assay; DOC-RT, sodium deoxycholate-solubilized rabbit testis preparation; deg, degenerated.

Reproduced with permission from Naz et al (41).

less than that for embryos obtained from rabbits inseminated with control serum-treated sperm (41). Incubation of normal fertilized eggs in vitro with the antibodies did not affect development. Neither antiserum nor immune uterine fluid reacted with four-day blastocysts in the indirect immunofluorescence technique. All these results indicate that the antibodies to GA-1 are not directly embryotoxic. The antibodies have to carried into the oocyte at the time of fertilization to exercise their effects. Thus, the active immunization with GA-1 results in a postfertilization reduction in fertility of rabbits and mice by inhibiting early embryonic development. The GA-1 may constitute an extranuclear sperm signal required for the normal development of an embryo.

Fertilization Antigen (FA-1)

The ideal antigen for the development of an antisperm contraceptive vaccine should be involved in the fertilization process (rather than embryonic development) for it to be widely acceptable. We have obtained such an antigen, FA-1, through the use of a sperm-specific monoclonal method (MCA-24) raised against human spermatozoa.

Characteristics of the MCA-24

Of 40 putative clones produced against human spermatozoa, MCA-24 appeared to be interesting and thus was selected for detailed analysis (37). The MCA-24 clone was determined to be of the 2a subclass of immunoglobulin G (IgG_{2a}) by the agar-gel double diffusion technique. This clone was negative in the agglutination (10) and the immobilization assays (15), two methods that are used to test for antibodies to sperm in serum of humans with involuntary infertility. MCA-24 was germ-cell-specific and did not bind to lymphocytes, erythrocytes, fibroblasts, lymphoblastoid cells, K562 leukemia cells, or other somatic cells by the indirect immunofluorescence technique (IFT) (37). The results of IFT were further checked in the absorption assay and in a test of residual activity in the enzyme-linked immunosorbent assay (ELISA) after the antibodies had been incubated with various somatic tissues.

The MCA-24 clone showed binding to the postacrosomes, midpieces, and tails of viable and methanol-fixed human spermatozoa by the IFT (Fig. 30.1). It showed the same binding pattern with capacitated, noncapacitated, and reduced (swollen-head) sperm treated with dithiothreitol and Triton X-100. It also reacted with murine, rabbit, bull, and rhesus monkey sperm. It did not react with vasectomized human seminal plasma. Thus, MCA-24 is specifically directed to an intrinsic membrane protein of sperm and not to one absorbed onto the sperm surface from accessory gland fluids.

The tissue-specific, but not species-specific, MCA-24 completely blocked binding and penetration by human sperm in zona-free hamster ova. MCA-24 also significantly inhibited fertilization in vitro of mouse eggs by murine sperm. It also blocked fertilization in female rabbits when they were artificially inseminated with sperm that had been treated with MCA-24. These results have been

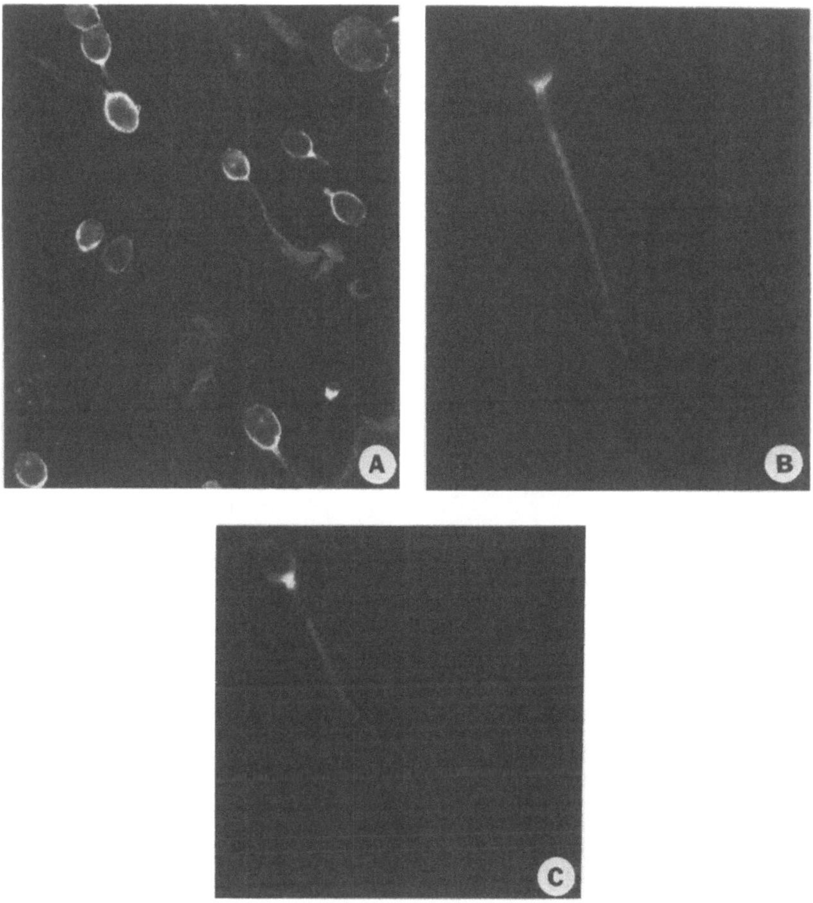

Fig. 30.1. Localization of the FA-1 on human sperm by indirect immunofluorescence technique. FA-1 is localized on the postacrosomal, tail and (to a limited extent) on mid-piece regions of sperm. Same pattern was noted in sperm that have been reduced and swollen (**A**), methanol-fixed (**B**), or capacitated in vitro (**C**). Magnifications: a,b,c × 1000. (Reproduced with permission from Naz et al (37))

summarized in Table 30.3. In the Western blot enzyme immunobinding assay, the MCA-24 clone reacted with a single 23 kDa-protein band (monomer) and/or 47–50 kDa-protein (dimer) when a detergent-solubilized membrane preparation of human testis was used.

Characteristics of FA-1

We have purified FA-1 from human as well as from murine male germ cells. The antigen isolated from either source is homologous. The antigen isolated from lithium diiodosalicylate (LIS)-solubilized murine testes has been characterized (40). The FA-1 was recovered at low (2.8) pH using stepwise washing of the

Table 30.3. Effects of MCA-24 on fertilization in various species of animals.*

Ascites fluid	Assay (No.)	Ova examined (No.)	Ova penetrated by sperm† (%)
1. Mouse			
In vitro fertilization (IVF)			
Control ascites	4	113	61[a] (69/113)
MCA-24‡ ascites	10	415	15[b] (61/415)
In vivo fertilization			
Control ascites	2	41	63[a] (26/41)
MCA-24‡ ascites	3	54	30[b] (16/54)
2. Rabbit			
In vivo fertilization			
Control ascites	2	25	76[a] (19/25)
MCA-24‡ ascites	2	32	28[b] (9/32)
3. Zona-free hamster ova penetration by human sperm			
Control ascites	2	34	88[a] (30/34)
MCA-24 ascites	2	33	0[b] (0/32)

*A portion of these data have been reproduced with permission from Naz et al (37).
† Significance tested by chi-squared analysis, a versus b, $P < .001$.
‡ Increasing the concentration of MCA-24 ascites fluid produced almost 0% penetration rates.

unbound proteins from the MCA-24 immunoaffinity column at pH 8 and pH 4. The FA-1 showed a single band of 47 kDa when analyzed by $NaDodSO_4$ /PAGE and silver staining. Following removal of the detergent and extensive dialysis at pH 5.8 or treatment with 0.15 mol/L NaCl, even in the presence of detergent, a monomer of 23 kDa was detected. Two-dimensional PAGE of FA-1 showed four or five polypeptides in the 47 kDa or 23 kDa range (Fig. 30.2). The dialyzed FA-1 contained a major 23 kDa and a minor 48 kDa band when separated on both sucrose and cesium chloride gradients. High-performance liquid chromatography (HPLC), size-exclusion, showed a major peak at 23 kDa and a minor peak at 50 kDa. Further analysis of the 23 kDa peak by reverse-phase chromatography resolved the antigen into three peaks (Fig. 30.3), giving two-dimensional gel patterns similar to the native FA-1. This indicates that the different polypeptides are the charge species of the same antigenic molecule. FA-1 has been purified from testis where this protein is being synthesized. The antigenic molecule having the same polypeptide composition at various stages of maturation or modification (glycosylation) or degradation is expected to be present. Mature sperm do contain the antigen, but in low quantity as compared with the testis, where it constitutes approximately 3.5% of the LIS-solubilized proteins. Lectin-affinity chromatography on a lens culinaris column demonstrated that a part of the antigen was bound to the lectin, while the rest was not. The FA-1 revealed a positive reaction with anthrone-sulfuric acid and periodic-Schiff reagent and contained glucose and mannose, which together constituted 18.8% of the total antigen mass. Amino acid analysis showed a high percentage of aspartic acid, glutamic acid, serine, and glycine.

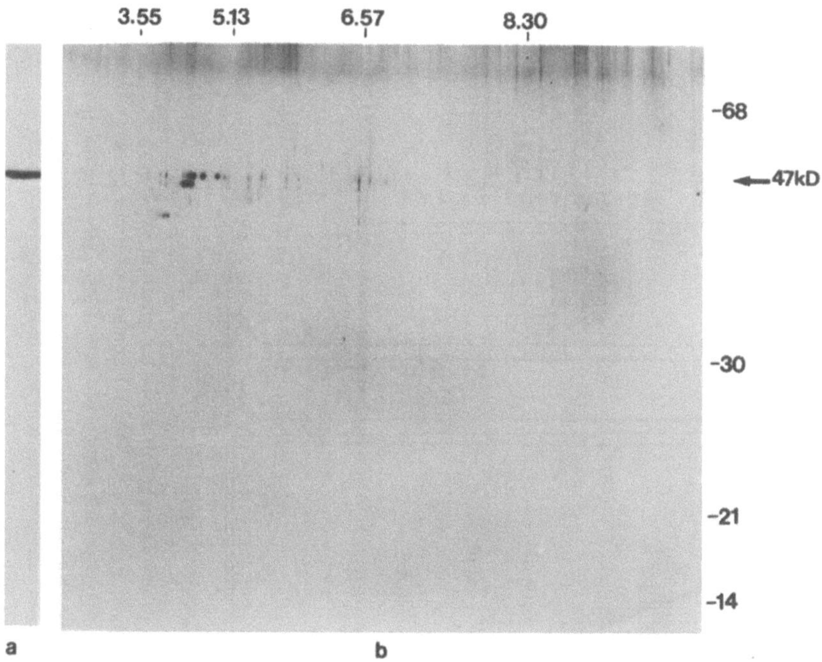

Fig. 30.2. The molecular identity of the reduced FA-1 as revealed by (**a**) ID/NaDodSO_4/PAGE and (**b**) 2D/PAGE. The antigen showed a single band of 47–50 kDa in ID/NaDodSO_4/PAGE and four or five polypeptides in the same molecular weight range in 2D/PAGE. Molecular sizes (kilodaltons) are indicated at the left, and the pH values of isoelectric points are indicated at the top of the gel. (Reproduced with permission from Naz et al (40)).

FA-1 Causes a Reduction in Fertility in Actively Immunized Animals

Recently, female rabbits were actively immunized against the purified and well-characterized FA-1 isolated from LIS-solubilized murine testes. There was significant ($P<.001$) reduction in fertility after immunization with FA-1 (Table 30.4). A majority of the immunized animals showed a complete block of fertility (36). Antisera collected from FA-1-immunized rabbits were negative in the agglutination and immobilization techniques and demonstrated modal titers of ≥1:2,560 in ELISA using FA-1. Antisera were tissue specific and showed binding to the specific protein bands in the Western blot procedure. Ova collected from rabbits inseminated with sperm that had been treated with antiserum from immunized rabbits showed reduced fertilization rates. Thus, the active immunization with FA-1 results in a tissue-specific immune response that causes a reduction of fertility in rabbits by a mechanism(s) involving an inhibition of the fertilization process (36).

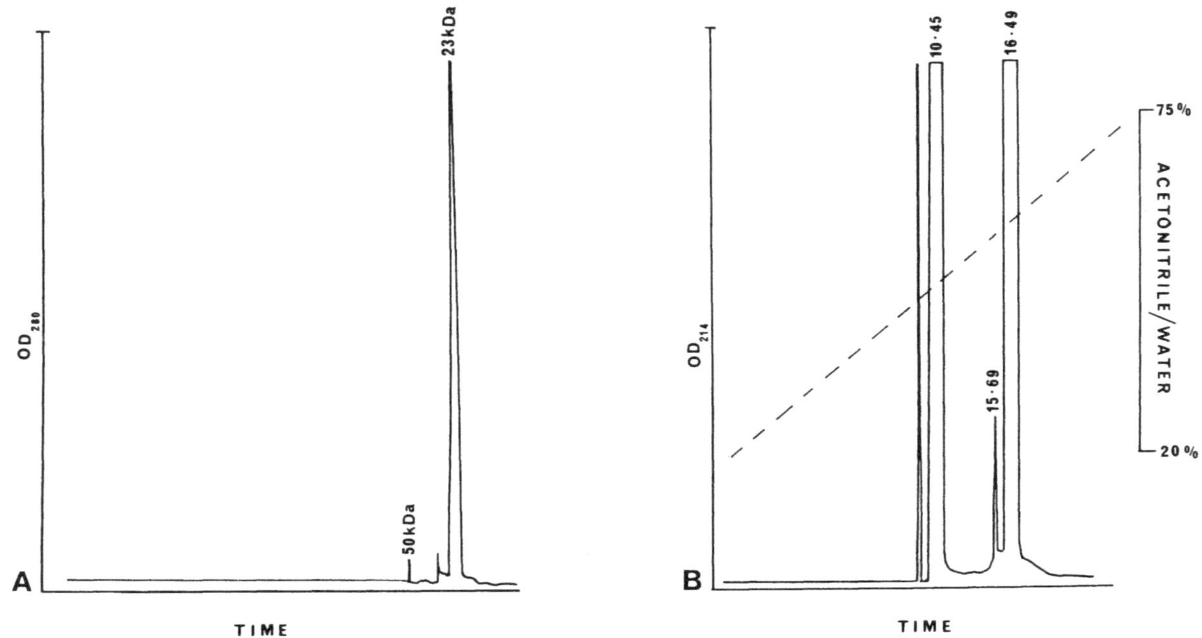

Fig. 30.3. (**A**) High-performance liquid chromatographic (HPLC) pattern of the FA-1. When FA-1 was run through TSK 3000 size exclusion column, two peaks were eluted at 50 kDa and 23 kDa. (**B**) Running the 23 kDa peak through reverse-phase liquid chromatography (RPLC) column revealed three peaks at 10.45, 15.69, and 16.49 minutes on a 20%–75% (vol/vol). acetonitrile/water gradient containing 0.1% trifluoroacetic acid. The 2D analysis of these peaks revealed a polypeptide pattern similar to that of the original FA-1. These results indicate that these peaks are the charged species of the same antigenic molecule. (Reproduced with permission from Naz et al (40)).

Table 30.4. Fertility at 8 days postinsemination of dose immunized against the fertilization antigen (FA-1).

Treatment	Rabbits (No.)	Viable implants/CL (No.)	Viable implants/CL (%)
FA-1	3	5/34	14.70[a]
Adjuvant	3	31/33	93.90[a]

[a] Mean differs significantly from control mean ($P < .001$).
CL, corpora lutea.
Reproduced with permission from Naz et al (36).

FA-1 Is Involved in Involuntary Immunoinfertility in Humans

Even if an antigen shows a complete block of fertility in animals, its utility for the development of an antisperm contraceptive vaccine in humans still remains to be determined. The involvement of the antigen in involuntary infertility in humans may indicate indirectly that (a) the antigen is immunogenic in humans and that (b) it may cause infertility in humans if sufficient antibody titers are present. Since most infertile couples are healthy individuals without any disease concomitant with infertility, the involvement of an antigen in involuntary infertility may be indicative, though not confirmative, of its tissue specificity.

FA-1 is involved in immunoinfertility in humans. FA-1 has been isolated from the human sperm cell plasma membranes (37) and is homologous with the antigen isolated from murine germ cells (40). FA-1 cross-reacts with sera from immunoinfertile patients (Fig. 30.4). Recently, we screened 30 more immunoinfertile sera and found that approximately 45% of them reacted strongly (SD units ≥ 2) with purified FA-1 in an ELISA. None of the sera from fertile humans reacted with FA-1 (SD units < 2).

The IVF performed in an immunoinfertile couple who demonstrated antibodies to FA-1 showed a complete inability of sperm to fertilize the oocytes. Fertilization failure was not due to loss of motility or agglutination of spermatozoa. Sperm maintained good motility and bound to the zona pellucida of oocytes, although in a small number. The penetration of sperm into the oocytes was completely blocked, and none of the oocytes retrieved were fertilized. Both partners were found to have antibodies to FA-1. In the man, they were present in serum, seminal plasma, and bound on the sperm surface. In the woman, they were found in serum and in follicular fluid. We are finding that many other couples who showed fertilization failure have high titers of anti-FA-1 antibodies in their serum. In our program, we generally observe ≥74% fertilization rates in other infertile patients (39).

Mechanism(s) of Action of Antibodies

The monoclonal antibodies, MA-24, and monospecific polyclonal antibodies developed against FA-1 inhibit zona-free hamster ova penetration by human

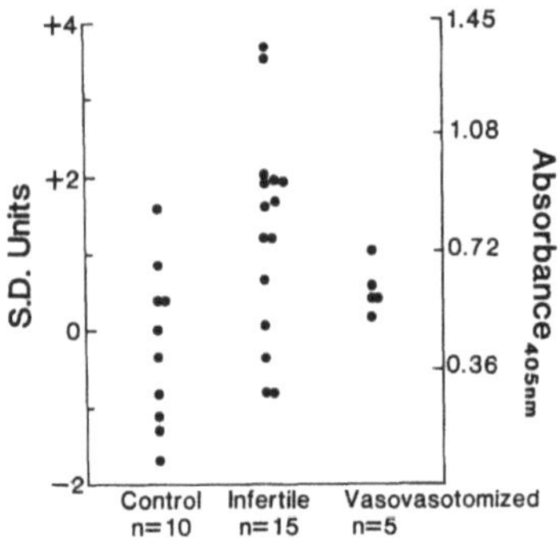

Fig. 30.4. Reaction of FA-1 with sera from fertile, infertile, and vasovasotomized patients have sperm antibodies by ELISA. The sera from some infertile patients showed a strong reaction with FA-1 (SD units > 2). (Reproduced with permission from Naz et al (37).

sperm, murine IVF and in vivo fertilization in rabbits. The preliminary data from immunoinfertile patients indicate that antibodies to FA-1 may be capable of interfering in the fertilization in vitro in humans. In most of these systems, sperm-zona pellucida binding is inhibited but penetration is completely blocked. The inhibition in fertilization is not due to agglutination or immobilization sperm, at least in in vitro systems. MCA-24 does not bind to ova, which indicates the sperm specificity of antibody. In the murine fertilization assay, performed in vitro, MCA-24 inhibited fertilization only in ova with intact zonae and not when zonae were removed, indicating that the inhibition occurs during interactions between sperm and zona. However, in the heterogametic zona-free hamster ova penetration assay, MCA-24 completely blocked binding and penetration of human sperm. This assay has been reported to measure capacitation indirectly (13). Thus, these antibodies may be directed to a sperm component (enzymatic or nonenzymatic) that is vital for capacitation.

Cleavage Signal (CS-1)

We have purified another antigen, the cleavage signal (CS-1), from human and rabbit male germ-cell plasma membranes. The antibodies to CS-1 do not inhibit fertilization. The cleavage of the fertilized zygotes is delayed or completely blocked and, whenever it occurs, is incomplete. The fertilized oocytes undergo a transition from G- to S-phase, expand to divide but never complete their division. Presently, we are further characterizing the CS-1 and its antibodies and investigating the potential utility of CS-1 as an antisperm contraceptive vaccine. It will be interesting to investigate whether these antibodies inhibit some mitotic signal by interacting with a component of the cytoskeleton necessary for cytokinesis. This antigen will also elucidate whether sperm plasma membrane provides an extranuclear signal(s) for the cleavage of fertilized oocyte. Antibodies to GA-1, FA-1, and CS-1 block fertility by causing inhibition at different points of the fertilization process (Fig. 30.5).

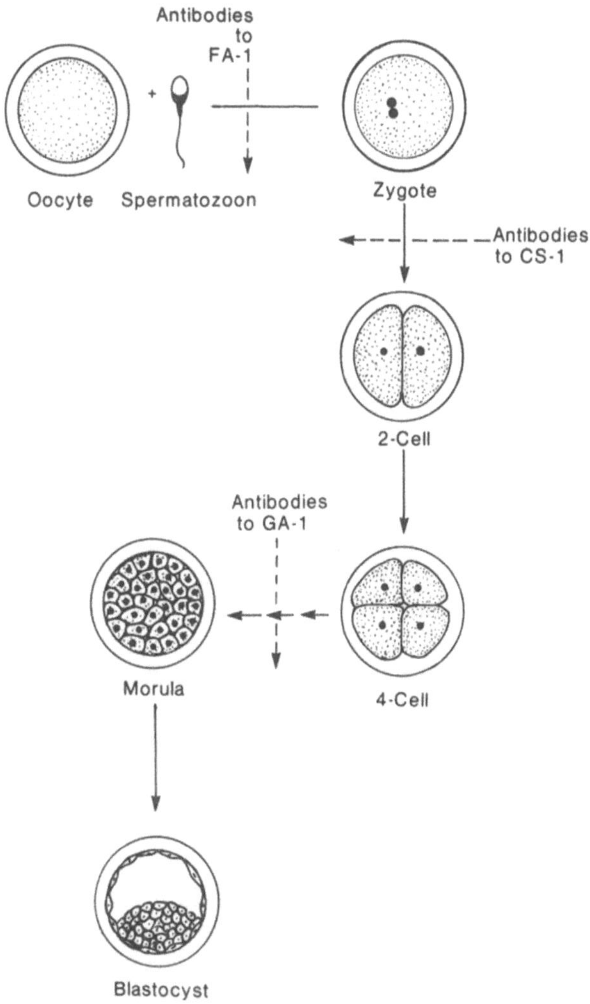

Fig. 30.5. Dissecting the fertilization process using antibodies to various defined sperm antigens. Antibodies to FA-1 cause a block in fertilization process, antibodies to CS-1 block cleavage without affecting the fertilization process, and antibodies to GA-1 show effects at the postfertilization preimplantation embryonic development stage at >16-cell stage. The antibodies (monoclonal or polyclonal) to GA-1 are not directly cytotoxic to the normal developing embryos. Antibodies have to be carried through sperm at the time of fertilization to demonstrate their effects during embryonic development after a few normal cleavages.

Conclusions

At the present time, FA-1 appears to be the most promising candidate for the development of an antisperm contraceptive vaccine and for the diagnosis and treatment of immunoinfertility in humans. The sperm-specific FA-1 is an interesting molecule because: (a) active immunization of female animals with FA-1 causes a significant reduction of fertility, with a complete block in the majority of the animals, (b) antibodies (monoclonal or polyclonal) to it can block fertilization in vitro and in vivo, and (c) it is involved in involuntary immunoinfertility, indicating, indirectly, its immunogenicity and effectiveness in humans.

We need a large quantity of FA-1 in homogeneous form for large-scale practical use. Initially, we purified FA-1 using MCA-immunoaffinity chromatography, which gave us enough purified antigen to do pilot studies. To obtain a large quantity of the antigen, we are currently employing a molecular biological approach to obtain the intact molecule and a protein synthetic approach to synthesize the antigenic fragments of the FA-1. Whether or not the synthetic antigenic fragments are as bioeffective as the whole molecule for the development of a contraceptive vaccine is currently being explored in our laboratory, as well as in others. The data emerging from phase 1 of clinical trials on two vaccines based on β-subunit of human chorionic gonadotropin (βHCG), i.e., Talwar's vaccine (52), based on the whole βHCG molecule to retain the native conformational integrity, and Steven's vaccine (50), based on C-terminal synthetic polypeptide, will help us understand the immunobiology and relative efficacy of the two approaches.

Acknowledgments. I wish to thank Karen Wayns-Lee, Anne Dullea, and Suresh Gupta for technial assistance rendered during the course of this study and Drs. Terry Phillips and Barnett Rosenblum for help in various immunochemical techniques. This project is supported by the Program for Applied Research on Fertility Regulation (PARFR) of Northwestern University, under a Cooperative Agreement with the US Agency for International Development (DPE-0546-A-00-1003-00). The financial assistance provided by Dr. Donn Duncan and John Payne is gratefully acknowledged.

References

1. Alexander NJ, Sampson JH, Fulgham DL: Pregnancy rates in patients treated for antisperm antibodies with prednisone. *Int J Fertil* 1983;28:63–67.
2. Baskin MJ: Temporary sterilization by injection of human spermatozoa: a preliminary report. *Am J Obstet Gynecol* 1932;24:892–897.
3. Bishop DW, Schrank WW, Musselman AD, et al: Testis sorbitol dehydrogenase (SDM) and activity changes during induced aspermatogenesis and cryptoorchidism. *Fed Proc* 1967;26:646a.

4. Chaffee JK, Schachner M: NS-6 (nervous system antigen–6): A new cell surface antigen of brain, kidney and spermatozoa. *Dev Biol* 1978;63:173–184.
5. doAmaral EB: Temporary sterilization of women by biologic means (injection of spermatozoa). *Rev Ginecol Obstet (Sao Paolo)* 1936;30:886–889.
6. Escuder CJ: Temporary biologic sterility of women produced by human sperm. *Am Fac Med Montevideo* 1936;21:889–937.
7. Edwards RG: Immunological control of fertility in female mice. *Nature* 1964;203: 50–53.
8. Feuchter FA, Vernon RB, Eddy EM: Analysis of the sperm surface with monoclonal antibodies:topographically restricted antigens appearing in the epididymis. *Biol Reprod* 1981;24:1099–1110.
9. Freund J, Thompson GE, Lipton MM: Aspermatogenesis, anophylaxis and cutaneous sensitization induced in the guinea pig by homologous testicular extract. *J Exp Med* 1955;101:591–603.
10. Friberg J: A simple and sensitive micromethod for demonstration of sperm-agglutinating activity in serum from infertile men and women. *Acta Obstet Gynecol Scand* 1974;74:21–29.
11. Fuchs EF, Alexander NJ: Immunologic considerations before and after vasovasostomy. *Fertil Steril* 1983;40:497–499.
12. Goldberg E: Current status of research on sperm antigens: potential applications contraceptive vaccines, in Zatuchni I (ed): *Research Frontiers in Fertility Regulation*, Chicago, Program for Applied Fertility Regulation, 1983, pp 1–11.
13. Gould JE, Overstreet JW, Yanagimachi H, et al: What functions of the sperm cell are measured by *in vitro* fertilization of zona-free hamster eggs? *Fertil Steril* 1983;40: 344–352.
14. Hendry WF, Stedronska J, Hughes L, et al: Steroid treatment of male subfertility caused by antisperm antibodies. *Lancet* 1979;ii:498–501.
15. Isojima S, Li TS, Ashitaka Y: Immunological analysis of sperm-immobilizing factor found in sera of women with unexplained infertility. *Am J Obstet Gynecol* 1968;101:677–683.
16. Jacob F: Mouse teratocarcinoma and embryonic antigens. *Immunol Rev* 1977;33: 1–32.
17. Jorgensen OS, Moller M: A testis antigen related to the brain D2 adhesion protein. *Dev Biol* 1983;100:275–286.
18. Katsh S: Infertility in female guinea pigs induced by injection of homologous sperm. *Am J Obstet* 1959;78:276–278.
19. Kerek G: Distribution of the blood group antigens A and B on human spermatozoa. *Int J Fertil* 1974;19:181–191.
20. Kiddy CA, Rollins MA: Infertility in female guinea pigs injected with testis. *Biol Reprod* 1973;8:545–549.
21. Kummerfeld HL, Foote RJ: Infertility and embryonic mortality in female rabbits immunized with different sperm preparations. *Biol Reprod* 1970;14:303–305.
22. Landsteiner K: Zur Kennetris der speziofisch auf Blutkorperchen Wirkenden Sera. *Zbl Bakt* 1899;25:546–549.
23. Lee GYS, Huang YS, Huant CH, et al: Monoclonal antibodies to human sperm antigens. *J Reprod Immunol* 1982;4:173–181.
24. Lee C-YG, Wong E, Teh C-Z, et al: Generation of mouse ooctye monoclonal isoantibodies:their effects and those of antisperm monoclonal antibodies on *in vitro* fertilization. *J Reprod Immunol* 1985;7:3–13.

25. Mancini RE, Andrada JA, Saraceni D, et al: Immunological and testicular response in man sensitized with human testicular homogenate. *J Clin Endocrinol Metab* 1965;25: 859–875.
26. Masson PL, Heremans JF, Dive CH: An iron-binding protein common to many external secretions. *Clin Chim Acta* 1966;14:735–739.
27. Mathur S, Goust J-M, Williamson HO, et al: Cross-reactivity of sperm and T-lymphocyte antigens. *Am J Reprod Immunol* 1981;1:113–118.
28. Menge AC: Immune reactions and infertility. *J Reprod Fertil Suppl* 1970;10:171–186.
29. Menge AC: Clinical immunological infertility: diagnostic measures, incidence of antisperm antibodies, fertility and mechanisms, in Dhindsa DS, Schumacher GFB (eds): *Immunological Aspects of Infertility and Fertility*. New York, Elsevier-North Holland, 1980, pp 205–225.
30. Menge AC, Fleming CH: Detection of sperm antigens on mouse ova and early embryo. *Dev Biol* 1978;63:111–117.
31. Metchnikoff E: Etudes sur la resorption de cellule. *Ann Inst Pasteur* 1899;13: 737–779.
32. Metz CB: Role of specific sperm antigens in fertilization. *Fed Proc* 1973;32:2057–2064.
33. Morton DB, McAnulty PA: The effect on fertility of immunizing female sheep with ram sperm acrosin and hyaluronidase. *J Reprod Immunol* 1979;1:61–72.
34. Myles DG, Primakoff P, Bellve AR: Surface domains of the guinea pig sperm defined with monoclonal antibodies. *Cell* 1981;23:433–438.
35. Naz RK: Toward the development of an antisperm contraceptive vaccine, in Talwar GP (ed): *Immunological Approaches to Contraception and Promotion of Fertility*. New York, Plenum Press, 1986, pp 183–190.
36. Naz RK: The fertilization antigen (FA-1) causes a reduction of fertility in actively immunized female rabbits. *J Reprod Immunol*, in press.
37. Naz RK, Alexander NJ, Isahakia M, et al: Monoclonal antibody against human germ cell membrane glycoprotein that inhibits fertilization. *Science* 1984;225:342–344.
38. Naz RK, Bradford CR, Menge AC: Isoantigenicity of rabbit sperm, testis and their extracts as demonstrated by Western blot enzyme immunobinding procedure. *Arch Androl* 1985;13:279–287.
39. Naz RK, Littman B, Stillman R: Successful human pregnancy following *in vitro* fertilization using frozen semen. *J IVF ET* 1985;2:143–145.
40. Naz RK, Phillips TM, Rosenblum BB: Characterization of the fertilization antigen-1 for the development of a contraceptive vaccine. *Proc Natl Acad Sci USA* 1986;83: 5713–5717.
41. Naz RK, Poffenberger RJ, Menge AC: Reduction of fertility in female rabbits and mice actively immunized with a germ cell antigen (GA-1) from the rabbit. *J Reprod Immunol* 1986;9:163–173.
42. Naz RK, Rosenblum BB, Menge AC: Characterization of a membrane antigen from rabbit testis and sperm isolated by using monoclonal antibodies and effect of its antiserum on fertility. *Proc Natl Acad Sci USA* 1984;81:857–861.
43. Naz RK, Saxe J, Menge AC: Inhibition of fertility in rabbits by monoclonal antibodies against sperm. *Biol Reprod* 1983;28:249–254.
44. O'Rand MG, Irons GP: Monoclonal antibodies to rabbit sperm antoantigens. II. Inhibition of human sperm penetration of zona-free hamster eggs. *Biol Reprod* 1984;30:731–736.

45. Rose NR, Hjort T, Rumke P, et al: Techniques for detection of iso- and auto- antibodies to human spermatozoa. *Clin Exp Immunol* 1976;23:175–199.
46. Saling PM, Irons G, Waibel R: Mouse sperm antigens that participate in fertilization. I. Inhibition of sperm fusion with the egg plasma membrane using monoclonal antibodies. *Biol Reprod* 1985;33:515–526.
47. Saling PM, Lakoski KA: Mouse sperm antigens that participate in fertilization. II. Inhibition of sperm penetration through the zona pellucida using monoclonal antibodies. *Biol Reprod* 1985;33:527–536.
48. Saling PM, Raines LM, O'Rand MG: Monoclonal antibody against mouse sperm blocks a specific event in the fertilization process. *J Exp Zool* 1983;227:481–486.
49. Schmell ED, Yuan LC, Gulyas BJ, et al: Identification of mammalian sperm surface antigens. 1. Production of monoclonal anti-mouse sperm antibodies. *Fertil Steril* 1982;37:249–257.
50. Stevens VC: Use of synthetic peptides as immunogens for developing a vaccine against human chorionic gonadotropin, in: *Synthetic Peptides as Antigens*. Ciba Foundation Symposium, 1986, vol 119, p 200.
51. Syner FN, Kuros BS, Moghissi KS: Active immunization of female rabbits with purified rabbit acrosin and effect on fertility. *Fertil Steril* 1979;32:468–473.
52. Talwar GP: Structural vaccines to control fertility and communicable diseases. *Crit Rev Trop Med* 1984;2:245.
53. Tsukui S, Noda Y, Yano J, et al: Inhibition of sperm penetration through human zona pellucida by antisperm antibodies. *Fertil Steril* 1986;46:92–96.
54. Tung KSK: Immunopathologic aspects of anti-sperm immunity. *Reproductive Immunology*. New York, Elsevier Science Publishers, 1986, pp 143–151.
55. Tung KSK, Goldberg EH, Goldberg E: Immunobiological consequence of immunization of female mice with homologous spermatozoa:induction of infertility. *J Reprod Immunol* 1979;1:145–158.
56. Wartman J: Vasectomy–what are the problems? *Popul Rep* [D] 1975;2:31-32.
57. Wolf DP, Sokoloski JE, Dandekar P, et al: Characterization of human sperm surface antigens with monoclonal antibodies. *Biol Reprod* 1983;29:713–723.
58. Zaneveld LJD, Polakoski KL, William WL: Properties of a proteolytic enzyme from rabbit sperm acrosome. *Biol Reprod* 1972;6:30–39.

CHAPTER 31

Origin of Zona Pellucida Antigens and Its Relevance to the Development of Contraceptive Vaccine

Shakti N. Upadhyay

The zona pellucida (ZP) is an extracellular glycoprotein matrix that closely surrounds the ovum and plays a major role in the process of sperm-egg interaction in mammalian species. In fact, it is the structure to which sperm attach and bind during initial stages of fertilization. Considering its important functional role, the ZP has been recognized as an ideal target for immunocontraception (1,5).

In recent years ZP glycoproteins have been isolated, characterized, and used for immunization studies (6). While the antifertility effect has been observed in ZP-immunized animals of various species, the adverse effect on the ovarian follicles has been a major concern for further investigations (4,7). The follicular atresia noted in ZP-immunized animals may be due to cross-reaction of ZP antibodies to follicular components other than oocyte, especially the granulosa cells, or it could also reflect the "granulosa cell" origin of ZP preparations used in such investigations. The origin of ZP glycoproteins could be an important consideration in selecting the antigen for ZP immunization.

The origin of ZP has been a subject of controversy. Granulosa cells or oocyte or both are believed to be involved in the formation of zona material (2,3). Recently, the evidence for involvement of the oocyte, at least in part, in synthesis and deposition of ZP was reported in a study on the differentiation and development of germ cells in the mouse adrenal glands (8,9). It was demonstrated that during embryonic development of the mouse, some of the primordial germ cells (PGC), migrating from the yolk sac to the genital ridges, become entrapped within the primordium of the developing adrenal gland. These ectopic PGCs do not undergo degeneration but instead continue to differentiate and develop normally, in spite of being located outside the gonads.

The study included adrenal glands from 39 male and female swiss albino mice from day 12½ of intrauterine life to postnatal day 21. Of the 39 animals, 26 contained germinal cells at various stages of differentiation in their adrenal glands. In early embryos, the germinal cells in the adrenal had typical features of primordial germ cells that differentiated into oogonia synchronously with those present in the gonads of the same animal. Beginning at day 17 of gestation, the oogonia entered meiotic prophase and by term had reached diplotene and differentiated into oocytes in both male and female fetuses. In postnatal animals

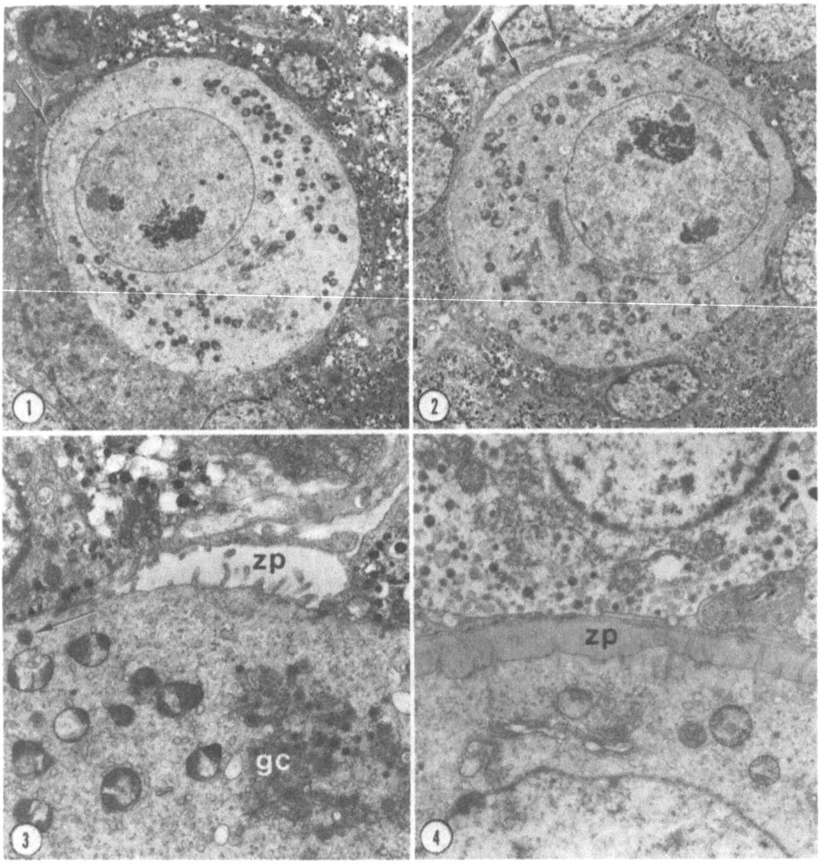

Fig. 31.1 (1,2). Electron micrographs illustrating zona pellucida deposition (arrows) around oocyte in the adrenal medulla of 13-day-old male (**1**) and 14-day-old female mouse (**2**) (× 4,000, shot at 64%.) (**3**) Electron micrograph of adrenal of 13-day-old female mouse. Note the deposition of zona pellucida (ZP) occurring in patches and resulting in focal separation of the oocyte from the surrounding adrenal medullary cells. Also note the microvilli developing on the surface of the oocyte, prominent Golgi complex (gc), and cortical granule (arrow). (× 20,000, shot at 64%.) (**4**) Electron micrograph of adrenal medulla of a 21-day-old female mouse. Note the continuous zona pellucida (zp) surrounding the oocyte and a prominent Golgi complex in the oocyte cytoplasm. (× 15,000, shot at 64%.)

all oocytes continued to develop through postmeiotic phase of growth exactly as they normally do in the ovarian follicles. In 2- and 3-week-old animals, the oocytes displayed features identical to those exhibited by oocytes in the large antral follicles. In some oocytes, synthesis and deposition of zona were noted initially in patches (Fig. 31.1 [1–3]) which in later stages appeared to coalesce and form a continuous layer (Fig. 31.1 [4]). The deposition of zona was closely associated with the presence of prominent of Golgi complex at the peripheral ooplasm (Fig. 31.1 [3 and 4]). Animals, both male and female, at around 3 weeks of age thus had fully grown oocytes surrounded by a continuous zona pellucida within their adrenal glands.

These observations unquestionably demonstrate that at least a part of zona is of "oocyte" origin and thus provides a lead for further identification and characterization of the zona glycoproteins synthesized by the oocyte. Immune response to ZP antigens of "oocyte" origin could be more specific in blocking sperm-egg interaction and may not have deleterious effect on the ovarian follicles.

Acknowledgment. The study reported here was carried out at the Department of Pathology, Harbor-UCLA Medical Center, Torrance, and University of California School of Medicine, Los Angeles, in collaboration with Dr. L. Zamboni.

References

1. Dunbar, BS: Antibodies of zona pellucida antigens and their role in fertility, in Wegmann T, Gill T, Cumming C, Nisbet-Brown E (eds): *Immunology of Reproduction*. Oxford, England, Oxford University Press, 1983, pp 506–534.
2. Guraya SS: Biology of Ovarian Follicles. Berlin–Heidelberg, New York, Springer-Verlag, 1985, pp 1–320.
3. Kan YH: Development of zona pellucida in the rat oocyte. *Am J Anat* 1974;139: 535–565.
4. Mahi-Brown CA, Yanagimachi R, Hoffmann JD et al: Fertility control in the bitch by active immunization with porcine zonae pellucidae: use of different adjuvants and patterns of estradiol and progesterone levels in estrous cycles. *Biol Reprod* 1985;32:761–772.
5. Sacco AG: Immunocontraception: Consideration of the zona pellucida as a target antigen, in Wynn RM (ed): *Obstetrics and Gynecology Annual vol 10*, 1981, pp 1–26.
6. Sacco AG, Subramanian MG, Yurewicz EC, et al: Antigenic profile and impact of immunization with zona pellucida antigens in primates, in Talwar GP (ed): *Immunological Approaches to Contraception and Promotion of Fertility.* New York, Plenum Press, 1986, pp 277–290.
7. Skinner SM, Mills T, Kirchick HJ, et al: Immunization with zona pellucida proteins results in abnormal ovarian follicular differentiation and inhibition of gonadotropin induced steroid secretion. *Endocrinology* 1984;115:2418–2432.
8. Upadhyay SN, Zamboni L: Ectopic germ cells: Natural model for the study of germ cell sexual differentiation. *Proc Nat Acad Sci USA* 1982;79:6584–6588.
9. Zamboni L, Upadhyay SN: Germ cell differentiation in mouse adrenal glands. *J Exp Zoo* 1983;228:173–193.

CHAPTER 32

Disillusions and Hopes About Zona Pellucida Glycoproteins

Shobha Sehgal and S.K. Gupta

Today there are serious concerns about the doubling time of the world's population, which is getting progressively shorter (64). On the Indian scene, the continent has only 2.4% of world's land area to support 14% of the population. The physical quality of life index is barely 40, which is far below the rest of the developed world (95 for the United States, 96 for Japan), and despite desperate efforts, the birth rate stands at 33 per 1,000. It is estimated that as many as 108 million couples are in the reproductive age group. There are 63,000 births against 25,000 deaths every day, and 13 million newborns added each year. The department of Science and Technology, Ministry of Health has rightly identified contraception as a thrust area. Over the last two decades several products have been floated for temporary control of fertility. These have largely been synthetic hormones but major drawbacks are that they have to be administered frequently over prolonged periods and have unacceptable side effects. It is apparent that the conventional modes of antifertility measures have been inadequate and impractical. Since vaccination has been an accepted mode for prevention of infectious diseases, it was deemed logical to devise vaccines for birth control as well. Theoretically it is possible to intercept pregnancy right from the hypothalamus to pituitary, gonads, and gametes (63). Contraceptive vaccines using β-subunit of human chorionic gonadotropins (HCG) and C-terminal peptide have already entered phase-I clinical trials (59,65). There are indications that gametes may also offer suitable modalities for birth control.

Zona pellucida is the glycoprotein coat that ensheaths oocytes of all mammals and maintains their integrity. It has been the subject of intensive investigation since the early seventies. One of the first descriptions in the literature refers to immunization with ovarian homogenates and studying the impact on the recipient animals. Table 32.1 illustrates the species of animals investigated using ovarian antigens. It became evident that the major antigen of significance originated from zona pellucida as the sera gave a unique zona precipitation reaction.

Preparation of Antigen

In all initial studies antisera were raised against ovarian homogenates, unfertilized eggs, or preimplantation embryos. The resultant inhibition of fertilization

Table 32.1. Effect of passive immunization with ovarian homogenates/isolated zonae on fertility.

Series No.	Source of zonae/eggs	Recipient	Method*	Fertilization (%) Control/test	References
1	Hamster ovary	Hamster	ZP ZD	48/0	Shivers et al (52)
2	Hamster ovary	Hamster	ZP ZD	100/0	Garavagno et al (13)
3	Hamster ovary	Hamster	ZP ZD	100/0	Oikawa et al (30)
4	Mouse	Mouse	ZP ZD	97/0	Jilek & Pavlok (23)
5	Rat	Rat	– –	75/4	Tsunoda & Chang (66)
6	Hamster	Hamster	ZP –	–	Gwatkin et al (19)
7	Mouse	Mouse	ZP ZD	79/0	Tsunoda & Sugie (68)
8	Marmoset	Marmoset	IF –	–	Shivers et al (55)
9	Mouse eggs	Mouse	ZP –	94/0	Tsunoda & Chang (69)
10	Mice	Mice	ZP –	–	Sacco & Shivers (40)
11	Mouse zonae	Mouse	ZP ID	86/28	Sacco (41)
12	Mouse zonae	Rat	ZP –	69/16	Tsunoda & Chang (69)
13	Mouse eggs	Hamster	ZP –	–	Tsunoda & Chang (69)
14	Cow zonae	Cow	CIE –	83/0	Tsunoda et al (72)
15	Bitches	Bitches	– –	–	Mahi-Brown et al (25)

*ZP, zona precipitation reaction; ZD, zona dissolution assay; CIE, counter-immunoelectrophoresis; IF, immunofluorescence.

in vitro and in vivo following either antiserum treatment of eggs or passive immunization was explained on the basis of zona precipitation reaction, which prevented sperm attachment/penetration (23,30,66). But these sera required various absorptive procedures to eliminate antibodies against contaminating-tissue components. Sacco and Palm (36) carried out heteroimmunization with isolated pig zonae and successfully raised monospecific antisera. Various techniques were devised from time to time to isolate zonae. Oikawa (31) published data on large-

Table 32.2. Effect of heteroimmunization with ovarian homogenates/isolated zonae on fertility (in vivo studies).

References	Source of immunogen	Species tested	% Efficiency
Shivers et al (52)	Hamster	Hamster	46
Garavagno et al (13)	Hamster	Hamster	100
Oikawa et al (30)	Hamster	Mouse	100
Jilek & Pavlok[a] (23)	Mouse	Mouse	100
Tsunoda & Chang (66)	Mouse	Mouse	100
Tsunoda & Chang (67)	Rat	Rat	40
Gwatkin et al (19)	Hamster	Mouse	93
Tsunoda & Sugie (68)	Mouse	Mouse	63
Gwatkin & Williams (20)	Bovine	Rabbit	75
Mahi-Brown et al (28)	Porcine	Canine	0/5[b]
Aitkin & Richardson (1)	Mouse ovary	Rat	96

[a] Histological evidence.
[b] Pregnant/mated.

Table 32.3. Effect of hetero-alloimmunization using oocytes zonae or HSZP.

References	Source of immunogen	Species tested	% Efficacy
Gwatkin, Anderson & Williams (21)	Porcine zonae	Rabbit	88
Wood et al (76)	Porcine zonae	Rabbit	100
Sacco et al (42)	Porcine zonae	Mice	Insignificant
Shivers et al (56)	Pig	Dogs	2/3
Sacco et al (43)	Pig	Mice	No effect on fertility
Drell et al (6a)	Pig	Rats	-do-
Gulyas et al (15)	Pig	Monkeys	90
Sacco et al (46)	Porcine	Squirrel monkey	Sperm binding
Skinner et al (57)	Porcine	Rabbit	+++
Mahi-Brown, Huang & Yanagimachi (26,27)	Porcine	Dogs	100
Gupta & Sehgal (16)	Porcine	Rabbit	++
Gupta et al (17)	Porcine	Monkeys	*
Suman et al (62)	Porcine	Monkeys	+

+++ = depressed ovulation
++ = low oocyte reserve
\+ = infertility at high antibody titers
* = acyclicity in monkeys

scale preparation of zonae in hamsters. Sacco et al (42) used either a meat grinder or different assortment of blades to homogenize ovaries and collected eggs using a stereomicroscope. Isolated zonae were then obtained mechanically by micropipette technique. Dunbar et al (8,9) used a special grinding machine with circular blades that substantially increased the speed of zonae preparation (12). She also modified the method for screening through nylon meshes of various sizes and a 75 μm mesh was found most suitable for final purification of zonae. Subsequently, experiments were conducted either with whole oocyte, whole zonae, or heat-solubilized zonae protein (HSZP) to corroborate earlier impressions that immunization with zona pellucida proteins prevented sperm-egg interaction and produced infertility in different species combinations (Tables 32.2 and 32.3). It is now possible to prepare large quantities of zonae from a large pool of frozen ovaries. This has facilitated characterization of zonae antigens from different species and demonstration of specific antibodies.

Methods of Antibody Detection

The studies using complete Freund's adjuvant (CFA) generated high titer antibodies, indicating that zonae proteins were reasonably immunogenic in a heterologous system. The earliest method used to detect antibodies was zona precipitation reaction in which the antibodies to zonae formed a dense precipitate around the zonae; this was better appreciated using phase-contrast studies (32,65). Immunoelectrophoresis, counter electrophoresis, and electroimmunodiffusion (17,47,75) indicated that the amount of antibody generated was very high indeed. Sacco et al (35) used indirect immunofluorescence to detect tissue-specific antigens in the ovary.

Further refinement of methods included radioimmunoassays (14,33,60,61,75) and immunofluorescence assays (71). Sacco et al (42) indicated that the end titers in radioimmunoassay were approximately 200 times greater than the immunofluorescence titers.

Palm et al (33) stressed that despite increased sensitivity, no cross-reactions could be observed with 22 porcine tissues. Gerrity and Dunbar (14) corroborated findings of Palm et al (33). More recent tests, e.g., enzyme-linked immunosorbent assay (ELISA), avidin biotin methods, have an edge over radioimmunoassay as the reagents are more stable and the use of sophisticated equipments can be obviated. Gupta et al (16,17) could detect high titers of antibodies both in monkeys and in rabbits immunized with porcine zona pellucida mixed with CFA.

Subunits. Zona glycoproteins are highly complex molecules depicting extensive charge heterogeneity. On two-dimensional gel electrophoresis it exhibits three to four families referred to as ZP1, ZP2, ZP3, and ZP4, depending upon the species studied (10,39,60,75,78). The deglycosylated proteins are also good immunogens (50). Skinner and Dunbar (58) used high resolution two-dimensional gel electrophoresis and confirmed reactivity of different glycoprotein families with antizona antisera using Western blot technique. The antibodies are known to cross placental barrier (44) and are of IgG type (17).

Efficacy Studies

The experiments in rodents, primates, bovines, and canines have been well planned and meticulously performed (see Tables 32.2 and 32.3) and excellent reviews are available in the literature (11,12,24,53). Studies indicate beyond doubt that zonae antigens are effective in intercepting pregnancy. However, the mode(s) of action is far more complicated than previously envisaged. It is now recognized that besides zona precipitation reaction, which makes zona resist enzymatic digestion and thus seriously interfere with sperm-egg interaction and fertilization (2,7,18,25,52,70), other factors may be equally important.

Species Specificity and Cross-Reactivity

Sacco clearly revealed that immunization in an allogenic system in mice failed to produce a contraceptive effect. He further showed that porcine zonae were also ineffective in inducing infertility in mice (43). Different authors used different combinations in the last decade (Table 32.4). The most important inference was that antigenic differences and antigenic cross-reactions are both necessary prerequisites for such an effect. Total antigenic identity in the homologous system fails to induce significant antibodies, but once the tolerance is broken to self-

Table 32.4. Species cross-reactivity of zona pellucida.

Tsunoda & Chang (66)	Rat/mice
Tsunoda & Chang (67)	Hamster/rat/mouse
Gwatkin et al (19)	Mouse/hamster/squirrel
Sacco (37,38)	Pig/man/marmoset/chimp
Shivers & Dunbar (54)	Man/pig
Gwatkin et al (19,21)	Mice/hamster & bovine/canine
Gwatkin & Williams (20)	Bovine/rabbit
Shivers et al (55)	Human/marmoset
Sacco (45), Wood & Dunbar (75)	Porcine/rabbit/monkey
Drell et al (6a)	Rabbit/porcine
Gupta & Sehgal (16)	Rhesus monkey/porcine
Suman et al (62)	Monkey/porcine

zonae, cross-reacting antibodies generated against the host zonae antigens damage the oocyte.

It has been documented by Yanagimachi (77) that zonae are the prime barrier for interspecies fertilization. However, it is now evident that zonae has both species-specific and species-nonspecific sites recognized by spermatozoa (4,34).

Dunbar (7,8) reiterated that in some animals, despite high titers of antibodies, fertility was unchecked; however, considering the vast reserve of oocytes, it is possible that immature oocyte escape attack by the antibodies. It is also evident that while the antibodies appear fairly early during the course of immunization, it requires at least 5 to 6 months in rabbits before fertility is drastically affected (16,57). Similarly in humans, antibodies may be present without infertility (41).

Mahi-Brown, Huang, and Yanagimachi (26) using a small number of bitches, made a pertinent observation: "Although our results suggested that anti-ZP immunization promises to be a successful mean for fertility control in the bitch, obviously there is much more that needs to be learnt especially as to the nature of the unexpected alterations in oestrogen cycles and the duration of infertility." The authors also admitted that they could not establish with certainty whether the infertility resulted from specific interference with fertilization, as in vitro, or from aberrations in ovarian function or both.

Effect of Immunization on the Ovary

Taking this as the lead, Skinner et al (57) conducted an excellent systematic study that clearly revealed aberrations in follicular maturation in rabbits immunized with porcine zonae. Hormonal assays conducted in these animals showed high levels of luteinizing hormone and follicular-stimulating hormone, with lowering of progesteron levels. Wolgemuth et al (73) substantiated in vivo binding of anti-zona antibodies using immunohistochemistry. The studies have been confirmed by Gupta and Sehgal (16) who showed a drastic fall of oocyte reserve in rabbits immunized with heat-solubilized zonae. Long-term follow-up of these animals

Table 32.5. Ovarian weights of control *v* immunized.

Group-1 (controls)		Group-2 (CFA)			Group-3 (alum)		
Rabbit No.	Weight (mg)	Rabbit No.	Weight (mg)	Duration (months)	Rabbit No.	Weight (mg)	Duration (months)
RC 0	520.0	R 2	293.0	6½	R 17	430.0	11
RC 1	643.0	R 3[a]	306.0	9	R 18	405.0	10
RC 2	620.5	R 5	181.0	8.24	R 19	555.0	10
RC 3	555.0	R 0	295.0	23	R 20	430	11
RC 5	696.5	R 9	245.0	23			
		R 10	285.0	7			
Mean±SD 607.0±70.3			267.5±47.29	14.36±8.4		455.0±67.7	10.5±0.5
			P<.001			*P*<.001	

[a] Left ovary at 24 months weighed 861 mg.

indicated that these changes were long lasting and irreversible. In normal rabbit ovary there are generally more than 100 primordial, 30 to 50 primary, and 12 to 26 secondary and mature follicles. Ovulation sites range from 4 to 10. In rabbit immunized with HSZP and CFA, the values declined as the titers increased. The details of ovarian weights and morphology with respect to oocyte content in control versus immunized animals are illustrated (Tables 32.5 and 32.6; Figs. 32.1 and 32.2). The mean ovarian weights were significantly lower in the immunized animals. They also used alum and glutardaldehyde as adjuvants and observed that alum induced significant titers but not comparable to those obtained with CFA, which generated precipitating antibodies as early as 4 weeks. Boosters were required for alum, and peak titers appeared in 3 to 4 months (51a).

Mahi-Brown et al extended studies pertaining to steroidal patterns (27) and ovarian histology (28) in immunized bitches. The authors noticed a sustained rise of estradiol even during estrus while progesteron remained very low. The authors concluded that ovulation failed in these animals. They also reported a high incidence of follicular cysts lined by theca interna. Suman et al (62) immunized

Table 32.6. Quantification of follicles: Control *v* immunized.

Treatment	Primordials	Primary	Secondary	Antral	CL[a]
Group 1					
Controls	>100	37.6±7.3	17.2±4.8	19.8±9.9	7.8±2.5
Group 2 (CFA)					
Up to 4 months	64.0±41.0	30.0±16.0	10.0±9.0	7.0±9.0	4.0±3.0
6 to 9 months	20.2±9.6	11.8±11.3	1.3±1.1	3.2±3.5	0.7±0.5
10 to 24 months	6.7±6.9	3.7±3.7	0.3±0.5	1.2±1.2	0.8±2.0
Group 3 (alum)					
10 to 11 months	11.3±10.8	5.8±4.8	0.8±0.8	1.5±1.1	0.0±0.0

[a] Corpora lutea.

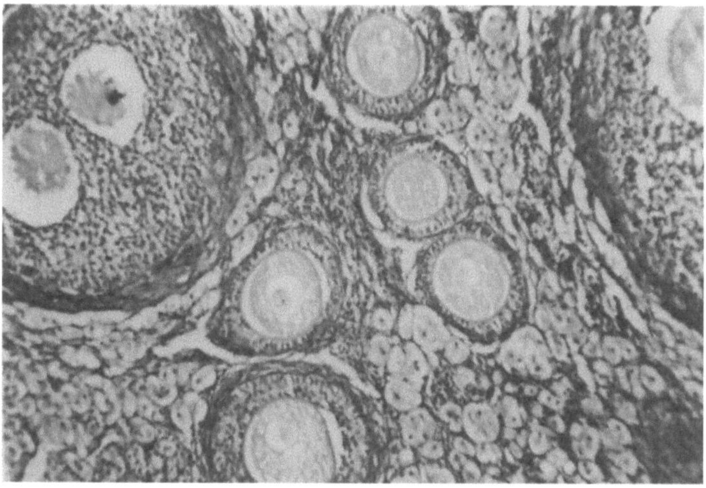

Fig. 32.1. Histological sections of ovary from a normal rabbit superovulated with HCG. Note abundant follicles in different stages of maturation (hematoxylin and eosin).

monkeys with whole oocytes and showed that monkeys regained fertility while the titers declined. However, animals with high antibody titers showed an abnormal follicular pattern. In a joint Indo-US study by Sehgal and Dunbar (51b), a drastic decline of oocyte reserve, lowering of ovarian weight and acyclicity have been reported in rhesus monkeys immunized with a single dose of 250 μg of

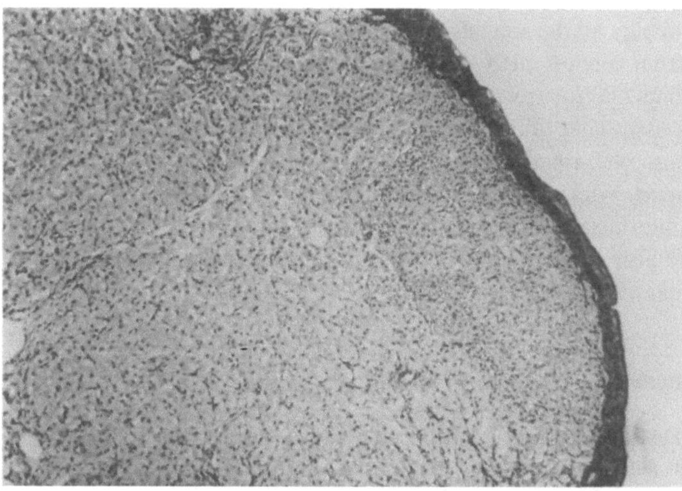

Fig. 32.2. Low-power histological sections from ovary of a rabbit immunized with HSZP + CFA and killed after 11 months. Note atrophy of cortex and lack of follicles (hematoxylin and eosin).

HSZP. Sacco et al (49) immunized 56 squirrel monkeys with 100 to 400 μg of a purified zona fragment (48), a 55 K molecule (ZP3) possessing sperm receptor activity. The laparoscopic oocyte recovery was drastically reduced in the immunized animals (0.1 versus 1.8 oocytes per animal). The histological examination was available in two animals.

The key question today is how to intercept sperm-egg interaction without altering the ovarian environment, and to answer this, a deep insight into the receptor ligand systems of zona pellucida and sperm is imperative. Various sugar molecules (e.g., *N*-acetyl-D-glucosamine, β-D-galactose, and, to some extent, α-D-galactose, α-D-glucose, and α-D-mannose, etc.) have been recognized on the surface of zonae with the help of fluorescein-labeled lectins (9,29,74). One of the major receptor-ligand combinations is galactosyl transferase on the sperm surface; although *N*-acetyl-D-glucosamine reside on the zona but they are not the crucial molecules for sperm interaction.

Monoclonal antibodies form the most vital tool to dissect antigenic determinants. Drell and Dunbar (6b) raised monoclonal antibodies to porcine zona but none of the monoclonals could inhibit sperm-egg interaction. Bamezai et al (3) raised monoclonals to zonae that interacted with a 55 kDa molecule in Western blot assays, but no passive immunization experiments were conducted. Isojima et al (22) raised monoclonals to HSZP and identified a 92 kDa molecule and claimed that this could be the sperm receptor, as antibodies could block human in vitro fertilization with goat antimouse γ-globulin.

Dean et al (5) published elegant immunoscintiphotographs to indicate high specificity of monoclonal antibodies to mouse zonae; two reacted with ZP2, one with both ZP2 and ZP3, and one with only ZP3. Passive immunization with 250 μg of antibody resulted in infertility in mice and blocked formation of two-cell embryos. ZP2 antibodies were equally effective, and in vitro 20 ng/mL of ZP2 and 185 ng/mL of ZP1 produced 50% inhibition. The block lasted 44 to 117 days and eventually all the animals had litters. The authors also constructed a cDNA library from PolyA+ mRNA from ovarian tissues and revealed that the DNA in and around ZP3 locus was hypomethylated in ovarian tissue but not in nonexpressing tissues like brain. Contrary to the earlier concept, homology of the ZP3 clone was established between human and mouse using cross-hybridization experiments. With such lead studies in biotechnology relating to zona, a wealth of new information awaits to be unfolded. Use of such refined epitopes holds great promise not only for fertility control but also for answering some fundamental questions relating to fertility.

References

1. Aitkin RJ, Richardson DW: Immunization against zona pellucida antigen, in Heap JP (ed): *Immunological Aspects of Reproduction and Fertility Control*. Lancaster, UK, MTP Press Ltd, 1980, pp 173–201.
2. Aitkin RJ, Richardson DW: Mechanism of sperm binding inhibition by antizona antisera. *Gamete Res* 1981;4:41–47.

3. Bamezai A, Talwar GP: Personal communication, 1986.
4. Bedford JM: Sperm/egg interaction: The specificity of human spermatozoa. *Anat Rec* 1977;188:477–488.
5. Dean J, Ringuette M, Sobieski DA, et al: Molecular genetics of the mouse zona pellucida: Implications for fertilization and early development, in Talwar GP (ed): *Immunological Approaches to Contraception and Promotion of Fertility.* New York, Plenum Press, 1986, pp 241–249.
6a. Drell DW, Wood DM, Bundman D, et al: Immunological comparison of antibodies to zonae pellucidae in rats and rabbits. *Biol Reprod* 1984;30:435–444.
6b. Drell DW, Dunbar BS: Monoclonal antibodies to rabbit and pig zonae pellucidae distinguish species specific as well as shared antigenic determinants. *Biol Reprod* 1984;30:445–457.
7. Dudkiewicz AB, Noske IG, Shivers CA: Inhibition of implantation in the golden hamster by zona precipitating antibody. *Fertil Steril* 1975;26:686.
8. Dunbar BS, Raynor BD: Characterization of porcine zonae pellucida antigen. *Biol Reprod* 1980;22:941–954.
9. Dunbar BS, Wardip MD, Hedrick JL: Isolation, physiochemical properties and macromolecular composition of zona pellucida from porcine oocytes. *Biochem* 1980;2:356–365.
10. Dunbar BS, Liu C, Sammons DW: Identification of the three major proteins of porcine and rabbit zonae pellucida by high resolution two dimensional gel electrophoresis: Comparison with serum, follicular fluid and ovarian cell proteins. *Biol Reprod* 1981;24:1111–1124.
11. Dunbar BS: Morphological, biochemical and immunochemical characterization of the mammalian zona pellucida, in Hartman J (ed): *Mechanism and Control of Animal Fertilization.* New York, Academic Press, 1983, p 139.
12. Dunbar BS, Wolgemuth DJ: Structure and function of the mammalian zona pellucida: a unique extra cellular matrix. *Mod Cell Biol* 1984;3:77–111.
13. Garavagno A, Rosada J, Barros C, et al: Some characteristics of the zona pellucida antigen in the hamster. *J Exp Zool* 1974;189:37.
14. Gerrity M, Niu E, Dunbar BS: A specific radioimmunoassay for evaluation of serum antibodies to zona pellucida antigens. *J Reprod Immunol* 1981;3:59–70.
15. Gulyas BJ, Gwatkin RBL, Yuan LC: Active immunization of cynomolgus monkeys (Macaca fascicularis) with porcine zona pellucida. *Gamete Res* 1983;7(4): 299–307.
16. Gupta SK, Sehgal S: Immunocontraception in rabbits using porcine zona pellucida antigens. *Bull Post-Graduate Institute*, Chandigarh, India, 1985;19,2. 54–58.
17. Gupta SK, Bhatnagar P, Dunbar BS, et al: Immunohistological studies during immunization with porcine zona pellucida, in Talwar GP (ed): *Immunological Approaches to Contraception and Promotion of Fertility.* New York, Plenum Press, 1986, pp 269–275.
18. Gwatkin RBL, Williams DT: Receptor activity of the hamster and mouse solubilized zona pellucida before and after the zona reaction. *J Reprod Fertil* 1977;49:55–59.
19. Gwatkin RBL, Williams DT, Carlos DJ: Immunization of mice with heat-solubilized hamster zonae: Production of anti-zona antibody and inhibition of fertility. *Fertil Steril* 1977;28:871–877.
20. Gwatkin RBL, Williams DT: Immunization of female rabbits with heat-solubilized bovine zonae: Production of anti-zona antibody and inhibition of fertility. *Gamete Res* 1978;1:19–26.

21. Gwatkin RBL, Anderson OF, Williams DT: Large scale isolation of bovine and pig zonae pellucidae: Chemical, immunological, and receptor properties. *Gamete Res* 1980;3:217.
22. Isojima S, Koyama K, Hasegawa A, et al: Monoclonal antibodies to porcine zona pellucida antigens and their inhibitory effects on fertilization. *J Reprod Immunol* 1984;6(2):77.
23. Jilek F, Pavlok A: Antibodies against mouse ovaries and their effect on fertilization in vitro and in vivo in the mouse. *J Reprod Fertil* 1975;42:377–380.
24. Jones WR: Immunization against the oocyte, in: *Immunological Fertility Regulation*. London, Blackwell Scientific Publication, 1982, pp 147–173.
25. Mahi CA, Yanagimachi R: Prevention of in vitro fertilization of canine oocytes by anti-ovary anti-sera: A potential approach to fertility control in the bitch. *J Exp Zool* 1979;210:129–135.
26. Mahi CA, Huang TF, Yanagimachi R: Infertility in bitches induced by active immunization with porcine zonae pellucida. *J Exp Zool* 1982;222:89–95.
27. Mahi-Brown CA, Yanagimachi R, Hoffman JC, et al: Fertility control in the bitch by active immunization with porcine zona pellucidae. Use of different adjuvants and patterns of oestradiol and progesteron levels in estrous cycles. *Biol Reprod* 1985;32:761–777.
28. Mahi-Brown CA: Prospects for control of fertility in the female dog by active immunization with porcine zona pellucida proteins, in Talwar GP (ed): *Immunological Approaches to Contraception and Promotion of Fertility.* New York, Plenum Press, 1986, pp 301–309.
29. Nicolson GL, Yanagimachi R, Yanagimachi H: Ultrastructural localization of lectin binding sites on the zonae pellucidae and plasma membranes of mammalian eggs. *J Cell Biol* 1975;66:263–274.
30. Oikawa T, Yanagimachi R, Nicolson GL: Species differences in the lectin-binding sites on the zona pellucida. *J Reprod Fertil* 1975;43:137.
31. Oikawa T: A simple method for the isolation of a large number of ova from pig ovaries. *Gamete Res* 1978;1:265.
32. Ownby CL, Shivers CA: Antigens of the hamster ovary and effects of anti-ovary serum on eggs. *Biol Reprod* 1972;6:310–318.
33. Palm VS, Sacco AG, Syner FN, et al: Tissue specificity of porcine zona pellucida antigen(s) tested by radio immunoassay. *Biol Reprod* 1979;21:709–713.
34. Peterson RN, Russell LD, Bundman D, et al: Sperm egg interaction: Evidence for boar sperm plasma membrane receptor for porcine zona pellucida. *Science* 1980;207:73–74.
35. Sacco AG, Shivers CA: Localization of tissue-specific antigens in rabbit ovary, oviduct and uterus by the fluorescent antibody technique. *J Reprod Fertil* 1973;32:415.
36. Sacco AG, Palm VS: Heteroimmunization with isolated pig zonae pellucidae. *J Reprod Fertil* 1977;51:165–168.
37. Sacco AG: Antigenicity of the human zona pellucida. *Biol Reprod* 1977;16:158.
38. Sacco AG: Antigenic cross reactivity between human and pig zonae pellucidae. *Biol Reprod* 1977;16:164–173.
39. Sacco AG: Immunological specificity of anti-zona binding to zona pellucida. *J Exp Zool* 1978;204:181–198.
40. Sacco AG, Shivers CA: Immunologic inhibition of development, in: *Methods in Mammalian Reproduction*, vol 9. New York, Academic Press, 1978, pp 203–228.

41. Sacco AG: Inhibition of fertility in mice by passive immunization with antibodies to isolated zonae pellucidae. *J Reprod Fertil* 1979;56:533–537.
42. Sacco AG, Subramanian MG, Yurewicz EC: Application of a radioimmunoassay (RIA) for monitoring immune response to porcine zonae pellucidae (41172). *Proc Soc Exp Biol Med* 1981;1967:318–326.
43. Sacco AG, Subramanian MG, Yurewicz EC: Active immunization of mice with porcine zona pellucida: Immune response and effect on fertility. *J Exp Zool* 1981;218:405–418.
44. Sacco AG, Subramanian MG, Yurewicz EC: Passage of zona antibodies via placenta and milk following active immunization of female mice with porcine zona pellucida. *J Reprod Immunol* 1981;3:313–322.
45. Sacco AG, Yurewicz EC, Subramanian MG, et al: Zona pellucida composition: species cross reactivity and contraceptive potential of antiserum to a purified pig zona antigen (PPZA). *Biol Reprod* 1981;25:997–1008.
46. Sacco AG, Subramanian MG, Yurewicz EC, et al: Heteroimmunization of squirrel monkeys (Saimiri Sciureous) with a purified porcine zonae antigen (PPZA): Immune response and biologic activity of antiserum. *Fertil Steril* 1983;39(3):350–358.
47. Sacco AG, Yurewicz EC, Zhang S: Immunoelectrophoretic analysis of the porcine zone pellucida. *J Reprod Fertil* 1983;68:21–31.
48. Sacco AG, Subramanian MG, Yurewicz EC: Association of sperm receptor activity with a purified pig zona antigen (PPZA). *J Reprod Immunol* 1984;6:89–103.
49. Sacco AG, Subramanian MG, Yurewicz EC, et al: Antigenic profile and impact of immunization with zonae pellucida antigens in primates, in Talwar GP (ed): *Immunological Approaches to Contraception and Promotion of Fertility.* New York, Plenum Press, 1986, pp 277–290.
50. Sacco A: 1986, Personal communication.
51a. Sehgal S, Gupta SK: Long term effects of immunization with porcine zona pellucida in rabbits, 1987, in press.
51b. Sehgal S, Bhatnagar P, Vasishta RKV, et al: Immunization of rhesus monkeys with porcine zona antigen: Effect on cyclicity. (Manuscript in preparation.)
52. Shivers CA, Dudkiewicz AB, Franklin LE, et al: Inhibition of sperm egg interaction by specific antibody. *Science* 1972;178:1211–1213.
53. Shivers CA: Immunological interference with fertilization, in Diczfalusy E (ed): *Immunological Approaches to Fertility Control.* Stockholm, Karolinska Institutet, 1974, p 223.
54. Shivers CA, Dunbar BS: Autoantibodies to zona pellucida. A possible cause for infertility in women. *Science* 1977;197:1087–1090.
55. Shivers CA, Gengozian N, Franklin S, et al: Antigenic cross reactivity between human and marmoset zonae pellucidae, a potential target for immuno countraception. *J Med Primatol* 1978;7:242.
56. Shivers CA, Sieg PM, Kitchen H: Pregnancy prevention in the dog: Potential for an immunological approach. *J Am Anim Hosp Assoc* 1981;17:823–828.
57. Skinner SM, Mills T, Kirchick HJ, et al: Immunization with zona pellucida proteins result in abnormal ovarian follicular differentiation and inhibition of gonadotropin induced steroid secretion. *Endocrinology* 1984;115(6):2418–2432.
58. Skinner SM, Dunbar BS: Species variation in the zona pellucida, in Talwar GP (ed): *Immunological Approaches to Contraception and Promotion of Fertility.* New York, Plenum Press, 1986, pp 251–268.
59. Stevens VC: Current status of antifertility vaccines using gonadotropin immunogens. *Immunol Today* 1986;7(12):369–374.

60. Subramanian MG, Yurewicz EC, Sacco AG: Specific radioimmunoassay for the detection of a purified porcine zona pellucida antigen (PPZA). *Biol Reprod* 1981;24:933–943.
61. Subramanian MG, Yurewicz EC, Sacco AG: Studies on the stability of 1251-purified porcine zonae antigen (PPZA) obtained by chloramine-T Glucose oxidase with lactoperoxidase and N-bromosuccinimide iodination procedures. *Gamete Res* 1985;11:283–288.
62. Suman Bamezai AK, Das C, Talwar GP: Effect of immunization of primates against porcine zonae on fertility and hormone profiles, in Talwar GP (ed): *Immunological Approaches to Contraception and Promotion of Fertility.* New York, Plenum Press, 1986, pp 291–299.
63. Talwar GP: Immunology in reproduction. *J Reprod Med* 1979;22(2):61–76.
64. Talwar GP: Immunological approaches to contraception: The need, basic promise and overview, in: *Immunological Approaches to Contraception and Promotion of Fertility.* New York & London, Plenum Press, 1986, pp 17–24.
65. Talwar GP: Phase I clinical trials using β-HCG vaccine, 1986, personal communication.
66. Tsunoda Y, Chang MC: Effect of anti-rat ovary anti-serum on the fertilization of rat, mouse, and hamster eggs in vivo and in vitro. *Biol Reprod* 1976;14:354–361.
67. Tsunoda Y, Chang MC: In vivo and in vitro fertilization of hamster, rat and mouse eggs after treatment with anti-hamster ovary antiserum. *J Exp Zool* 1976;195: 409–416.
68. Tsunoda Y, Sugie T: Inhibition of fertilization in mice by antizona pellucida serum. *Jpn J Zool Sci* 1977;48:784.
69. Tsunoda Y, Chang MC: Effect of anti-sera against eggs and zonae pellucidae on fertilization and development of mouse eggs in vivo and in culture. *J Reprod Fertil* 1978;54:233–237.
70. Tsunoda Y, Chang MC: The suppressive effect of sera from old female mice on in vitro fertilization and blastocyst development. *Biol Reprod* 1979;20:355–361.
71. Tsunoda Y, Sugie T, Mori J: Quantitative determination of titres of antizona serum. *J Exp Zool* 1979;207:315.
72. Tsunoda Y, Soma T, Sugie T: Inhibition of fertilization in cattle by passive immunization with anti-zona pellucida serum. *Gamete Res* 1981;4:133–138.
73. Wolgemuth DJ, Celenza J, Bundman DS, et al: Formation of the rabbit zona pellucida and its relationship to ovarian follicular development. *Dev Biol* 1984;106:1–14.
74. Watanabe M, Muramatsu T, Shirne H, et al: Discrete distribution of binding sites for dolichos biflorus agglutinin (DBA) and for peanut agglutinin (PNA) in mouse organ tissues. *J Histochem Cytochem* 1981;29:779–790.
75. Wood DM, Dunbar BS: Direct detection of two cross reactive antigens between porcine and rabbit zona pellucida by radioimmunoassay and immunoelectrophoresis. *J Exp Zool* 1981;217:423–433.
76. Wood DM, Celilia L, Dunbar BS: Effect of alloimmunization and heteroimmunization with zonae pellucidae on fertility in rabbits. *Biol Reprod* 1981;25:439–450.
77. Yanagimachi R: Specificity of sperm-egg interaction, in Edidin M, Johnson MH (eds): *Immunobiology of Gametes*. London and New York, Cambridge Univ Press, 1977, pp 187–207.
78. Yurewicz EC, Sacco AG, Subramanian MG: Isolation and preliminary characterization of a purified pig zona antigen (PPZA) from porcine occytes. *Biol Reprod* 1983;29:511–523.

Part VII
Basic Considerations in Immunological Approaches to Control of Fertility

CHAPTER 33

New Developments in Epitope Analysis

J. Griffiths, N.A. Mitchison, D.B.G. Oliveira,
and P. Salgame

We are witnesses to a great surge of activity in epitope analysis. Powerful new motivations are joining together to drive powerful new technologies. The field is exploding with new ideas and new experiments. We aim here to survey this excitement to guess how the field will develop, and to place our own work within the context of these developments.

New Motivations

With the end of colonialism we have seen a new and better relationship grow between the developing and the developed world. The common struggle against ill-health reflects this new relationship. The development of new vaccines is a key element in this struggle and has been largely responsible for the current upsurge of interest in epitope analysis. The overseas aid programs of the richer countries have been an important source of support for this development, either directly or through the World Health Organization. Support of a similar nature has come from a multitude of other agencies, some private and some public. For instance, one of the authors of this chapter is an Indian who receives support from the Wellcome Foundation for research in the United Kingdom, and another is British who is supported by the World Health Organization for coming to this meeting.

We emphasize that financial support is not all that matters. In the world as it now is we take it as a matter of course that a scientist from the developing world should feel free to draw on the developed world in whatever way will best solve her indigenous problems, and that her colleagues in the developed world should feel bound to help. Without this spirit none of the networks of WHO would matter. We feel confident that the spirit of international solidarity is here to stay and to grow.

New Technologies

Epitope analysis enters vaccine development at the point where second-generation vaccines begin to replace the older first-generation ones. The first generation consists of infectious agents themselves, in attenuated or killed form, or of products derived from them such as toxoids. Many of them date from the first 50 years of immunology, during the colonial era of tropical disease research. Second-generation vaccines are products of protein fractionation, peptide, or oligosaccharide synthesis, or recombinant DNA. Their essential feature is that they include an element of human design. It is assumed that antigens, as they occur in nature, either have parts that make them work badly as vaccines, and that therefore need to be cut off, or parts that need to be added. To design vaccines rationally we need to understand how the various parts or "epitopes" of a macromolecule work in stimulating an immune response, hence epitope analysis.

Exactly the same principles apply to the design of birth control vaccines, that is, vaccines against naturally occurring components of the reproductive system that are designed to lower fertility.

At this point it is only fair to confess that no second-generation vaccine has yet reached general use. At the start of this year the first synthetic vaccine (actually a conjugate of a synthetic peptide with a naturally occurring protein) entered a phase I clinical trial, after prolonged animal tests. It is an antihuman chorionic gonadotrophin vaccine designed for birth control, developed by the Human Reproduction Program of WHO. In the immediate future the fear of using human blood products and other safety considerations, combined with the difficulty of growing infectious agents such as *Mycobacterium leprae*, will provide the main driving force for second-generation vaccines, rather than any grand immunological strategy. But deep immunological issues are at stake and the introduction of the new vaccines will provide valuable opportunities to explore them through epitope analysis.

In all these developments recombinant DNA provides the central technology. This has become the easiest way to acquire sequence information and competes with peptide synthesis to provide peptides for evaluation. For instance in epitope analysis of the main 65KD protein of *M leprae*, within the network of the WHO special program in Tropical Disease Research, a peptide synthesis approach is competing directly with a minilibrary prepared in λgt11 bacteriophage and neither method seems to have a decisive advantage. Whatever one may believe about the relative importance of proteins and carbohydrates as antigens, and the recent syntheses of phenolic-glyco-lipid I of *M leprae* elegantly illustrating how well carbohydrate chemistry is progressing, proteins offer a far more versatile array of structures. And finally, recombinant DNA is well adapted to biotechnological production scale-up, as well as to cheap production in recombinant vaccinia virus or wounded bacteria.

New Ideas: Naive Structuralism Versus Subtle Evolutionism

Very broadly, immunologists are pursuing two alternative lines of approach to epitope analysis. One school starts from antigenic structures and searches empirically for parts that serve as T or B epitopes. It tends to examine proteins more or less in order as their structure becomes known, without paying much attention to their evolutionary relationship with the immune system. This approach reveals for instance that surface loops, rather than helices, attract the attention of B cells; hence the value of computer-aided modeling of three-dimensional protein structure such as that of β-human chorionic gonadotropin (HCG) recently completed for WHO/HRP (S. Cooper and T. Blundell, personal communication). This approach has two recent triumphs with T cells: the identification of epitopes and agretopes. Maybe an amphipathic helix is not the final answer for epitopes (1, J. Rothbard personal communication). The point is that the search for common features defined by a structural algorithm is now accepted as valid. Agretopes now have a firm footing (2,3), with the implication emphasized by Sercarz that the pairing of an epitope with a particular agretope can determine whether T-helper (Th) or T-suppressor (Ts) cells become activated.

The alternative school focuses on the evolutionary relationship between antigens and the immune system. Any attempt to predict the distribution of epitopes on a protein lays emphasis on the following four considerations.

Self-Tolerance: Clone Elimination

Self-macromolecules can be expected to purge the T-helper compartment of reactive clones. For monomeric proteins the concentration in body fluids needed for purging is known with some precision both from repeated-injection experiments and from studies on naturally occurring proteins, and it is low (~ 10^{-9} mol/L). The B and T-suppressor compartments are relatively refractory to purging, and accordingly "intermediate"-concentration proteins effectively define a set of suppressor (and B) epitopes. When an intermediate-concentration protein also expresses an allo-epitope recognized by helper T cells, this has the effect of dividing it into distinct Th and Ts domains (4,5).

Self-Tolerance: Suppression

Self-macromolecules that for any reason tend to slip through the net of clone elimination must select for immune-suppression (Is) genes and express corresponding Ts epitopes. This kind of evolution is hard to demonstrate in action, but a comparison between F liver protein and the complement component C5 is suggestive. Mice do not normally contain *activated* suppressor T cells specific for F (4), whereas for C5 they do so (Y. Borel, personal communication). The reason for this difference is not understood, but for present purposes the point is that self-proteins evidently vary in the expected manner.

Immunopathology Selects for Ts Epitopes

We have advanced the argument that chronic infection and its attendant immunopathology may be the main force in selecting for Is genes and their corresponding Ts epitopes. We regard immunopathology essentially as a deterrent generated by parasites and other agents of chronic infection and designed to enforce this selection, which in turn benefits the parasite (6). If this is so (and like most evolutionary arguments it may be correct in theory but of minor importance in practice), then infection-related proteins will define a major set of Ts epitopes.

Control of the Receptor Repertoire Through Connectedness

Connectedness (i.e., idiotype-anti-idiotype binding of lymphocyte receptors and, to an extent, also the binding of these receptors to other self-macromolecules) is a powerful force moulding the repertoires. Among T cells we have recently reviewed the evidence that this applies particularly to Ts (3). Among B cells, the prevalent idea is that activated B cells in the spleen of young mice comprise a population that is highly connected, autoreactive, multispecific ("sticky," so that one antibody can bind several antigens), and that expresses predominantly the two 3́ VH gene families and belongs to the CD5 (Lyl) subset (7,8). Most of the B cells that are used for epitope analysis, being present in the adult and $CD5^-$, do not belong to this subset; nevertheless, the repertoire that they express seems to be more or less under its control. So B epitopes must also fall under its control, to an extent that is as yet unclear.

The references cited above make it clear that the authors of this chapter belong to the second school and that we hold in scorn the naive structuralist who argues of epitopes that "if it occurs in lysozyme and galactosidase and then turns up in cytochrome C, it must be true." The weakness of our position is that it does not yet help make definite predictions about candidate vaccine proteins, whereas the structuralists can already do so. Nevertheless, we are convinced that when all the facts are revealed, it will be the evolutionists who will provide the deeper and intellectually more satisfying view.

Our Own Collection of Proteins for Epitope Analysis

We have selected certain proteins for intensive epitope analysis on the grounds that they are likely to take us in interesting directions. At the head of our list is F liver protein, chosen because the way that it is recognized by helper T cells is profoundly influenced by self-tolerance. As this chapter makes clear, we are following a standard procedure for determining primary structure and should soon be able to make an accurate assignment of the Th-defined alloepitope. Next comes a screening enterprise carried out on *M leprae*. Here our aim is to discover how far we can get with a T-cell screen as applied to the entire set of 4,000 proteins that this organism's genome is believed to encode. The work of others suggest that the response to this entire organism is extraordinarily selective, with just five proteins dominating both T and B responses. Our screen will test this suggestion. Another feature of this response is the dramatic immunopathology of leprosy, particularly in the borderline tuberculoid form of the disease. This, we believe, may have selected for Ts epitopes. Because of the extreme diversity of serotypes, we believe that rhinovirus may be good material for testing the hypothesis that T as well as B epitopes become polymorphic in response to immunoselection. This hypothesis is not as obvious as it sounds: one has to ask why the nucleoprotein of influenza virus is not as polymorphic as the hemagglutinin.

F Liver Protein

F is a liver protein of unknown function found in all vertebrates so far studied. In the best studied animal, the mouse, there are two allotypes, F_1 and F_2. Mice do not respond to self-F, and only certain strains (haplotype k) will respond to immunization with allo-F antigen. However, antibodies, once produced, recognize both self- and allo-F antigen, indicating that whereas B-cell determinants are shared, the (probably single) Th epitope is allovariable. Low responsiveness to allo-F is due to suppression in certain strains and hybrids; these Ts cells can also be activated by self-F antigen, indicating that the Ts epitope(s) is also shared (for general review of the F antigen system, see ref. 9). Our aim is to dissect those areas of the F-antigen molecule responsible for activating these different types of T cells.

Using a monoclonal antibody, we have been able to produce large amounts of the protein. High-pressure liquid chromatography (HPLC) separation of a pepsin digest generated a series of peptides. Using in vitro T-cell proliferation assays, a subset of these peptides has been identified as containing the T-proliferative (and therefore probably helper) epitope; further fractionation of this subset is in progress.

The amino acid sequence of a short peptide was determined as a means of constructing a DNA oligomer for screening a mouse liver cDNA library. Three different oligomers were used for screening: a 12mer, a 48mer, and three overlapping 12mers. All three approaches proved unsuccessful, and so the alternative approach of screening an expression library (λgtll, cDNA) with antibodies was employed (10).

A polyclonal antiserum to murine F protein was raised in a rabbit by subcutaneous injection of an F/CFA (complete Freunds adjunct) emulsion. When expressed in λgtll, F is in the form of a galactosidase fusion protein. The feasibility of using the antiserum for detection of denatured F protein was ascertained by Western blotting onto nitrocellulose filters. These were screened with rabbit antiserum using a biotinylated-antirabbit, streptavidin-horseradish peroxidase detection system. A band of 41 to 42K molecular weight was detected in samples of murine and human F proteins and a crude murine liver extract. The antiserum did not bind significantly host cell lysate (*Escherichia coli* Y1090), nonspecific λgt11 lysogen or actin (mol wt 43K).

This antiserum was therefore used to screen a mouse liver λgtll library and a human fetal liver library. In both cases 1×10^5 phage clones were screened, yielding approximately 50 positives each (i.e., 1/200). Several of these were selected, replated at low density, and rescreened so that single positive clones

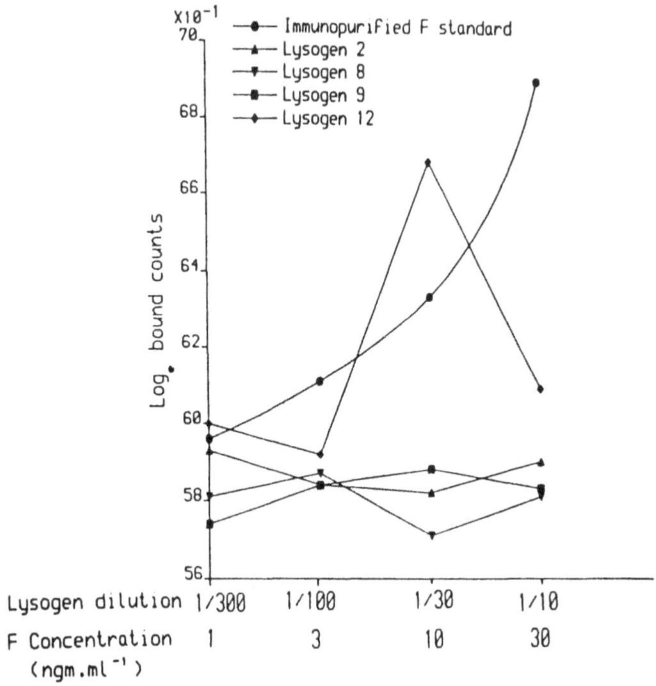

Fig. 33.1. Capture radioimmunoassay for F protein (polyclonal anti-F protein antibodies on solid phase; captured material detected by binding of ^{125}I labelled monoclonal anti-F protein antibody) showing cross reactive material present in lysogen no. 12.

could be isolated for amplification and sequencing. To ensure that selected clones contained F protein DNA, lysogens of each were prepared and tested in a highly specific radioimmunoassay (11) (incorporating a monoclonal anti-F second layer). At least one of the lysogens scored positive in this assay (Fig. 33.1). They were also checked by Western blotting for the presence of F protein and β-galactosidase with their respective antibodies, and again scored as positive.

Insert purification and subcloning into vectors suitable for sequencing is currently in progress. The inserts will then be labeled and used to rescreen the libraries at the DNA level until the entire sequence of the genes has been determined.

The difference between the sequences for the two murine F allotypes should reveal the Th epitope. Elucidation of Ts epitopes will be more difficult, but there are a number of possible strategies. Once identified the effect of grafting these epitopes onto other carriers can be tested.

Leprosy

M leprae is the etiologic agent of leprosy and remains the only major bacterial pathogen of man that has not yet been cultivated. As mentioned earlier, DNA recombinant technology offers a means to obtain, in large quantities, all the protein antigens of a pathogen encoded in linear segments of the pathogen's genome. Such a strategy has been successfully used to construct a genomic *M leprae* library in λgtll that expresses the foreign insert DNA under the control of prokaryotic gene signals in *E coli* (12).

Using antibody probes and T-cell clones, various groups have picked individual recombinant clones. Of particular interest are the clones expressing a 65 Kd and 18 Kd protein antigen of *M leprae*. However, such a screening procedure brings in a degree of bias, and hence our approach has been to use a more general and random screening procedure of the library for T-cell reactive epitopes. The feasibility of this approach is important to determine, as it provides a larger selection window for T-cell reactivity.

We have developed a technique for partially purifying the fusion proteins expressed in the bacterial lysates and absorbing them onto nitrocellulose. The fusion protein was partially purified by extractions with the weak detergents deoxycholate and Triton-X, solubilized in 1% sodium dodecyl sulfate (SDS), absorbed onto nitrocellulose discs, and put directly into the assay. From this, adequate signals for screening purposes have been obtained from T cells of mouse lymph nodes primed with whole killed *M leprae* (*M leprae* was obtained from J. Colston, Mill Hill, London).

The sensitivity of the assay was tested by a simple titration experiment whereby the number of positive clones required in a pool of negative clones to give a good signal was determined. To increasing numbers of λ gtll, constant numbers of a positive clone were added and the antigens expressed and assayed for proliferation using the above method. As observed in Table 33.1, at antigen concentrations of 0.1 μg, a good signal was obtained at a ratio of 1:25 of positive to negative clones. Even at a 1:200 ratio there was a positive signal. With

Table 33.1. Stimulation index.

Ratio of +ve to −ve	Partially purified lysate (μg/mL) 0.01		0.1		10	
	Experiments					
	1	2	1	2	1	2
Only +ve	1.4	0.8	2.7	1.8*	18	8.6
1:1	1.4	0.7	0.9	1.26	14	16.7
1:10	4.9	1.4	2.8	2.5*	25	27.6
1:25	1.9	1.1	2.2	1.5*	23	30
1:50	2.6	1.0	1.6	2.6†	23	52
1:100	2.0	1.2	2.0	1.9*	10	10
1:200	1.6	0.7	1.8	1.7*	22	25.9
1:500	1.5	0.6	1.3	0.8	15	21.2
1:1000	0.7	0.6	1.5	0.5	8	16.1
Only −ve	1.4	0.7	1.1	0.8	10.5	6.6

+ve clone, clone Y.3155 picked from the *M tuberculosis* library by a monoclonal antibody TB 68.
−ve clone, λgt11 containing no insert.
P values obtained by student's t-test compare test cpm responses to λgt11 (−ve clone only) cpm responses. Levels of significant differences are $*P<.05$; $\dagger P<.01$.

Table 33.2. Fractionation of 10,000 phages randomly picked from the *M leprae* library.

1st tract	A	B	C	D	E	F	G	H	I	J
2nd tract	1-10	11-20	21-30	31-40	41-50	51-60	61-70	71-80	81-90	91-100
					Original pool: 2.0 (0.1 μg/mL)					
	1	2	3	4	5	6	7	8	9	10
A	0.9	1.7	1.3	1.4	0.8	0.6	1.6	11.1	1.2	2.1
B	2.0	2.5	1.6	1.2	0.5	2.4	1.7	1.8	1.6	3.0
C	1.3	1.4	1.1	0.8	3.3	6.2	6.7	2.5	6.2	4.4
D	5.0	1.8	1.3	3.6	4.6	8.6	8.2	3.0	2.8	1.8
E	5.6	7.3	2.1	1.6	2.8	2.3	1.0	0.6	0.5	0.5
F	4.6	7.0	0.8	1.2	–	–	4.0	3.2	0.6	0.4
G	2.4	2.2	3.7	1.5	–	3.5	1.6	2.0	4.4	4.1
H	3.7	4.0	4.0	7.7	3.6	4.1	6.0	5.6	2.9	10.6
I	2.1	1.9	2.1	2.1	1.5	2.4	7.8	4.4	1.3	1.8
J	1.3	1.1	2.9	3.2	9.0	0.9	1.9	4.3	1.3	2.2

Data have been represented as stimulation index (cpm test/cpm control). The circled numbers are the positive pools. SI of λgt11 lysate was 2.1 and pools having an SI twice that of λgt11 have been considered as positive pools. cpm of *M leprae* (5×10^8/mL) stimulated culture was 19,889 ± 5,894, ConA (2.5 μg/mL) 134,376 ± 35,014; λgt11 lysate 8,083 ± 1,213 and media alone was 3,780 ± 1,180.

increase in antigen concentration, the detection limit went up; however, the sensitivity of the assay decreased.

Using the above system we proceeded to fractionate a randomly selected aliquot from the *M leprae* library and screen for T-cell proliferation. A total of 10,000 phages were randomly picked from the library and fractionated to ten equal pools. Each of the ten pools was further subfractionated to ten minipools. A total of 111 pools were obtained. Each pool was amplified before fractionation. Crude lysate expressing fusion proteins was prepared from each pool, partially purified and absorbed onto nitrocellulose discs placed in 96 well plates, in four concentrations ranging from 0.01 to 10 μg protein/ml. The plates were sterilized by exposure to ultraviolet rays for 20 minutes. The wells were rinsed twice with media and each well received 4×10^5 lymph node cells from *M leprae* immunized mice. The antigens were found to absorb well onto nitrocellulose even in the presence of SDS, as the washings from each well were negative when tested for protein. With antigen concentrations of 0.1 μg and 1 μg protein/ml, 23 and 30 positive pools, respectively, were obtained. As the antigen concentration was increased, the sensitivity of the assay decreased (Table 33.2). The experiment has been duplicated with all the positive clones and one negative clone (A3) with similar results.

We are aware that we may have picked some false positives at this stage, largely because we have not used purified fusion proteins. However, the first round of screening does eliminate a large number of negative pools. With the titration experiments we obtained a positive signal even at 1:200 ratio of positive to negative clones; hence a negative signal with pools of 100 phages indicates that there is no positive clone in this pool, as it falls within the sensitivity limits of our assay. Also, it is likely that we are missing clones with a low proliferative signal. Dilution analysis of *M leprae* primed cells and their reactivity toward the antigens present in the positive pools will select the true positives.

At this point it is important to determine whether the antigens picked by our screening belong to the five antigens that have been selected by the antibody probes. We are in the process of immunoscreening our positive pools with a mixture of *M leprae* monoclonal antibodies.

Acknowledgment. Dr. P. Salgame is the recipient of a Wellcome Trust Research Fellowship.

References

1. Berkower I, Buckenmeyer GK, Berzofsky JA: Molecular mapping of a histocompatibility-restricted immunodominant T cell epitope with synthetic and natural peptides: Implications for T cell antigenic structure. *J Immunol* 1986;136:2498–2503.
2. Buus S, Colon S, Smith C, et al: Interaction between a "processed" ovalbumin peptide and Ia molecules. *Proc Natl Acad Sci USA* 1986;83:3968–3971.
3. Mitchison NA: Suppressor T cells in the network. *Nature* 1985;316:676–677.
4. Mitchison NA, Lukic ML, Nardi N: Proteins at the margin between being tolerated and being noticed. In *Antibodies: Protective, Destructive and Regulatory Role 9th Int Convoc on Immunol.* Milgrom F, Abeyounis CJ, Albini B, Karger S (eds); Karger, Basel, 1985, pp. 241–248.
5. Mitchison NA: Repertoire purging by medium-concentration self-macromolecules is the major factor determining helper and suppressor repertoire differences. In *Haematology and Blood Transfusion Modern Trends in Human Leukaemia VI.* Neth, Gallo, Greaves, Janka (eds); Berlin, Springer-Verlag, 1985, pp. 3–6.
6. Mitchison NA, Oliveira DBG: Chronic infection as a major force in the evolution of the suppressor T-cell system. *Parasitology Today* 1986;2:312–313.
7. Martinez C, Bernabe RR, de la Hera A, et al: Establishment of idiotypic helper T-cell repertoires early in life. *Nature* 1985;317:721–723.
8. Hayakawa K, Hardy RR, Herzenberg LA: Progenitors for Ly-1 B cells are distinct from progenitors for other B cells. *J Exp Med* 1985;161:1554–1568.
9. Griffiths JA, Mitchison NA, Nardi N, et al: F protein. In *Immunogenicity of protein antigens.* Sercarz EE, Berzofsky J (eds); Boca Raton, 1986 (in press).
10. Huynh TV, Young RA, Davis W: Constructing and screening cDNA libraries in gt10 and gtll. *DNA Cloning, a Practical Approach.* Glover DM (ed); Vol 1, Chapter 2, ED IRL Press, 1985, pp. 49–78.
11. Oliveira DBG: Bulk immunopurification and development of a radioimmunoassay for human and murine F liver protein. *J Immunol Methods* 1986;91:99–105.
12. Young RA, Mehra E, Sweetser D et al: Genes for the major protein antigens of the leprosy parasite *Mycobacterium leprae. Nature* 1985;316:450.

CHAPTER 34

Immunological Difficulties and Requirements for Immune Manipulations in the Field of Reproduction

Guy André Voisin

Introduction

Immune manipulations in the field of reproduction have two potential applications aiming at two opposite goals: (a) immunotherapy for sterility and repeated abortions and (b) anticonceptional vaccination for control of fertility.

Both of these potential domains are based on a fact and an idea. The *fact* is that gametes, embryos, and gestation involve particular substances that may behave an antigens, that is, they may be able to give rise to immune responses. The *idea* is that immune reactions against these antigens are bound to have adverse effects on their targets, with resulting impairment of either fertility or early gestation—whether spontaneous (immunopathology of sterility and repeated abortions) or deliberately provoked (anticonceptional vaccination)—and that, as a corollary, an absence of immune reaction to these antigens is necessary for normal reproductive functions. Whereas the preceding fact is correct, the preceding idea is partially erroneous, as we shall see later.

Clearly, anticonceptional vaccination is an old idea. It has been proposed, experimented, and even clinically tried long ago (30,37,39). There is not yet, however, a definitive immunological method available for common use, whereas hormonal methods more recently proposed are widely applied. On the other hand, the role of immune reactions in some types of sterility has been known for many years and has long been suspected in unexplained repeated abortions. Still, we are only at the beginning of some tentative immunotherapy.

Another paradoxical situation is that of spermatozoa and their immunological status: they are among the most autoantigenic cells in all species studied. Spontaneous immunization against them with pathological consequences such as infertility is, however, unfrequent in spite of the frequent presence of so-called natural antispermatozoa autoantibodies. Immunization of female guinea-pigs with guinea-pig spermatozoa autoantigens may result in a state of immunity with a complete array of humoral and cellular reactivity and with antibodies even in the genital tract. Still, the observed effect on fertility and gestation is minimal, if any (18). It is well known from old works, including Mitchison's and ours, that immunizing a pregnant female with paternal antigens prepares her to reject, in an accelerated way, a graft from the father (or even from the fetus) but has no dramatic action on the in situ fetus.

Finally, is it not paradoxical that the same type of manipulation, namely, immunization, will try to inhibit fertility or early gestation on the one hand–meaning that this immunization will have adverse effects on spermatozoa (or ovum or early embryo)–and, on the other hand, will try to cure repeated abortions–meaning that this immunization will have favorable effects on the conceptus survival?

These various paradoxical situations must be understood and the potent immunoregulatory mechanisms that they imply must be overcome if one is to obtain a successful anticonceptional vaccine. These constitute the main prerequisites for successful immune manipulations in the fields of reproduction.

There are other well-known prerequisites that will not be considered here: they go from specificity to reproductive target (which is present only on gametes or early embryo or substances necessary only to reproduction) to great simplicity and cheapness and pass by extreme percentage of efficacy, absence of noxious side effects, reversibility after a given time, and reproducibility thereafter.

As far as the immunological prerequisites are concerned, they result from two crucial aspects. The first one is concerned with the nature of an immune reaction to an antigen-bearing cell and the consequences for that cell. The second one resides in the local immunological status of the genital tract.

Nature of an Immune Reaction to an Antigen-Bearing Cell

Driving Forces During Evolution and Final (Present) Selection

Besides the mechanisms (molecular idiotypic network, cellular T lymphocyte circuitry, cell messengers) through which the immune reaction and its regulation operate, it is crucial to imagine what the driving forces that have ultimately led to the selection of the immune system, its agents, and mechanisms have been.

It appears to me axiomatic that the essential driving force for the selection of the immune agents, and for the responsible immune system and corresponding genes, has to be the final fate of the antigen-bearing target. Any target that is dangerous to the survival of the individual and the species must be eliminated or neutralized if the species is to survive and, eventually, to continue its evolution. The relevant targets are bacteria, viruses, fungi, and parasites, as well as allografted and, possibly, some tumoral cells. This is at the origin of the reaction of immune elimination or immune rejection reaction (RR), which means the selection of its immune mechanisms and agents. On the other hand, targets that are necessary to the survival of the individual and the species (such as autoantigen-bearing organs and alloantigen-bearing embryos and fetuses) must not be threatened by such an RR or be protected against it. The selected reaction that attains that goal, and is more subtle and more difficult to uncover, is what we have named the immune facilitation reaction (FR). It includes what may be called active tolerance. In addition, another superimposed driving force, common to all physiological reactions, is the necessity of a regulation to prevent the immune RR toward invading foreign agents to reach a level and a duration that would have immunopathological consequences and preempt the immune system, preventing its reaction against other new invaders. This regulation also is ensured by the immune agents of the FR. In other words, active tolerance is the extreme result of a strong immune down-regulation. It has been and is our concept to consider that the "raison d'être" of the FR is to maintain the immune reaction in the proper limits, reaching a well-balanced stage when foreign invaders are attacked, neutralized, and/or rejected while substances and cells from the individual itself (and its conceptus) are protected against the induction and/or the effects of such an immune RR. The same sets of mechanisms of the FR may, according to the conditions and its intensity, go from a simple down-regulation of an ongoing reaction (and its eventual termination) to a state of active immune tolerance (51,60). This is observed and obtained in highly evolved animals.

Responsible Immune Agents of RR and FR

These two opposite aspect of the immune reaction are both mediated by antibodies and cells. The immune agents of the RR are mainly the following ones: cytotoxic T cells, delayed hypersensitivity T cells, helper T cells, activating macrophages, and natural killer (NK) cells through lymphokines such as macrophage-activating factor (MAF) and γ-interferon. Antibodies involved in rejection are those able to activate killer (K) cells through the antibody-dependent cell-mediated cytotoxicity (ADCC) phenomenon, or to activate complement component 1 (C1), resulting in opsonization or, more seldom, destruction. The immune agents of the FR are suppressor or regulatory cells, mainly suppressor T cells specific of antigen or of idiotype and components of the suppressor circuit. They also are "facilitating" antibodies (i.e., facilitating the survival of the antigen-bearing target), including classic enhancing antibodies (nonactivating C1 or K cells) or complexed-to-soluble antigens (blocking factors) and anti-idiotypic. There are strong relations between regulatory antibodies and regulatory cells.

Milestones of the Immune Reaction During Evolution (55)

1. From the unicellular protozoa, a primitive cell-surface recognition seems necessary to allow nutrition from exterior medium and to preserve the cellular individuality. This apparently can be bypassed by intracellular grafting.
2. In multicellular, cell-colony-like (with minimal cell differentiation) animals such as sponges and also in primitive metazoans such as coelenterates, transplantation studies show the presence of a self/nonself system of recognition shown by the nonacceptance of cells grafted from foreign colonies (Theodore, personal communication, November 1972). Such a simple binary system is not compatible with sophisticated recognition between various nonself, i.e.,

immunological specificity, with immunological memory, and with high degree of cell differentiation.

3. In differentiated protostoma such as annelids and earthworms, transplantation studies point to the individual recognition of different foreign colonies accompanied by specific second set rejection reactions (17,23), indicating the presence of an immunological memory, which is a correlate of specific recognition. This accelerated rejection can be adoptively transferred by cells of the coelom, but not passively by its fluid. This shows the early role of immune cells in the RR. In addition, there are indications that specialized cells recognize and remove altered cells from the body.
4. An interesting milestone in the fields of immune reaction and reproduction is the primitive urodele amphibian *Salamandra salamandra*. Transplantation shows the elements of a classic alloimmune RR, mainly cellular, but apparently also with a component of cytotoxic antibodies and with some primitive type of complement. The RR is becoming more sophisticated. *Salamandra salamandra* also represents one of the first attempts of tetrapods at viviparity. Actually it is an ovoviviparous species, i.e., without placenta, and its gestation is of particular interest. Embryos are numerous (up to over 40). They may stay intramaternal until full maturity or be rejected while larvae. Spleen cells from the pregnant mothers can specifically kill in vitro cells from their embryos (up to 95%). This "immunity" increases with the number of gestations (and embryos). Correlatively the sera of the very same pregnant mothers can prevent this in vitro cytotoxicity. This protective capacity also increases with the number of gestations (16). It is due to two factors: the first is specific of the embryos and specifically absorbed by them and localized in the IgM fraction (the only Ig apparently present on that species); the second factor is nonspecific, acting on the maternal spleen cells and possibly anticipating the mammalian pregnancy-associated factors, while the specific component anticipates the mammalian regulatory FR. No suppressor cells, however, were detected in these reactions (1,16). This model, therefore, represents an interesting attempt but is still very imperfect.
5. The last and highly evolved landmark is the human – or the mouse (the latter being reasonably close to the former for both its immune and its reproductive systems). For experimentation, one prefers mice for more than one reason. In that species the immune system, its mechanisms, and its immune agents are better known than in any other. RR and FR have been mainly defined in it. Reproductive phenomena are close to the human ones. Any knowledge on immunology of reproduction acquired in studying the mouse is likely to be useful in further active intervention in the human species.

Immunology of Gestation and Role of Placenta

We have demonstrated that during gestation, the balance of RR and FR is shifted from a predominance of the RR, as is observed in a conventional allograft, to a predominance of the FR, with obviously an importantly decreased overall immune reaction. To summarize the essential facts, one may conclude the following:

Pregnant Females Elaborate a Weak RR Toward the Conceptus Paternal Antigens. The alloreaction may be detected through signs of delayed hypersensitivity (44,66), priming for a secondary type of mixed lymphocyte reaction (MLR) (8) or through signs of graft-versus-host reaction (61), and is adoptive transfer of accelerated rejection in some cases (11). This weak rejection does not go, however, to the point of elaborating the effector agents of the RR. In the mouse the antipaternal immune agents are not C′-activating and NK cell-activating IgG2 antibodies (3,11,58), neither are they cytotoxic T cells (14).

Pregnant Females Elaborate a (Relatively) Stronger FR Toward the Conceptus Paternal Antigens. This reaction is essentially made of enhancing antibodies and suppressor T cells. Both are better (but not only) seen in multiparous gestations. IgG1-enhancing antibodies specific of paternal antigens are fixed on the placenta (spongiotrophoblast) (11,58,63). Anti-idiotypic antibodies to antipaternal antibodies may be found (12) and are expected to play a role. Suppressor cells are found initially in the para-aortic lymph nodes, then in the spleen where they can be studied. In vivo they are able to transfer some degree of facilitation (11), and in vitro they can suppress a corresponding MLR (8) and the generation of cytotoxic T cells (14), genetically restricted and active through suppressor factors (9,10). The presence of immune agents of FR (enhancing antibodies and suppressor cells) toward paternal antigens—together with the absence or extreme poorness in immune agents of the RR (cytotoxic T cells and cytotoxic antibodies) against antigens of the same origin—in pregnant mice clearly points to an immune deviation from a predominant RR (as it is usually observed in experimental or clinical allografts) to a predominant FR. The hypothesis was put forward that the conceptus itself, through "immunomodulatory" placental substances, would signify to the mother to react to the accompanying conceptus antigens in the direction of predominant regulatory and protective immune agents rather than aggressive and cytotoxic ones.

Placental Extracts Injected Together With Limited Doses of Allogeneic Cells Induce an Immunodeviated Response Mimicking the One Induced by a State of Gestation. Recipient mice (CBA for instance) injected with 5×10^5 to 10^6 allogeneic donor cells (A/J for instance), together with CBA placental extracts or fractions, will not reject A/J sarcoma 1 cells ($3\text{–}5 \times 10^6$) in an accelerated way, as do control animals (in which placental extracts have been replaced by RPMI or liver extracts), but will keep them for a longer time than naive mice and sometimes permanently (13,20). Looking at the spleen cells of these modified recipient mice, the following observations were made:

1. The spleen cells are less reactive than controls in graft-versus-host reactions (62), in MLR (13), and in cell-mediated lympholysis (CML) (19).
2. They are suppressive in vivo and in vitro. When transferred in vivo to CBA mice, they slow the rejection of A/J tumor cells, whereas the control spleen cells accelerate it (13,20). In vitro they are suppressive of a corresponding relevant MLR (13) and CML (19). The responsible cells were found to be IJ^+ T cells (19).

3. Looking now at the anti-A/J serum antibodies in the CBA alloimmunized, placenta-treated CBA animals, the general level was not very modified by placental treatment but the isotypes were shifted, with less and less rapidly appearing cytotoxic (IgM and mainly IgG2) antibodies and more and more rapidly appearing anaphylactic ones (IgE and mainly IgG1). The in vivo correlate was the enhancing activity of these antibodies favoring the take and growth of SA 1 A when injected simultaneously to new CBA recipients.

   In brief, the immunological behavior (both cellular and humoral) of mice moderately alloimmunized in the presence of placental extracts or fractions mimicks the one of pregnant mice and is similar to the immunodeviation characteristics of the regulatory FR (53).
4. The responsible substances are found in the spongiotrophoblast, absent in the labyrinth, but also present in the decidua either as a contaminant or because of a physiological migration and a possible participation of the decidua itself in response to placental messages. Several substances exist endowed with different immunoregulatory properties. Modulation of alloimmune reaction was found in concanavalin A retained fractions and in sodium deoxycholate solubilized placenta cell membranes (27). Modulation of antisheep red blood cell antibody-forming cells was found in two fractions of opposite action: one depressive, of 40 kDa and one stimulatory, of 60 kDa (5) acting on T suppressor cells and on B cells, respectively (6). Obviously more has to be known of the biochemical nature and mode of action of each of the participating substances, the interaction of which results in the described immunodeviation.

The Conceptus and Paternal Antigens Responsible for the Maternal Immune Response Are Not Completely Known. It is, nevertheless, of crucial importance to know these antigens if one is going to attempt immune manipulation against them (fertility control) or in favor of them (prevention of habitual abortions). These antigens are of two types: allogenic and organ-specific. *Allogenic antigens* involved are polymorphic and of three kinds: major histocompatibility complex (MHC)-coded, MHC-linked, and MHC-independent. Concerning MHC-coded antigens, class II Ia antigens are absent from the trophoblast interface, but anti-Ia maternal reactions may be induced by fetal cells passing the placenta. Class I antigens are expressed at the interface but not every where and not in a classical way. Actually MHC-linked allotypic antigens involved in gestation might be modified class I antigens, still allotypic but much less polymorphic. This might be the case for the TA antigens of the described trophoblast/lymphocyte cross-reactive (TLX) antigens in humans (24) and the Pa antigens in the rats (25). The recognition of such class I antigen by the mother seems essential to the building of a favorable immune reaction to the conceptus, which seems necessary to the good development of the latter. This is the basis for the "vaccination" against abortion. However, the lack of precise knowledge of the involved antigens does not allow us to decide between paternal or panelled leukocytes (and which precise cells) to compose the vaccine. MHC-independent allotypic antigens are the minor histocompatibility antigens that may play an important role in gestation

(61,62). *Organ-specific (monomorphic ?) antigens* are even less known on the placenta. Nevertheless, there is evidence that they exist (4,57).

Placental substances stimulate a regulatory network that prevents the conceptus to induce and undergo a detrimental immune reaction—an RR—of the pregnant mother and that helps to induce a favorable reaction to the recognized antigens—an FR: a protective reaction as well as, apparently, an immunostimulation of the antigen-bearing target, possibly by growth factor-like lymphokines (43,64). This regulatory network, however, is not conceived to prevent the artificial induction of an antipaternal RR by a strong allogeneic stimulus. As a matter of fact, such an RR can be induced if the stimulus is adequate, with the elaboration of the corresponding immune agents, active on a test graft. In spite of that, the conceptus remains unaffected, clearly indicating that some local protective effector mechanism is taking place, as part of a particular immunological status of the genital tract, especially the female one, which is analyzed later.

Immunology and Immunopathology of Spermatozoa

Spermatozoa are extremely autoantigenic, possibly because of their late appearance in individual life. Still, infertility linked to auto- or isoimmune reactions against them is unfrequent.

Spermatozoa Autoantigens Are Numerous (52,56). Besides spermatozoa-coating antigens (SCA), the main categories are the ones directly accessible to immune agents, i.e., plasma membrane antigens such as the guinea-pig T autoantigen, which has recently been shown to be of a family of nine immunoprecipitable molecules (46). There are also those that play a crucial role in the process of fecundation—the acrosomal substances (such as guinea-pig S and P autoantigens), cytoplasmic and mitochondrial enzymes such as mouse and human lactic dehydrogenase (LDH) C4 (26), and nuclear antigens such as sperm-specific protamines (32). These autoantigens give rise to autoantibodies that are able to affect the spermatozoa functions, at least in vitro. Injected in vivo, they give rise to several immune reactions with consequences on fertility.

Biological Actions of Antispermatozoa Antibodies. These have been studied mainly in vitro:

- Spermatozoa mobility (and survival in presence of complement) is affected by anticytoplasmic membrane autoantibodies such as anti-T.
- Cervical mucus passage is often impaired in the presence of antibodies of male or female origin giving rise to the "shaking phenomenon" (33).
- Capacitation and acrosome reaction are affected by autoantibodies against S and T autoantigens, which modify their localization during the process (38).
- Binding to zona pellucida is inhibited by antispermatozoa antibodies, especially by auto-anti-T (40).
- The fusion of acrosome-reacted guinea-pig spermatozoa with zona-free oocytes is prevented by anti-T, partially inhibited by anti-P, unaffected by anti-S (40).

In Vivo Antisperm Immunization and Consequences on Fertility. Autoimmunization of male animals with spermatozoa or purified autoantigens leads to strong immune reactions and to an autoimmune aspermatogenic orchitis (AIAO). Although identical at the mature stage, the disease is induced by different immunopathological mechanisms, depending on the autoantigen involved, as shown for S, P, and T in the guinea pig (47,59). S-induced AIAO is due to a delayed type of hypersensitivity shown cutaneously, at the level of the epididymis and by transfer, which is possible by mononucleated cells but not by serum. P-induced AIAO results from an auto-Arthus reaction with typical infiltrates in the epididymis; it is passively transferable by anti-P IgG1 antibodies containing a small amount of IgG2, which is the best situation to transfer Arthus reactivity (35). T-induced AIAO is due to and transferable by IgG2 spermotoxic antibodies. Although interesting in the field of immunopathology, this type of infertility-inducing sperm autoimmunity is inapplicable to man for contraceptive purpose in view of the existence of orchitis lesions. The role and choice of the adjuvants may be of crucial importance.

Sperm isoimmunization of females is more logical given the absence of corresponding antigens from the female organism. The observed results of decreased fertility are far from homogeneous, varying according to the species and the authors. The best results described were obtained in rabbits with whole spermatozoa [for a summary see Voisin (54)]. With purified preparations the best results were obtained with LDH C4. However, this does not lead to complete infertility. In our personal experience immunization of female guinea-pig with whole spermatozoa, whole spermatozoa extracts, or isolated S, P, or T autoantigens led to a complete array of systemic specific reactions, cellular and humoral, with specific antibodies even in the uterovaginal secretions. The fertility, however, was not much impaired except for a delay in fertilization (the consequences of which were easily overcome by keeping longer the females with the males for longer time periods) and a higher rate of intrauterine deaths, raising the problem of common antigens (18). By flushing the uterine horns 72 hours after mating, it was seen that anti-P immunization induced a significant decrease in the percentage of cleaved eggs (67% decrease) and in anti-T, a borderline one (30% decrease) (40).

The presence of "natural" antispermatozoa antibodies is often considered as extremely frequent (with a low titer) and without any detrimental consequences on fertility. These antibodies are presumably induced by bacteria with which they give cross-reactions. It has been argued that they might play a role in preventing spontaneous active immunization against spermatozoa (56).

The antigenic substances (and cells) necessary to reproduction, essentially the gamete autoantigens, and the conceptus alloantigens are protected, especially in the female, against detrimental reactions. They, themselves, initiate a rather protective reaction; on the other hand, their environment is such that they are protected against the effector agents of an artificially induced detrimental reaction. This is related to the particular immunological status of the reproductive tract.

Immunological Status of the Reproductive Tract

This status is particular in several respects. It involves anatomical and physiological barriers or rather filters, phenomena and agents of local suppression, phenomena and agents of local immunity. The last two phenomena go in opposite directions but both point to a relative autonomy.

Barriers and Filters

From an immunological point of view these barriers are mainly selective filters reducing quantitatively and modifying qualitatively the cells and molecules that can be exchanged in both directions: antigens in one, immune agents in the other.

In the nonpregnant female the gametes and their antigens travel in the lumen of the uterus, its horns, and the oviducts. As in the case of the intestine, there is a highly restricted communication with the systemic circulation, with the possible exception of the time of menstruation. As in the case of the respiratory lumen, there are on the Fallopian tubes epithelial cell microcilia transporting the cells and particles present in the lumen.

In pregnant females the barrier built between the embryo or fetus and the mother and its immune system are trifold at the tissular, cellular, and molecular levels. At the tissular level the most important barrier is, of course, the separation of maternal and fetal circulations, which prevents uncontrolled exchanges between the two. At the cellular level, the continuous trophoblastic layer forms the real interface at the placenta and fetal envelops level. The presence of intercellular tigh junctions leads to an obligatory control over cellular and molecular fetomaternal traffic. At the molecular interface level, several substances are bound to play a role: a fibrin layer, more or less important depending on the species and the stage of gestation, forms a dense network that, even if not continuous, undoubtedly also decreases the cellular traffic. The neuraminic acid-rich sialomucin coating the trophoblast microvilli at their apical (i.e., maternal) poles confers a highly negative charge to the cell surface, which could serve to repel maternal lymphocytes. These indeed have a negative charge, which is accentuated during gestation, possibly through the action of human chorionic gonadotropin (HCG) (65).

As a result the traffic is reduced to the essential necessary for the good development of the embryo and fetus. An example of selective traffic is given by the transtrophoblastic passage of antibodies: IgG and only IgG are transported (actively) and only if they are not directed against fetal antigens present on the trophoblast.

In the male the main barrier is the blood-testis barrier, which greatly isolates the germinal cell (especially the more evolved ones) from the rest of the organism after meiosis, when most autoantigens appear. This barrier is made by the seminiferous tubule basement membrane (the most obvious anatomically), by the Sertoli cells and their tigh junctions (the most efficient physiologically), and, in the efferent ducts of the epididymis, by the epithelial cells and their tigh junctions. However, this barrier has weak points, more permeable (for instance to Evans Blue fixed on albumin): the rete testis and the following efferent tubules. These are precisely the places where immunopathological lesions due to autoimmune reactions to spermatozoa autoantigens begin before becoming extended to the rest of the testis.

These are the main barriers of the reproductive tract. They would be, however, insufficient to prevent many auto- and isoimmune aggressions. Another important aspect follows.

Local Immunosuppression

Uterus in nonpregnant females in the absence of decidualization has no immunological privilege: a skin allograft implanted in a normal uterus is rejected as elsewhere (2). *In nonpregnant but decidualized uterus*, whether decidualization is induced by natural or artificial means, a skin allograft lying on the decidual bed will enjoy a prolonged survival (2). It is interesting to note in that connection that a uterine cell able to block the maternal cytotoxic T cell response has been described in the uterus of pseudopregnant mice of two to three days postmating. This is a large Thy 1^+, Lyt 2^+ lymphocyte. It is hormone-dependent and might stimulate macrophages to release prostaglandin E_2 (PGE_2), blocking the production of interleukin 2 (IL2) (7).

In pregnant uterus a skin allograft implanted close to the embryo's bed is accepted (2). The decidua itself contains suppressor cells able to block the maternal cytotoxic T cells response. These are small, granulated, Fc receptor-bearing, non-T lymphocytes under the influence of placental substances for their recruitment and acting through a ~ 100 kDa factor that is able to block IL2 action (45). As for the placenta, it contains and releases immunoactive substances, in addition to the suppressor cell recruiting factor. Some are able to block the cytotoxic efferent immune agents: cytotoxic T cells, K cells, ADCC, NK cells, and even C′-dependent lysis (15,22). Some block the induction of the immune response (13,21) and some act as immunomodulators, as reminded earlier. One must also bear in mind the immunosuppressive capacity that has been advocated for many hormones such as progesterone and proteins, especially placenta proteins. The richness of protective mechanisms at the pregnant uterus levels is therefore remarkable.

In the male genital tract immunosuppressive substances able to inhibit mitogen-induced proliferation of lymphocytes and MLRs have been described in the seminal fluid (42). A high molecular weight component obtained by high-speed centrifugation or gel filtration on Sephadex G 100 column of human seminal plasma was found to be inhibitory of rosette formation, MLR, and complement activity. The first Sephadex peak was found to inhibit phytohemagglutinin (PHA)-induced proliferation and NK cell activity, it had affinity to cells, including human red blood cells. The authors suggest that the responsible substance may be operational in protecting the spermatozoa against an immune attack (36). One must also keep in mind the presence of prostaglandins in seminal plasma.

At the opposite side of this local suppression, there is the possibility of local immunity in the reproductive tract.

Local Immunity

In the female the local immune reaction seems to be mainly localized at the cervical mucosa level. Subepithelial plasma cells synthetize IgG and mainly IgA (dimeric IgA). The latter, going first between and then inside the epithelial cells, combine with the secretory component and are excreted in the lumen with the cervical mucus. There is an increase in subepithelial IgA-containing plasma cells in the endocervix of infertile women (29). This local immune apparatus is functional, as shown by pathological infections (*Neisseria*, *Trichomonas*, or *Candida*) and by experimental immunization. Local immunization with horseradish peroxidase resulted in the production of IgA and IgG antibodies locally (plus IgG in the serum) whereas systemic immunization induced only IgG antibodies in the serum and in the genital tract (34). This local apparatus does not seem responsible for the demonstrated presence of immunoglobulins (including antispermatozoa antibodies) in oviductal and follicular fluid. [For a summary of uterine local immunity see Vaerman and Ferin (48), Jones (31).]

In the male the local production of autoantibodies to spermatozoa has long been debated. However, the described fact that IgA found in the seminal plasma is 60% to 90% secretory IgA indicates a local production (28). No data are available on conceivable local cellular immunity in the male and female genital tracts (except for new data disclosed at the 1986 New Dehli symposium by D. Anderson).

The presence of clear immunological barriers, of local suppression, and of local immunity point to the relative autonomy of the immune reactions in the sphere of the reproductive tract. This autonomy, together with the systemic immunoregulatory phenomena we have described, in turn points to the difficulty inherent in attempts at immunological manipulations either with fertility control views or with therapeutic goals in infertility or habitual abortions.

Can One Overcome the Difficulties?

There are reasons to believe that the preceding difficulties may eventually be overcome or bypassed. These reasons are of two types: factual and theoretical.

Factual Data

In the field of therapeutics of sterility and primary habitual abortion, a few successes have been published on recovered fertility by immunosuppressive therapy (corticoids) in males infertile with antispermatozoa antibodies; the antibody level drop being a prerequisite. Of course, one would prefer a more specific (and more efficient) treatment. Primary habitual abortions have often yielded to immunization by leukocytes from the father or from a panel of donors.

In the field of fertility control, the encouraging experimental results obtained with LDH-C4 and even the clinical ones with βHCG indicate that the way may be followed with hope and confidence. At the same time the results obtained in the preceding fields indicate those current limitations that have to be surmounted.

Theoretical Aspects

Mastering the immune regulation mechanisms goes further than the tolerance breaking "cliché." It implies several works in depth. Among them are:

1. A deeper knowledge of the regulatory mechanisms, their immune agents, and the interplay of the latters.
2. A precise knowledge of the antigen sequences, structures, and epitopes; especially the immunogenic and tolerogenic ones, depending of the type of lymphocyte targeted. The type of work made by Elie Sercarz (41) on lysozyme is exemplary. It is precisely at this point that genetic engineering may be particularly helpful in realizing the ideal molecule for a given purpose.
3. Correlatively a mastering of the processing and presentation of these epitopes.
4. Playing with the idiotypic network by means of monoclonal antibodies might also help.

Overcoming the "autonomous" local status of the reproductive tract may be considered from several points of view. Considering the genital tract antigens, one should either bypass the barriers, proceed to local immunization, which is not an easy task, or overflow the defenses by a strong reaction with overwhelming titers of antibodies (cellular reactions are rather to be avoided in view of their immunopathological potential).

But the preceding difficulties may be avoided if one targets a molecule necessary to the fertility and/or the early gestation and not sheltered from circulating antibodies. Is it necessary to say that these (almost) ideal molecules are the hormones necessary to the reproductive phenomena: HCG at the first place, ovine luteinizing hormone and follicle-stimulating hormones, luteinizing hormone-releasing hormone. Of course, necessary chemical manipulations and controls will avoid undue cross-reactivity.

The problem is then of another type: *increasing and homogenizing the level of reactivity*, that is, to obtain high titers of antibodies in all (or almost all) immunized subjects. It is classic to obtain these results in other fields of immunology by using a good carrier and a good adjuvant. The carrier must be innocuous and, if possible, useful; hence the recommended use of a toxin such as tetanus, diphtheria, cholera, or of bacillus Calmette-Guerin (BCG) vaccine. Also, two vaccines may be used as reciprocal carriers. The chemical link and its length may be of importance.

The adjuvant is not an easy problem. The present choice is rather limited, Freund's adjuvant being eliminated: muramyl-dipeptide (MDP) or analogs (but they need to be incorporated in water-in-oil emulsion), the old but efficient alum, lipopolysaccharide (sodium phthalyl derivative of lippopolysaccharide), and a few other ones, the ideal still being to be discovered.

Conclusions

During the length of evolution Nature has selected the most adapted ways to ensure effective reproduction of the species—and human beings can testify that it is a success—protecting the necessary structures against the simultaneously selected immune rejection reaction that is aimed at foreign invaders.

During the last decades Man has intervened and attempted to master the process in view of the exponential increase of the human population. Immunological intervention offers a promising possibility, with some results already in view. Improvements will be obtained when we have a better understanding of the immunoregulatory mechanisms at work in both systemic and local areas in order to bypass them, a better knowledge of the structure and epitope immunological properties of the antigens involved, the adjustment of a good carrier, an efficient possible association of antigens, and an efficient and innocuous adjuvant. Such are the immunological prerequisites to successful immune manipulations in the various fields of reproduction.

References

1. Badet MT, Chateaureynaud-Duprat P, Mouches C, et al: Séparation et étude de deux facteurs sériques responsables de l'inhibition de la réaction cytotoxique maternelle à l'égard des cellules embryonnaires, chez la salamandre gestante. *Ann Immunol (Paris)* 1980;131 D:57–69.
2. Beer AE: The paradox of foeto-placental units as allografts, in Talwar GP (ed): *Recent Advances in Reproduction and Regulation of Fertility.* Amsterdam, Elsevier/North Holland, 1979, pp 501–513.
3. Bell SC, Billington WD: Major anti-paternal alloantibody induced by murine pregnancy is non-complement-fixing IgG1. *Nature* 1980;288:387–388.
4. Billington WD, Davies M, Bell SC: Maternal antibody to foetal histocompatibility and trophoblast-specific antigens. *Ann Immunol (Paris)* 1984;135:331–335.
5. Bobe P, Doric M, Kinsky RG, et al: Modulation of mouse anti-SRBC antibody response by placental extracts. *Cell Immunol* 1984;89:355–364.
6. Bobe P, Kanellopoulos-Langevin C, Bleux C, et al: Modulation of mouse anti-SRBC antibody responses by placental extracts. II. Antigen specificity and regulatory role of B and T cell populations affected by two distinct placental fractions. *J Immunol* 1986;136:574–581.
7. Brierley J, Clark DA: Characterization of a hormone-dependent suppressor cell in the uterus of pregnant and pseudopregnant mice. *J Reprod Immunol* [Suppl] 1986;6:118.
8. Chaouat G, Voisin GA: Regulatory T cell subpopulations in pregnancy. I. Evidence for suppressive activity of the early phase of MLR. *J Immunol* 1979;122:1383–1388.
9. Chaouat G, Voisin GA: Regulatory T cells in pregnancy. IV. Genetic characteristics and mode of action of early MLR suppressive T cell subpopulations. *J Immunol* 1981;127:1335–1339.

10. Chaouat G, Voisin GA: Regulatory T cells in pregnancy. V. Allopregnancy induced T-cell-suppressor factor is H-2 restricted and bears Ia determinants. *Cell Immunol* 1981;62:186–195.
11. Chaouat G, Voisin GA, Escalier D, et al: Facilitation reaction (enhancing antibodies and suppressor cells) and rejection reaction (sensitized cells) from the mother to the paternal antigens of the conceptus. *Clin Exp Immunol* 1979;35:13–24.
12. Chaouat G, Kinsky RG, Duc HT, et al: The possibility of anti-idiotypic activity in multiparous mice. *Ann Immunol (Paris)* 1979;130:601–605.
13. Chaouat G, Chaffaux S, Duchet-Suchaux M, et al: Immunoactive products of mouse placenta: I. Immuno-suppressive effects of crude and water soluble extracts. *J Reprod Immunol* 1980;2:127–139.
14. Chaouat G, Monnot P, Hoffmann M, et al: Regulatory T cells in pregnancy. VI. Evidence for T-cell-mediated suppression of CTL generation toward paternal alloantigens. *Cell Immunol* 1982;68:322–331.
15. Chaouat G, Kolb JP, Riviere M: Role of the placental interface and of trophoblast/maternal tissue interactions in the survival of the murine foetal allograft. *Ann Immunol (Paris)* 1984;135:302–306.
16. Chateaureynaud P, Badet MT, Voisin GA: Antagonistic maternal immune reactions (rejection and facilitation) to the embryo in the urodele amphibian Salamandra salamandra Lin. *J Reprod Immunol* 1979;1:47–60.
17. Cooper EL, Rubilotta LM: Allograft rejection in Eisenia Foetida. *Transplantation* 1969;8:220–223.
18. D'Almeida M, Voisin GA: Resistance of female guinea pig fertility to efficient iso-immunization with spermatozoa autoantigens. *J Reprod Immunol* 1979;1:237–247.
19. Duc HT, Monnot P, Voisin GA: Inhibitory effects of placental extracts on the effector CTL, in vitro action. 5th Int. Congress of Immunology, workshop presentation 706.28, Kyoto, Japan, 1983, p 208.
20. Duc HT, Masse A, Bobe P, et al: Deviation of humoral and cellular alloimmune reactions by placental extracts. *J Reprod Immunol* 1985;7:27–39.
21. Duc HT, Monnot P, Righenzi S, et al: The in vitro immunomodulatory effect of placental substances on mixed lymphocyte reaction (MLR) and cell mediated lysis (CML). *Fed Proc* 1986;45:631.
22. Duc HT, Righenzi S, Monnot P, et al: Immunomodulatory effects of placental substances on mixed lymphocyte reaction (MLR) and cell mediated lysis (CML). *J Reprod Immunol* [Suppl] 1986;6:116.
23. Duprat P: Etude de la prise et du maintien d'un greffon de paroi du corps chez le lombricien Eisenia Foetida Typica. *Ann Inst Pasteur (Paris)* 1967;113:867–881.
24. Faulk WP, McIntyre JA: Immunologic studies of human trophoblast: markers, subsets and functions. *Immunol Rev* 1983;75:139–175.
25. Gill TJ: Speculations on the transplantation biology of the maternal-fetal interface. *Ann Immunol (Paris)* 1984;135:307–309.
26. Goldberg E: Effects of immunization with LDH-X on fertility, in Diczfalusy E (ed): *Immunological Approaches to Fertility Control*. Stockholm, Karolinska Institutet Publication, 1974, pp 202–222.
27. Gupta GS, Kinsky R, Duc HT, et al: Effects of placental extracts on the immune response to histocompatibility antigens: class deviation of alloantibody response and allograft enhancement. *Am J Reprod Immunol* 1984;6:117–123.
28. Hjort T: Immunological capacity of the male genital tract. A workshop report, in Boettcher B (ed): *Immunological Influence on Human Infertility*. New York, Academic Press, 1977, pp 115–117.

29. Hutcheson RB, Anderson TD, Holborow EJ: Cervical plasma cell population in infertile patients. *Br Med J* 1974;3:783–784.
30. Jarcho J: Artificial production of sterility with special reference to experimental temporary sterility biologically induced in the female. *Am J Obstet Gynecol* 1928; 16:813–827.
31. Jones WR: Immunological factors in male and female infertility, in Hearn JP (ed): *Immunological Aspects of Reproduction and Fertility Control*. Falcon-House, England, MTP Press, 1980, pp 105–140.
32. Kolk AHJ, Samuel T: Isolation, chemical and immunological characterization of two strongly basic nuclear proteins from human spermatozoa. *Biochem Biophys Acta* 1975;393:307–319.
33. Kremer J, Jager S, Kuiken J: The clinical significance of antibodies to spermatozoa, in Boettcher B (ed): *Immunological Influence on Human Fertility*. Sydney, Academic Press, 1977, pp 47–66.
34. McAnulty PA, Morton DB: *J Clin Lab Immunol* 1978;1:255.
35. Maillard JL, Voisin GA: Elicitation of Arthus reactions in guinea pigs by homologous $\gamma 1$ and $\gamma 2$ immunoglobulins. *Proc Soc Exp Biol Med* 1970;133:1188–1194.
36. Marcus ZH, Dondero F, Lunenfeld B, et al: Effect of the inhibitory material from male genital tract on natural killing activities. *Immunol Lett* 1985;10:1–5.
37. McCartney JL: *Am J Physiol* 1923;63:207.
38. Marquant-Le Guienne B, De Almeida M: Role of guinea-pig sperm autoantigens in capacitation and the acrosome reaction. *J Reprod Fertil* 1986;77:337–345.
39. Metalnikoff S: Etudes sur la spermotoxine. *Ann Inst Pasteur (Paris)* 1900;14:577–589.
40. Munoz de Vera G, Marquant-Le Guienne B, De Almeida M, et al: Role of guinea-pig sperm autoantigens in sperm binding to the zona pellucida and oocyte penetration. *J Reprod Fertil* 1986;77:347 353.
41. Oki A, Sercarz E: T cell tolerance studied at the level of antigenic determinants. I. Latent reactivity to lysozyme peptides that lack suppressogenic epitopes can be revealed in lysozyme-tolerant mice. *J Exp Med* 1985;161:897–911.
42. Prakash C, Coutinho A, Moller G: Inhibition of in vitro immune responses by a fraction from seminal plasma. *Scand J Immunol* 1976;5:77–85.
43. Prehn RT: The immune reaction as a stimulator of tumor growth. *Science* 1972; 176:170–172.
44. Rocklin RE, Zukerman JE, Alpert E, et al: Effect of multiparity on human maternal hypersensitivity to foetal antigen. *Nature* 1973;241:130–131.
45. Slapsys RM, Beeson JH, Clark DA: Localization of the trophoblast dependent decidua-associated suppressor cell. *J Reprod Immunol* [Suppl] 1986;6:121.
46. Tiffoche C, Kanellopoulos J, Voisin GA: Solubilization and partial characterization of guinea pig spermatozoal plasma membrane T autoantigen. *Fed Proc* 1985; 44:959.
47. Toullet F, Chadenier F, Maillard JL, et al: Cellular transfer of autoimmune aspermogenic orchiepididymitis (AIAO) by the i.v. route in the guinea pig. *Clin Exp Immunol* 1985;59:593–603.
48. Vaerman JP, Ferin J: Local immunological response in the vagina, cervix and endometrium, in Diczfalusy E (ed): *Immunological approach of fertility control*. Stockholm, Karolinska Institutet Publication, 1974, pp 281–301.
49. Voisin GA: Le phénomène de facilitation immunologique, concept élargi du "enhancement phenomenon," in Miescher PA, Grabar P (ed): *IVth International Symposium Immunopathology*. Bâle, Stuttgart, Schwabe & Co, 1966, pp 165–193.

50. Voisin GA: Immunological facilitation, a broadening of the concept of the enhancement phenomenon. *Prog Allergy* 1971;15:328–485.
51. Voisin GA: Immunity and tolerance: a unified concept. *Cell Immunol* 1971;2:670–689.
52. Voisin GA: Sperm specific autoantigens, in Shulman S, Dondero F, Nicotra M (eds): *Immunological Factors in Human Reproduction*. New York, Academic Press, 1982, pp 147–161.
53. Voisin GA: Enhancing antibodies and suppressor cells in pregnancy: role of the placentation, in Iojima S, Billington WD (eds): *Reproductive Immunology*. Amsterdam, Elsevier, 1983, pp 121–131.
54. Voisin GA: Spermatozoa autoantigens in animals. A possible way for immunological contraception, in: *Immunological Factors in Human Contraception*. Roma, ACTA MEDICA, 1983, pp 23–38.
55. Voisin GA: Antigen-bearing targets have selected the immune mechanisms and their regulation during evolution: the example of gestation. *Asian Pac J Allergy Immun* 1984a;2:293–300.
56. Voisin GA: Active immunization against sperm and sperm autoantigens, in Crighton DB (ed): *Immunological Aspects of Reproduction in Mammals*. London, Butterworths Scientific Ltd, 1984, pp 291–303.
57. Voisin GA: Levels and nature of immunoregulatory mechanisms in pregnancy. *Contrib Gynecol Obstet* 1985;14:36–43.
58. Voisin GA, Chaouat G: Demonstration, nature and properties of maternal antibodies fixed on placenta and directed against paternal antigens. *J Reprod Fertil* [Suppl] 1974;21:89–103.
59. Voisin GA, Toullet F: Autoimmune aspermatogenic orchitis (A.I.A.O.). A model for three possible mechanisms of autoimmune lesions. *Folia Allergol (Roma)* 1971;18: 310–316.
60. Voisin GA, Kinsky RG, Duc HT: Immune status of mice tolerant of living cells. II. Continuous presence and nature of facilitation-enhancing antibodies in tolerant animals. *J Exp Med* 1972;135:1185–1203.
61. Voisin JE, Monnot P, Voisin GA: Maternal alloimmune reactions towards the murine conceptus and graft-versus-host reaction (GVHR). I. Priming for antipaternal GVHR by gestation. *J Reprod Immunol* 1986;9:73–83.
62. Voisin JE, Kinsky RG, Voisin GA: Maternal alloimmune reactions towards the murine conceptus and graft-versus-host reaction (GVHR). II. Inhibition of priming by placental extracts. *J Reprod Immunol* 1986;9:85–94.
63. Wegmann TG, Singh B, Carlson GA: Allogeneic placenta is a paternal strain antigen immunoabsorbent. *J Immunol* 1979;122:270–274.
64. Wegmann TG: Foetal protection against abortion: is it immunosuppression or immunostimulation? *Ann Immunol (Paris)* 1984;135:301–351.
65. Wrezlewicz W, Rashkoff E, Lucas T, et al: Effects of human chorionic gonadotropin and anti-thymocyte globulin on the electrophoretic mobility of human lymphocytes. *Transplant Proc* 1977;9:1441–1445.
66. Youtananukorn V, Matangkasombut P, Osathanondh V: Onset of human maternal cell-mediated immune reaction to placental antigens during the first pregnancy. *Clin Exp Immunol* 1974;16:593–598.

CHAPTER 35

The Past Is Prologue: Implications of the Symposium

Egon Diczfalusy

In 1961 the late Prime Minister of India, Mr. Nehru, said, "the future belongs to science and to those who make friends with science." And to make friends with science also means to learn to live in a rational way in a seemingly irrational, very rapidly changing world where 5 billion human beings try to pursue their happiness today and 10 billion will do so tomorrow.

Indeed, there is broad agreement that the world population will continue to grow, albeit at a progressively slower pace, for another 100 to 150 years; it is expected to stabilize between 10 and 11 billion by the end of the 21st century, or so. The World Bank's present projections of the hypothetical size of stationary populations suggest that by that time, India will be the most populous country on earth, with 1.7 billion inhabitants, followed by China with 1.6 billion, and Nigeria the third, with more than half a billion.

One of our great human virtues is our ability to reach consensus, to agree; but our most distinctive characteristic is to disagree. The available information is frequently interpreted by us so as to suit our personal beliefs, our emotional background, and ethical and moral value systems. Indeed, as Mahatma Gandhi remarked 60 years ago, "to conquer the subtle passions seems to me to be harder far more than the physical conquest of the world by the force of arms." Hence it is not too surprising that, at least in the Western world, there is considerable disagreement today on the likely impact of the rapid population increase on the economic growth in the developing world and on the impact of family planning in slowing down population growth. There is agreement, however, that sustaining a global population of 10 or 11 billion human beings will require a high level of technical civilization and an intimately interconnected world economy. There seems to be also agreement that family planning alone cannot solve all the problems of the developing world; however, there is a growing consensus, at least in the developing countries, that without family planning it will be very difficult, or perhaps impossible, to solve their problems.

There is also general agreement that the impact of family planning goes far beyond the issues of population growth and development, since it is a major tool to improve the general health status of populations, especially in developing countries. The effective use of family planning will delay the age at first pregnancy and space further births; when integrated into primary health care, both of these factors will lower infant, child, and maternal morbidity and mortality and significantly reduce illegal abortions and their health hazards. The problem is, however, that available methods and services of family planning fail to meet, as yet, the demand of developing countries. This is suggested by the number of unwanted births, the high discontinuation rates with available methods, and the number of induced abortions.

Hence, what is urgently needed is *more research* to develop a much wider variety of safe and effective methods that will suit the individual situation, the socioeconomic condition, and the cultural and religious values of different couples. There is an equally urgent need for increased social, behavioral, and service research to better understand the factors influencing the acceptability of various methods and the complicated interactions between users and providers of services. This need is underlined by the major differences among developing countries in family formation patterns, attitudes, perceptions, and user preferences. In several countries, acceptability is frequently specific to a given locality and culture.

There is an old Latin proverb, "Tempora mutantur et nos mutamur in illis," times are changing and we are changing with them. True, but not with the same speed. The changes around us are so rapid that our perceptions of them are lagging more and more behind and might even become old-fashioned. In 1974 at the United Nations Conference on Population in Bucharest, the Western countries were those that were urging the then very suspicious and reticent developing countries to join a global effort and adopt family-planning measures. Only ten years later, in 1984 at the Mexico Conference, the roles were completely reversed. The developing countries were now urging all governments to provide more support to research on family planning, whereas certain Western governments showed a lukewarm interest, and others expressed doubts whether there really was such a great need for family planning, not to mention national population policies.

The Inaugural Symposium of the National Institute of Immunology provided ample evidence to show how the focus of research in general and on fertility regulation in particular is also gradually moving from developed to developing nations. In my view, this process is historically inevitable, because developing countries are simply more concerned and more committed. And I hope and believe that when the history of the 21st century will be written, frequent reference will be made to countries that then will be called "former developing countries." I also believe that by that time India, China and Thailand will be such "former developing countries" and that they will provide leadership and assistance to developing countries, which may then be at different levels of development.

An inaugural symposium is a very special and an auspicious occasion; it is like being present at the creation, to use an expression coined by Alphonso the Learned, king of Spain 700 years ago. We stop for a moment for reflection, to take stock of where we stand and how we intend to proceed. At the same time, we may also perceive the perspectives and future lines along which the next generation is likely to make its contributions. During the three days of this symposium we have covered an enormous territory, and it would be both pretentious, overambitious, and unwise of me to attempt to review and summarize here the highlights of each of almost 60 full-dress presentations. In addition, Professor Voisin gave a masterly summary of the state of the art. Indeed, we have covered a vast area. We have discussed demographic trends, population policies and strategies, and the specific needs of developing countries in research on human reproduction. We have also considered a number of recently developed long-acting methods of fertility regulation, such as implants, intrauterine devices, vaginal rings and pills; and it is entirely my own fault that no paper was presented on long-acting injectable contraceptives, an area in which the World Health Organization Special Programme on Human Reproduction has developed a wide variety of new and improved formulations.

Then, after an interesting session on the problems of neuroendocrine regulation, the stage was set for an in-depth review of the state of the art within the different areas of the immunological control of human fertility. It was most gratifying to see that after 15 years of developmental efforts, four different preparations of β-subunit of human chorionic gonadotropin (βHCG) reached the stage of clinical evaluation. Two of those were developed here by scientists at the National Institute of Immunology, one by the WHO Programme, and one by the Population Council. We have also seen several very promising new approaches under development by this Institute, which may reach the clinic in the early nineties. We then carefully considered the problems related to the safety of the vaccines developed and to the available methods for their preclinical assessment.

We have also covered the complex area of the immunobiology of pregnancy and noted the great potential in this field to provide entirely new approaches to the immunological control of human fertility in the not too distant future.

There has been major progress also in the use of gamete antigens, and we have noted with satisfaction that some of these have shown efficacy when tested in different primate species. We have also obtained new insights into a variety of immunological factors that appear to be instrumental in causing infertility. This may be very important, since the diagnosis, prevention, and treatment of infertility is an integral part of the WHO Human Reproduction Programme.

Last, but not least, we have examined some new concepts and fresh ideas, which are likely to direct our future efforts toward the development of a variety of novel, improved, and safe immunological methods for fertility regulation.

On the first day of this symposium, Professor Fathalla recalled that in May 1981, the World Health Assembly in Geneva had a unique and historic opportunity to listen to the Opening Address delivered by the late Indira Gandhi, Prime Minister of India. One of her statements is frequently quoted by scientists and statesmen alike: "Family planning programmes are awaiting a big breakthrough. Without a safe, preferably oral, drug which women and men can take, no amount of Government commitment and political determination will avail." And I repeat this statement here, since by now this became the universal message to and cultural heritage of individual scientists working in fertility regulation, not only here in India, but all over the world.

However, the late Indira Gandhi also said something else in the same speech in Geneva. She said, "Life is and perhaps always will be a struggle, although the nature of it keeps changing. To meet it we need vision, faith, courage and dogged perseverance. These are the qualities I admire in individuals and in organizations." This is then our message to research institutions and to their scientific leaders in our field, like Professor Talwar, and, again, not only all over India, but also in a global context.

I started this summary by quoting a sentence from a speech by the late Prime Minister, Mr. Nehru. Permit me now to quote it more fully:

> It is science alone that can solve the problems of hunger and poverty, insanitation and illiteracy, of superstition and deadening custom and tradition, of vast resources running to waste, of a rich country inhabited by starving people . . . Who indeed could afford to ignore science today? At every turn we have to seek its aid . . . The future belongs to science and to those who make friends with science.

And this is then the message of India to political leaders and decision-makers all over the world in both developing and developed countries. And I wish that those leaders would pay more attention to it, because research is cost effective, and to invest in research on human reproduction and fertility regulation means to invest into a brighter future for mankind.

Finally, there is also an individual message from India for all of us, the 5 billion individuals of today and the 10 billion human beings of tomorrow. It comes, again, from Mahatma Gandhi, and it is as fresh, relevant, and appropriate today as ever. It says, "I have nothing new to teach the world. Truth and non-violence are as old as hills."

I would like to close by expressing our deep appreciation and most sincere gratitude to Professor Talwar and all his colleagues for a most generous hospitality and for the exceptional privilege to be present at the creation of a new and most important instrument designed by the Government of India to improve the quality of life, not only in India, but in every corner of a troubled world. And – to use the 100-year-old words of Shelley – under "the gigantic shadows, which futurity casts on the present," all of us here want to tell you, Professor Talwar, and all members of this Institute that during many, many years to come, our high hopes and very best wishes will follow every step of you in your mission of critical importance, which is to improve the human condition.

Index

MIX
Papier aus verantwortungsvollen Quellen
Paper from responsible sources
FSC® C105338

If you have any concerns about our products,
you can contact us on
ProductSafety@springernature.com

In case Publisher is established outside the EU,
the EU authorized representative is:
Springer Nature Customer Service Center GmbH
Europaplatz 3, 69115 Heidelberg, Germany

Printed by Libri Plureos GmbH
in Hamburg, Germany